Advances in Intelligent Systems and Computing

Volume 1432

The series "Advances in Intelligent Systems and Computing" contains publications on theory, applications, and design methods of Intelligent Systems and Intelligent Computing. Virtually all disciplines such as engineering, natural sciences, computer and information science, ICT, economics, business, e-commerce, environment, healthcare, life science are covered. The list of topics spans all the areas of modern intelligent systems and computing such as: computational intelligence, soft computing including neural networks, fuzzy systems, evolutionary computing and the fusion of these paradigms, social intelligence, ambient intelligence, computational neuroscience, artificial life, virtual worlds and society, cognitive science and systems, Perception and Vision, DNA and immune based systems, self-organizing and adaptive systems, e-Learning and teaching, human-centered and human-centric computing, recommender systems, intelligent control, robotics and mechatronics including human-machine teaming, knowledge-based paradigms, learning paradigms, machine ethics, intelligent data analysis, knowledge management, intelligent agents, intelligent decision making and support, intelligent network security, trust management, interactive entertainment, Web intelligence and multimedia.

The publications within "Advances in Intelligent Systems and Computing" are primarily proceedings of important conferences, symposia and congresses. They cover significant recent developments in the field, both of a foundational and applicable character. An important characteristic feature of the series is the short publication time and world-wide distribution. This permits a rapid and broad dissemination of research results.

Indexed by DBLP, INSPEC, WTI Frankfurt eG, zbMATH, Japanese Science and Technology Agency (JST).

All books published in the series are submitted for consideration in Web of Science.

For proposals from Asia please contact Aninda Bose (aninda.bose@springer.com).

Subarna Shakya · Ke-Lin Du · Klimis Ntalianis
Editors

Sentiment Analysis and Deep Learning

Proceedings of ICSADL 2022

Volume 1

 Springer

Editors
Subarna Shakya
Department of Electronics and
Computer Engineering
Pulchowk Campus, Institute
of Engineering, Tribhuvan University
Lalitpur, Nepal

Ke-Lin Du
Department of Electrical and
Computer Engineering
Concordia University
Montreal, QC, Canada

Klimis Ntalianis
University of West Attica
Aigaleo, Greece

ISSN 2194-5357 ISSN 2194-5365 (electronic)
Advances in Intelligent Systems and Computing
ISBN 978-981-19-5442-9 ISBN 978-981-19-5443-6 (eBook)
https://doi.org/10.1007/978-981-19-5443-6

This Springer imprint is published by the registered company Springer Nature Singapore Pte Ltd.
The registered company address is: 152 Beach Road, #21-01/04 Gateway East, Singapore 189721,
Singapore

We, the organizers, dedicate this ICSADL 2022 proceeding to the worldwide community of artificial intelligence, information, and communication technologies [ICT] researchers. We also dedicate this proceeding to the authors and editorial team of the conference, who have highly discussed the solutions to problems that exist in the international research community.

Preface

The 2nd International Conference on Sentimental Analysis and Deep Learning [ICSADL 2022] aims to bring together the researchers from the relevant fields of artificial intelligence and other computing technologies to provide a broad forum to discuss the recent innovations in deep learning and behavioral analysis domain. A specific goal of ICSADL 2022 is to study how the proposed research idea of context may be understood and used to create fully interactive information and computing systems.

ICSADL 2022 has achieved this goal by inviting research contributions that approached intelligent information system contexts from many perspectives such as deep learning, social media mining, behavioral analysis, sentiment analysis, and its instances in social media analysis. We received 332 papers from various countries in total and accepted 75 of them. Further, the oral presentations were delivered over two days by organizing different parallel sessions along with a keynote session delivered by plenary speakers.

Collectively, the research works discussed at the conference contributed to our knowledge and basic understanding on the state-of-the-art mechanisms involved in the design, development, and analysis of the intelligent information systems. This knowledge will lead to progress in computing and communication research, as well as improved behavioral analysis and online data mining technologies. All that is evident when looking at the papers presented in the conference under four different categories: sentiment analysis, deep learning, big data analytics, and real-time applications.

We acknowledge the distinguished speaker of the conference: Dr. Manu Malek for the acceptance to deliver a keynote talk on the respective fields of expertise. We also thank all organizing and program committee members and national and international referees for their excellent work. Our high appreciation is to all the principal speakers at the conference event and also the participants for making this conference successful and enjoyable.

Prof. Dr. Subarna Shakya
Professor
Department of Electronics
and Computer Engineering
Pulchowk Campus, Institute
of Engineering, Tribhuvan University
Lalitpur, Nepal

Dr. Ke-Lin Du
Affiliate Associate Professor
Department of Electrical and
Computer Engineering
Concordia University
Montreal, QC, Canada

Dr. Klimis Ntalianis
Professor
University of West Attica
Aigaleo, Greece

Contents

About the Editors

Prof. Dr. Subarna Shakya is currently a Professor of Computer Engineering, Department of Electronics and Computer Engineering, Central Campus, Institute of Engineering, Pulchowk, Tribhuvan University, Coordinator (IOE), LEADER Project (Links in Europe and Asia for engineering, eDucation, Enterprise and Research exchanges), Erasmus Mundus. He received M.Sc. and Ph.D. degrees in Computer Engineering from the Lviv Polytechnic National University, Ukraine, 1996 and 2000 respectively. His research area includes E-Government system, Computer Systems and Simulation, Distributed and Cloud Computing, Software Engineering and Information System, Computer Architecture, Information Security for E-Government, and Multimedia system.

Dr. Ke-Lin Du has been a research scientist at Center for Signal Processing and Communications, Department of Electrical and Computer Engineering, Concordia University since 2001, where he became an Affiliate Associate Professor in 2011. He received a Ph.D. degree in electrical engineering from Huazhong University of Science and Technology in 1998. He has published five textbooks, 50 papers, and holds six granted U.S. Patents. Presently, he is on the editorial boards of IEEE Spectrum Chinese Edition, IET Signal Processing, Circuits Systems and Signal Processing, Mathematics, Scientific Reports, and AI. He served as program chair of dozens of conferences. His research area includes wireless communications, signal processing, and machine learning. He has been a Senior Member of the IEEE since 2009.

Dr. Klimis Ntalianis is a full professor at the University of West Attica, Athens, Greece. He has worked as Senior Researcher in multi-million research and development projects, funded by the General Secretariat of Research and Technology of Greece (GSRT), the Research Promotion Foundation of Cyprus (RPF), the Information Society S.A. of Greece and the European Union. He is also serving as a Senior Project Proposal Evaluator for GSRT, RPF, the European Union, the Natural Sciences and Engineering Research Council of Canada and the National Science Center of Poland. In parallel he is a member of several master theses and Ph.D. evaluation committees in Greece, Cyprus, Germany and India. He is also serving as promotion evaluator for Saudi Arabia and academic staff. He has served as general chair in several conferences (IEEE etc.). Dr. Ntalianis has published more than 160 scientific papers in international journals, books and conferences. His main research interests include social computing, multimedia analysis and information security.

Ranking Roughly Tourist Destinations Using BERT-Based Semantic Search

Myeong Seon Kim, Kang Woo Lee, Ji Won Lim, Da Hee Kim, and Soon-Goo Hong

Abstract With the development of the Internet, extensive online reviews have been generated in the tourism field. To utilize big data and avoid the shortcomings of traditional techniques, we designed a tourist destination ranking system by introducing a semantic search that extracts data related to the input query. For this system, we reviewed tourist spot reviews and pre-processed text reviews. Then, we embed the corpus and query using SBERT, measure their similarity, and leave data similar to the query above the threshold. By implementing a count-based ranking algorithm with the data within the boundary, the tourist destinations are derived in a semantically similar order to the query. We entered three queries to obtain the top 5 relevant tourist destinations. Although there are problems with optimal thresholds and imbalanced data, semantic search derives information of desired conditions and may be referenced in policymaking and recommendation systems.

Keywords Semantic search · BERT · Ranking system · Tourism destination

1 Introduction

With the growth of the Internet, there have been a larger number of online tourism reviews. Tourists' decision-making is influenced by the travel experiences of other individuals that are often form, online review comments [1]. Therefore, the analysis of online review data is essential. It can help tourism practitioners and stakeholders reflect results in the policymaking process and raise policy issues by identifying

M. S. Kim · D. H. Kim
Department of Computer Engineering, Dong-a Univ. 37, Nakdong-daero 550beon-gil, Saha-gu, Busan, South Korea

K. W. Lee · J. W. Lim
Smart Governance Research Center, Dong-a Univ, 225 Gudeok-ro, Seo-gu, Busan, South Korea

S.-G. Hong (✉)
Department of Management Information System, Dong-a Univ. 225 Gudeok-ro, Seo-gu, Busan, South Korea
e-mail: shong@dau.ac.kr

© The Author(s), under exclusive license to Springer Nature Singapore Pte Ltd. 2023
S. Shakya et al. (eds.), *Sentiment Analysis and Deep Learning*,
Advances in Intelligent Systems and Computing 1432,
https://doi.org/10.1007/978-981-19-5443-6_1

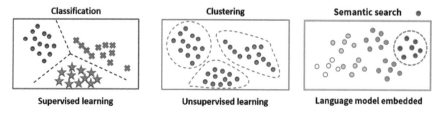

Fig. 1 Comparison of classification, clustering, semantic search [2]

which tourist destinations are popular and which are problematic. Alternatively, it may provide a practical recommendation system that helps tourists have a better travel experience.

Techniques for extracting meaningful information from big data already exist. For classification with these techniques, computers are employed to automatically classify input objects according to their content in predefined classes. This requires the use of labeled data to be used for training in supervised learning. Because it is difficult to obtain large datasets including labels, this method is both demanding and expensive. Clustering is a technique that identifies similarities between objects and groups them according to their characteristics. However, because clusters are ambiguous, the accuracy of these clustering results may be poor.

To avoid the shortcomings of traditional techniques, we introduced a semantic search to effectively extract the necessary information. When a query is entered, the semantic search returns a score for the semantic similarity between the query and corpus. By filtering data below a certain score, it is possible to coarsely define a boundary that contains data related to a common query. This study uses semantic search with this property to design a ranking system that shows tourist attractions that are related to a specific query in order as shown in Fig. 1.

2   Semantic Search with SBERT

2.1   *Semantic Search with SBERT*

Semantic search, which can be distinguished from lexical search based on literal matches without the semantic understanding of a query, denotes search with an understanding of the query meaning [3]. In traditional search engines, which only find documents based on lexical matches, semantic search engines can also find semantically similar sentences [4]. Therefore, it provides more significant search results by assessing and understanding the search phrase and discovering the most appropriate results in a website, database, or any other dataset [5].

Word embedding is word representation that allows words with similar meanings to have a similar representation. Using bidirectional encoder representations from transformer (BERT)-based word embedding, a sentence (or word) can be represented in the dense form of a vector. It allows the identification of sentences similar to a search query by measuring the distance in the vector space. Sentence-BERT (SBERT) is a modification of the BERT network using Siamese and triplet networks to improve its efficiency and suitability for semantic searches [6].

2.2 *Extracting a Boundary by Applying Different Thresholds*

To demonstrate the properties of semantic search based on semantic similarities, we extract a boundary of vectors by applying different threshold values. If the threshold is high, the boundary can omit target sentences (signals) that must be included, whereas non-target sentences (noise) are less likely to be included. On the other hand, if the threshold is low, the boundary contains most of the target sentences but may also include too many non-target sentences. The threshold is key to determining the signal-to-noise ratio. Therefore, it is important to select an appropriate threshold for the search task.

To provide an intuitive understanding of the change in boundary changes according to the threshold values, we applied a semantic search algorithm to a tourist destination, Gwangalli Beach [7]. The results are shown in Fig. 2. They show that the range of boundaries is formed by the query "It is a beautiful beach" with different 3 thresholds (0.7, 0.5, and 0.3). The review data on the beach were embedded in the SBERT. There are 1943 reviews from TripAdvisor [8], and they are labeled according to whether they are similar to the query. In the graph, the colored points indicate queries within the boundary. The blue line is the matched target (signal), and the red line is the non-matched target (noise). The figure shows that the boundary range increased as the threshold decreased.

For a more formal explanation, we measured the signal-to-noise ratio (SNR). The SNR is a measure that compares the detection of the desired signal with the detection of background noise [9]. Table 1 provides detailed information such as the number of signals, noise, and SNR. The SNR was the highest, 14.50, when the threshold was 0.7. The SNR is dramatically reduced to 0.29 and 0.13 when the SNR is set to 0.5 and 0.3, respectively.

3 System Architecture

Figure 3 shows the structure of the count-based ranking system implemented in this study. It consists of dataset construction, data pre-processing, semantic search, and ranking result derivation. We created a dataset by crawling review data for tourist destination on TripAdvisor [8] and performed pre-processing. Subsequently, a semantic search derives the semantic similarity between a specific query and each

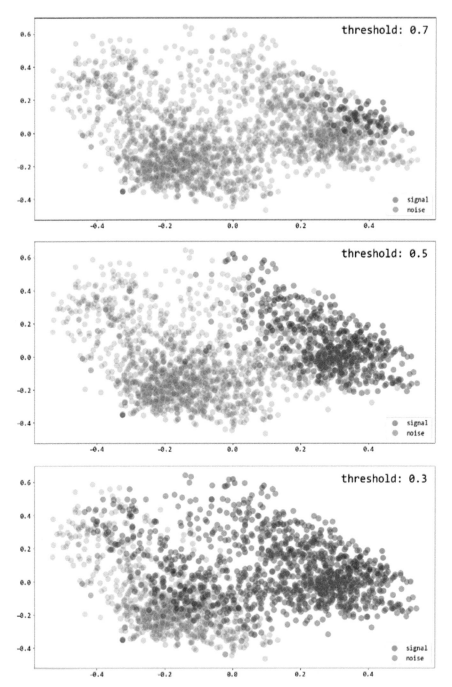

Fig. 2 The boundary at each threshold

Table 1 Numerical Information according to Threshold

Threshold	# of Signal	# of Noise	SNR	# of Signal not included
0.7	58	4	14.50	77
0.5	118	409	0.29	17
0.3	133	1010	0.13	2

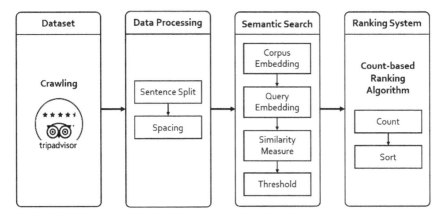

Fig. 3 Ranking system architecture using semantic search

data point in the dataset. Based on the results of the semantic search, tourist destinations with data similar to the query are ranked, making it possible to identify tourist locations suitable for a specific query.

3.1 Data

Reviews of tourist attractions in Busan, South Korea, were collected from TripAdvisor [8]. We scrapped the reviews of 201 tourist spots, including Haeundae, Gamcheon Culture Village, and Haedong Yonggungsa, and stored the data as a comma-separated value (CSV) file. The data consist of "names of tourist destination," "languages," "dates visited," "reviews," and "ratings" attributes. We used 11,751 reviews written in English between July 2007 and May 2021. One review consisted of approximately five sentences on average, and each sentence had an average consisted of approximately 15 words.

3.2 Text Cleaning and Pre-processing

Because reviews include opinions on various aspects of tourist destinations, they must be separated into individual sentences to interpret each opinion more clearly. Using the Natural Language Toolkit (NLTK) library [10] developed for English natural language processing (NLP), sentence segmentation was performed for all reviews. Then, data that had no actual meaning (were not composed of English characters and numbers) were removed. As a result, 58,175 sentences were extracted.

Some sentences had non-spaced phrases such as "activitiessceneryspiritual" and "housemysonwanted." As they are not separated word by word, they have the potential to lead to inaccurate results. Therefore, the sentences were spaced using the "Wordsegment" library [11] from Apache2.

3.3 Embeddings

In this study, the data were embedded using SBERT. It applies Siamese and triplet networks to the existing BERT to more efficiently express embeddings that reflect the meaning of sentence data and to quickly calculate similarity through cosine similarity. This makes it a suitable embedding model for semantic searches where similarity comparison is important for a large-scale dataset.

Sentences were vectorized by a pre-trained SBERT model named "all- MiniLM-L6-v2." It had sentence embeddings and semantic search performances of 68.06 and 49.54, respectively. This model has a speed of 14,200 sentences per second, which is five times faster while retaining good quality [4].

First, a corpus of review sentences was embedded in vector form. The query of interest is represented as a vector in vector space of the review sentences.

3.4 Similarity Measure

We were able to compute the similarity between the corpus and query by embedding them. Similarity is measured using cosine similarity, and the formula can be obtained as follows:

$$\text{similarity} = \cos(\theta) = \frac{A \cdot B}{A\,B} = \frac{\sum_{i=1}^{n} A_i B_i}{\sqrt{\sum_{i=1}^{n} A_i^2}\sqrt{\sum_{i=1}^{n} B_i^2}} \tag{1}$$

We set a threshold and assumed that sentences above the threshold value were semantically related to the query. In this study, 0.6 was rather arbitrarily set as the default threshold value. However, it should be noted that an appropriate threshold value can vary depending on the data.

3.5   Ranking Algorithm

The ranking algorithm returns a list of tourist destinations suitable for a specific query. A count-based ranking algorithm considers the number of tourist spots that semantically matched the query. We counted the number of review sentences that were above the threshold value for each tourist destination. Tourist attractions were sorted based on the obtained information.

Algorithm 1 Count-based Ranking Algorithm

1. *ranked_destination = dictionary{ }*
2. **for each** *tourist_destination* ∈ *dataset* **do**
3.   *ranked_destination['tourist_destination']* $+=1$
4. **end for**
5. *sort_by_value(ranked_destination)*
6. **return** *ranked_destination*

4   Result of Ranking System

4.1   Result of Ranking System

A ranking system with three queries was applied.

The first is a positive query "the night view is beautiful," and the tourist spots suitable for the query are obtained and ordered. The number of reviewed data points was 377. The ranking results of the queries are shown in Fig. 4. Gwangan Bridge [12], Gwangalli Beach [7], Haeundae Beach [13], Taejongdae [14], and Haedong Yonggungsa Temple [15] were placed in order for the query. All tourist destinations were located on the seaside. In particular, the Gwangan Bridge is known to have the best night view because different colors are produced using 7011 LED lights for each season.

The second query concerns the toilet facility, "Toilet is dirty." A total of 23 cases were included in our study. The ranking results of the queries are shown in Fig. 5. Haeundae Beach [13], Gamcheon Culture Village [16], Haedong Yonggungsa Temple [15], Beomeosa Temple [17], and Busan subway were placed in descending order. Because millions of tourists visit Haeundae Beach annually. Tourists on the beach often complain about the maintenance and cleanliness of toilet facilities.

The third query is concerned with the crowdedness of tourist destinations "it is very crowded." A total of 219 cases matched our queries. The ranking results of the queries are shown in Fig. 6. Haeundae Beach [13], Haedong Yonggungsa Temple [15], BIFF Square, Gamcheon Culture Village [16], and Gwangalli Beach [7] were ranked for the query. Unsurprisingly, Haeundae Beach is the most popular tourist destination in Busan, where millions of tourists visit each year.

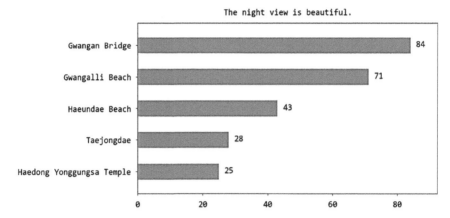

Fig. 4 Result of ranking system for "*The night view is beautiful*"

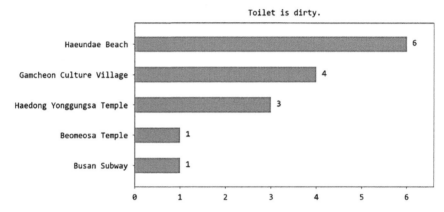

Fig. 5 Result of ranking system for "*Toilet is dirty*"

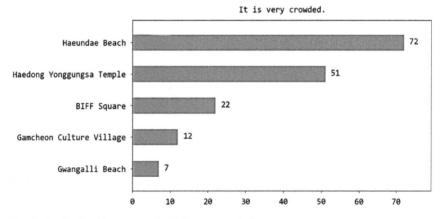

Fig. 6 Result of ranking system for "*It is very crowded*"

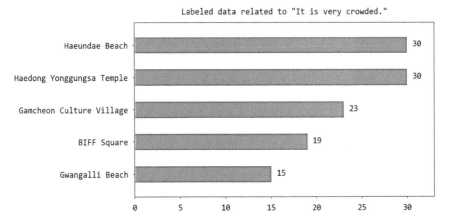

Fig. 7 Ranking for labeled data related to "*It is very crowded*"

4.2   Evaluation of Result

To evaluate the ranking algorithm, we directly labeled 58,175 sentences whether the data were semantically related to the third query on crowdedness. The tourist spots with data labeled as relevant to the query were ranked orderly in Fig. 7.

The top five tourist destinations are Haeundae Beach [13], Haedong Yonggungsa Temple [15], Gamcheon Culture Village [16], BIFF Square, and Gwangalli Beach [7]. Compared with Fig. 6, which is the result of the ranking algorithm, they yield the same top places, although in different rankings.

The labeling process guarantees accuracy but requires a lot of time and effort. The ranking algorithm produces similar results and is even extremely fast and convenient. The ranking algorithm of this study is competitive as an economical and efficient method to roughly derive tourist attractions related to a query.

5   Discussion and Conclusion

To extract meaningful information and provide insight from online tourism big data, this study ranks tourist destinations according to a specific topic using semantic search. For the three queries on night view, toilet facilities, and crowdedness, we derived a ranking of tourist destinations with high semantic relevance to the query.

This study extracts information limited to a particular subject (query), which facilitates access to the desired information. This makes it possible to intuitively rank and recommend tourist attractions according to the query. Because this recommendation is information generated based on the opinions of real tourists, not companies or governments, it is likely to be more reliable and appropriate. This might help tourists optimize their destination choices and find unexpectedly lovely places. From the point of view of a policymaker or relevant stakeholder, tourist sites with related issues can be recognized and reflected in policy decisions by using appropriate queries, and more strategic solutions can be devised by analyzing the properties of the obtained tourist destinations. For example, budgets can be assigned deliberately based on the ranking of tourist destinations. When planning a local tour product, they can find suitable places for the theme of the tour. Practitioners and stakeholders can refer to the results of tasks, such as policy decision-making, budget allocation, and new business development, allowing them to improve their services.

Some of the shortcomings of this study include optimal thresholds and imbalanced data. The appropriate threshold value is different for each query and depends on the corpus (data). Determining the appropriate threshold is a difficult and important task for achieving good performance. As the tourist destinations that appear frequently in the ranking of the results show, there is an imbalanced data problem, in which the review data of some tourist spots dominate. Among the data used in this study, Gamcheon Culture Village, Haeundae Beach, and Haedong Yonggungsa Temple, which commonly appear in the above results, have 7977 (39.7%), 6379 (31.7%), and 5431 (27.01%) sentences, respectively, which are large percentages of the total data. The imbalanced data make the count-based algorithm biased because tourist spots with more data are likely to have more data within the boundary.

In future research, we plan to develop algorithms other than the count-based ranking algorithm to solve the unbalanced data problem. We plan to develop an algorithm based on the average ratings and an algorithm that weighs the portion, which is relatively insensitive to the amount of data.

Semantic search, compared to classification, is economical and simpler but has relatively poor accuracy. Compared to the clustering method, this method is less ambiguous. Just as precise measurements are necessary for science, it is equally important to make rough estimates of quantities using rudimentary ideas and common observations. In practical terms, a semantic search is a useful tool for obtaining specific results, although the boundary is arbitrary.

Acknowledgements This study was supported by the Ministry of Education of the Republic of Korea and the National Research Foundation of Korea (NRF-2018S1A3A2075240).

References

1. Ye, Q., Law, R., Gu, B., & Chen, W. (2011). The influence of user-generated content on traveler behavior: an empirical investigation on the effects of e-word-of-mouth to hotel online bookings. *Computers in Human Behavior, 2*(27), 634–639.
2. Analyst Prep. https://analystprep.com/study-notes/cfa-level-2/quantitative-method/superv ised-machine-learning-unsupervised-machine-learning-deep-learning/
3. Wikipedia Semantic search. https://en.wikipedia.org/wiki/Semantic_search
4. SBERT. https://www.sbert.net
5. Roy, S., Modak, A., Barik, D., & Goon, S. (2019). An overview of semantic search engines. *Int. J. Res. Rev, 10*(6), 73–85.

6. Reimers, N., & Gurevych, I. (2019). Sentence-BERT: Sentence embeddings using Siamese BERT-networks. In *2019 Conference on empirical methods in natural language processing* (pp. 671–688). Association for Computational Linguistics, Hong Kong.
7. Wikipedia Gwangalli Beach. https://en.wikipedia.org/wiki/Gwangalli_Beach
8. Tripadvisor. https://www.tripadvisor.co.kr/
9. Wikipedia Signal-to-noise ratio. https://en.wikipedia.org/wiki/Signal-to-noise_ratio
10. Loper, E., & Bird, S. (2002). NLTK: The natural language toolkit. In *ACL-02 workshop on effective tools and methodologies for teaching natural language processing and computational linguistics* (Vol. 1, pp. 63–70).
11. Wordsegment. https://pypi.org/project/wordsegment/
12. Wikipedia Gwangan Bridge. https://en.wikipedia.org/wiki/Gwangan_Bridge
13. Wikipedia Haeundae Beach. https://en.wikipedia.org/wiki/Haeundae_Beach
14. Wikipedia Taejongdae. https://en.wikipedia.org/wiki/Taejongdae
15. Wikipedia Haedong Yonggungsa. https://en.wikipedia.org/wiki/Haedong_Yonggungsa
16. Wikipedia Gamcheon Culture Village. https://en.wikipedia.org/wiki/Gamcheon_Culture_Village
17. Wikipedia Beomeosa. https://en.wikipedia.org/wiki/Beomeosa

A New Image Encryption Technique Built on a TPM-Based Secret Key Generation

Pallavi Kulkarni, Rajashri Khanai, and Gururaj Bindagi

Abstract Present communication trend involves the exchange of information using multimedia data. Images are predominantly used as a primary source of information. The evolution of new technology brings the burden of an enhanced threat landscape. Increasing data breach incidences emphasize the need for a robust and reliable mechanism to secure the information. In our work, we propose a new image encryption algorithm that uses a shared secret key to generate an actual key of encryption. The secret key used in symmetric key encryption can be generated using a neural network. The exchange of a secret key over public channels built on neural synchronization using different learning rules provides an acceptable substitute to number theory-based cryptography algorithms. Neural networks with tree topologies are classified as the most secure and reliable network models. In the neural key exchange protocol, the two tree parity machines synchronize homogeneous weights and these weights are used to establish a key exchange protocol over a public channel for the secure transmission of confidential information. To improve the security of keys generated using neural key exchange protocol, in our work we propose to use two keys: one is a shared key between two communicating entities and the second one is generated using the first key. The generated new key is used for image encryption through the AES algorithm. To understand the strength of the new key and to evaluate the quality of encryption, we have carried out statistical and differential attack analyses. The comparative study shows that the proposed method accomplishes improved results in terms of efficiency and security and helps us achieve better integrity and privacy.

Keywords Tree Parity Machine (TPM) · Secret key · Image encryption · AES · Security

P. Kulkarni (✉) · R. Khanai
Department of Electronics and Communication Engineering, KLE Dr. MSSCET, Belgaum, India
e-mail: pallavik15@gmail.com

G. Bindagi
Accenture, Bengaluru, India

© The Author(s), under exclusive license to Springer Nature Singapore Pte Ltd. 2023
S. Shakya et al. (eds.), *Sentiment Analysis and Deep Learning*,
Advances in Intelligent Systems and Computing 1432,
https://doi.org/10.1007/978-981-19-5443-6_2

13

1 Introduction

Cryptography is the art of constructing and analyzing protocols to secure the data that is being shared between two or more entities. This involves secure communication, controlling the existence or effect of third parties and averting leakage of information to unauthorized parties. It deals with many facets of data safety, e.g., data confidentiality/integrity and authentication. These schemes need to be continually improved as there is a constant advancement in computation techniques and the algorithms involved are gearing up their speed. Encryption, Steganography, and Watermarking techniques are widely used to secure the data/multimedia, among which encryption is the most efficient and common method used. Encryption will control the illegal access of data. For secure, robust, and effective encryption, we need a strong key. But sometimes, the key itself can become the weakest link with respect to security.

Since the secure channel has an extensive range of applications, the need for fast, powerful, and reliable transmission protocols is in demand [6]. Diffie-Hellman key exchange protocol is proposed in the 1970s, which is susceptible to a man-in-the-middle attack [7]. The capacity to construct a reliable channel is one of the toughest tasks in modern communication. Subsequently, various techniques are proposed based on number theory, which needs a great amount of computational power. To overcome these disadvantages, several other concepts and techniques have been explored. Among them, the concept of neural networks can be considered to generate a symmetric key, and this can offer one of several possible solutions to this critical issue of key exchange.

An artificial neural network has an input, hidden, and output layer as shown in Fig. 1. The weight vectors connect the input and hidden layer. Based on this concept, various methods are developed. One of the techniques is the Tree Parity Machine (TPM) [2, 5] where we can generate a secret key using one of the following rules: Hebbian Rule, Anti Hebbian Rule, and Random Walk.

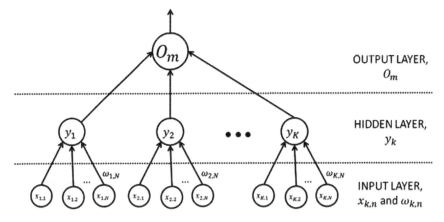

Fig. 1 The architecture of neural network

1.1 Neural Network and Tree Parity Machine (TPM)

A new method of computation is developed known as an artificial neural network which is greatly influenced by the human nervous system. The fundamental computing element is known as a neuron and has a similar kind of functionality as our nervous system. The concept of a neural network is applied in cryptography. A Tree Parity Machine, known as TPM, is a special type of feed-forward neural network [3]. As the information moves only in the forward direction, it is known as a feed-forward network. In such kinds of networks, there is no loop formation between the units.

X_{ij} is the input vector that can take the values from 1 to N, defining the input layer of TPM [10, 11].

Input and hidden layers are connected by the weight edges in a span of $-L$ to L. Sigma (σ) denotes the output of the hidden layer, expressed as in (1).

$$\sigma i = sgn \sum_{j=1}^{n} Wij * Xij \qquad (1)$$

Sgn(x) is known as signum function which returns the values $\{-1, 0, 1\}$, i.e., output of hidden layer neuron can take the values of -1, 0, and 1 as calculated by (2).

$$sgn(x) = \{-1 \text{ if } x < 0,\ 0 \text{ if } x = 0,\ 1 \text{ if } x > 1\} \qquad (2)$$

Neural network output is binary and denoted by tau (τ). It is calculated by using (3) as given below:

$$\tau = \prod_{k}^{i=1} \sigma i \qquad (3)$$

Though the TPM-based key generation is considered to be strong as compared to the traditional method, reference [4] suggests four different kinds of attacks that it is susceptible to—Brute force attack, Genetic algorithm for weight prediction, Man-in-the-middle interception attack, and Sign of weight classification using neural networks. Though the key generated using TPM is random in nature but is vulnerable to the above-said attacks. To enhance the security of the key generated by TPM, we propose a new method in which the actual key used for encryption is generated by the key exchanged between two parties. Initially, the secret key is shared between two communicating entities, and this key is given as input to the TPM. If at all the shared secret key is compromised, it is impossible to get the new key generated by TPM. In our work, we have used the new key generated by TPM for encrypting the images using the AES algorithm and carried out a detailed (SWOT) analysis.

2   Methodology

Encryption is the technique of transforming information into an incomprehensible form in order to offer security to sensitive data. Symmetric key encryption is the most widely used encryption scheme where the same key is used for both encryption and decryption. The one who possesses the "key" can decrypt the message successfully. Important security issues related to symmetric key encryption are:

1. Sharing the secret key between two parties.
2. Keeping the original key secure from the intruder so that there is no compromise in the communication.

Our research is focused on securing the key from the intruder. To overcome the above-mentioned security issues and to further enhance security, we propose a new key generation technique using TPM.

A flowchart of the proposed key generation technique is shown in Fig. 2. Key generation using neural key exchange protocol from the shared secret key is the differentiating novel factor as compared to other methods. Initially, both sender and receiver share the 128-bit key. In the proposed technique, a shared secret key of 128-bits is given as input to the Neural Key Exchange Protocol. Details of key generation using traditional Neural key exchange protocol and proposed technique are given in Sects. 2.1 and 2.2, respectively. We use the AES-128 for image encryption.

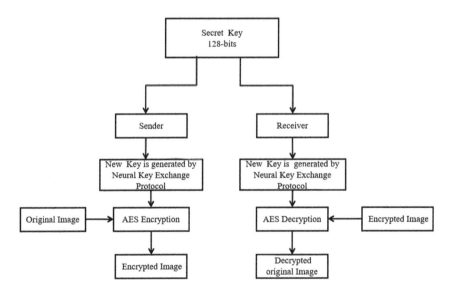

Fig. 2 Proposed methodology

2.1 Traditional Method of Generating Secret Key Using Tree Parity Machine

The parameters $I[n]$, H, M, and L are defined by sender and receiver.

I The input vector of size n.
H The number of hidden neurons.
M The number of input neurons connected to hidden neurons.
L Defines the range of each weight $\{-L, ..., 0, ..., +L\}$.
     Weights $W_{ij} = \{-L, ..., 0, ..., L\}$.

The steps involved in the secret key generation based on traditional TPM can be detailed below [9, 12, 13]:

1: Consider two Tree Parity Machines, TPM-X, and TPM-Y.
2: Initialize the network weights randomly in a range of $\{-L ..., 0, ..., L\}$.
3: Initialize the desired input vector on the input node.
4: Calculate sigma (σ) using the value of inputs and weights as given in Eq. (1).
5: Find the output tau of both TPMs using Eq. (3).
6: Compare the outputs of TPM-X and TPM-Y. If it is the same then go to Step 7 if not go to Step 2.
7: Update the weights of both TPMs using the Hebbian rule given in Algorithm 1.

Algorithm 1 Algorithm for updating the weights using the Hebbian rule

$$\text{Input: } \{W, I, , \tau_{1,} \tau_{2,1}\}$$
$$\text{for each } (i, j) \text{ in W}$$
$$\text{do}$$
$$W_{i,j} \Leftarrow W_{i,j} + I_{i,j} * \tau_1 * \theta(\sigma_i, \tau_1) * \theta(\tau_2, \tau_1)$$
$$W_{i,j} \Leftarrow f_{clip}(W_{i,j})$$
$$\text{end for}$$

$$\text{where, } \begin{cases} \theta = & 1 \text{ if } \tau_1 = \tau_2 \\ & 0 \text{ otherwise} \end{cases}$$

$$f_{clin} = \begin{cases} L & \text{if } W_{i,j} > L \\ -L & \text{if } W_{i,j} < -L \end{cases}$$

8: Continue Step 7 till weights on both the sides are same. When both TPMs synchronize to the same synaptic weight, that final weight is taken as a secret key.

2.2 Proposed Rule with Algorithm

In the proposed method, we use the secret key shared between two parties as to the input to the Tree Parity machine. This makes the key generation process more robust.
 The following section explains the steps involved in the proposed algorithm:

1: Let key1 be the key shared between two parties and key2 is the new key generated by the proposed algorithm.

$$key2 \rightarrow func(key1)$$

2: Define the structure of NN by choosing the number of neurons in the input and hidden layer.
    Basically, the structure of the neural network is defined by the length of the key shared between two parties. For example, if we take a 16-byte key, i.e., a 128-bits key we can have:
    64/32/16 neurons in the input layer, 2/4/8/ neurons in the hidden layer, and 1 neuron in the output layer, respectively.
3: Consider two Tree Parity Machines (TPM) to say, TPM-X and TPM-Y. Both TPMs need to agree on the structure of NN based on key1.
4: Get the ASCII equivalent of key1.
5: Apply the confusion and diffusion method on the ASCII values of key1 to get modified key1. This makes the relation between key1 and key2 more complex.
6: Expand modified key1 to the required key length (128,192, or 256-bits).
7: We use this modified key1 to build the random input vector of TPM as well as to initialize the random weights between the input and hidden layer.
8: Initialize the desired input vector on the input node.
9; Find the value of a hidden layer, i.e., sigma, using the value of inputs and weights as given in Eq. (1).
10: Find the output tao of both TPMs using Eq. (3).
11: Compare the outputs of TPM-X and TPM-Y. If it is the same, then go to Step 7; if not, go to Step 2.
12: Update the weights of both TPMs using the Hebbian rule given in Algorithm 1.
13: Continue Step 7 till weights on both the sides are same. When both TPMs synchronize to the same synaptic weight, that final weight is taken as the secret key.

3 Results and Discussion

In this segment, we discuss the experimental results which are divided into two categories—Performance parameters and Security parameters which are explained in detail in the following section. All the experiments are performed on Intel core i3 CPU 1.8 GHz, 4 GB RAM machine using MATLAB 2019b. For experimental analysis, we have used grayscale images with dimensions 128 × 128 pixels.

Table 1 Comparison of performance parameters

Parameter	Pepper		Camera man		Coins	
	AES	AES with NN key	AES	AES with NN key	AES	AES with NN key
Execution time(s)	8.84	9.04	8.94	8.85	8.85	8.85
CPU usage (%)	0.1004	0.1018	0.10126	0.0999	0.10026	0.10009
Throughput (Bytes/s)	128.68	125.84	114.04	115.08	114.78	114.79

3.1 Performance Parameters

The Performance parameters help us to analyze the performance of the proposed algorithm in terms of Execution time, CPU utilization, and Throughput. Execution Time, measured in seconds, is the measured CPU time required to execute the encryption algorithm. CPU Utilization shows Percentage processor utilization. Throughput is the number of bytes encrypted per second. Table 1 gives us the comparison of performance parameters.

3.2 Security Parameters

An efficient encryption technique should withstand all kinds of attacks. In our work, we have carried out statistical and differential attack analyses [8, 14]. The statistical analysis primarily contains histogram analysis, chi-square test, and adjacent pixel correlation analysis. The differential analysis mainly includes finding NPCR, UACI values, and information entropy analysis.

The analysis of the adjacent pixel's correlation

In an ordinary image, there is a high correlation between neighboring pixels. One of the purposes of encryption is to reduce the association between two pixels. A reduced correlation value suggests an improved encryption effect and better security [4]. The correlation coefficient is calculated using the equations given below:

$$E(x) = \frac{1}{N} \sum_{i=1}^{N} x_i \tag{4}$$

$$\text{cov}(x, y) = \frac{1}{N} \sum_{i=1}^{N} (x_i - E(x))(y_i - E(y)) \tag{5}$$

Table 2 Correlation values in a horizontal direction

Correlation coeff	Pepper	Camera man	Coins
AES	0.027145	0.004167	−0.01596
Ref. [1]	0.002	0.0002267	NA
Proposed algorithm	−0.0034225	−0.0248105	0.00918835

$$r(x, y) = \frac{cov(x, y)}{\left[\sqrt{D(x)}\sqrt{D(y)}\right]} \tag{6}$$

$$D(x) = \frac{1}{N} \sum_{i=1}^{N} (x_i - E(x))^2 \tag{7}$$

where

x and y values of the neighboring pixels
N Number of selected neighboring pixels.

The correlation coefficient can take the values between $+1$ to -1. Maximum correlation is reached when the value is $+1$, which indicates the image can be easily recognized. A good encryption algorithm should provide a very low /close to zero/ a negative value so that it is hard for the intruder to get the information. The correlation coefficient is calculated in a horizontal direction between pixels and is tabulated in Table 2.

Histogram analysis of the encrypted image

A histogram of pre- and post-encryption images is shown in Table 3. We notice that the histogram of the ciphered image is even and notably varies from that of the original. This makes the statistical analysis difficult. From the table below, it is observed that the original image histogram presents a single or few peaks.

Information entropy analysis

Information entropy measures the randomness of a source. The information entropy is symbolized as $H(X)$ and given by the equation:

$$H(x) = \sum_{i=1}^{n} p(x_i)\log_2 p(x_i) \tag{8}$$

where $p(xi)$ is the probability of occurrence of the pixel. The maximum value is entropy is 8. Powerful encryption technique results in high entropy values and can successfully resist the decipher attacks. Consequently, the proposed technique is assessed against entropy measure and results are tabulated in Table 4 and compared with that of [1] and with standard AES.

Table 3 Histogram analysis

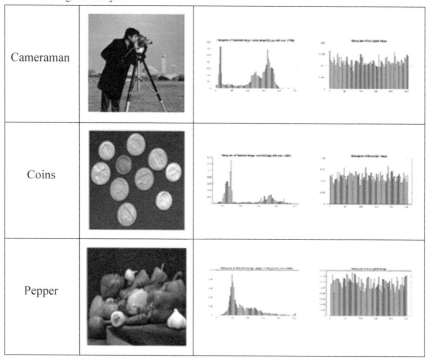

Cameraman		
Coins		
Pepper		

Table 4 The value of information entropy

Entropy	Pepper	Camera man	Coins
AES	7.45	7.89	7.62
Ref. [1]	7.9972	7.9971	NA
Proposed algorithm	7.915	7.925	7.992

Key sensitivity analysis

Key sensitivity analysis is highly conclusive for any cryptosystem. It is required to have a key that produces a completely different ciphered image by a slight variation in the key. The scenario of the key sensitivity analysis is shown in Fig. 3 as follows:

1. The original image I is encrypted with two different keys.
2. Key1 is the original key. Key2 differs Key1 only by single bit.
3. A new key is generated using Neural key exchange protocol.
4. I is encrypted using Key1 to obtain C1.
5. I is encrypted using new key Key2 to obtain C2.
6. Compare both encrypted images bit by bit.

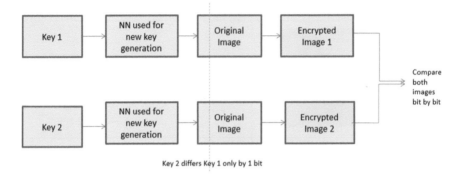

Key 2 differs Key 1 only by 1 bit

Fig. 3 Key sensitivity analysis (Key 2 differs Key 1 by a single bit)

Table 5 Number of bits changed after 1-bit change in key

Number of bits changed	Pepper	Camera man	Coins
AES	4098	4289	4647
Proposed algorithm	16,266	16,519	16,274

Referring to Table 5, we can conclude that a single bit change in the key is producing a huge change in the cipher image. This is also supported by the results in Table 6.

Hit Collision Analysis

An ideal image encryption technique should be sensitive to the secret key. The communication between two parties is negotiated if the key falls into wrong hands. The attacker may get the original key by a trial-and-error method by changing the bits in the key. In this type of attack, the attacker's goal is to get the secret key. "Hit" is the number of bits mapping between the two keys after changing the 1-bit, 2-bits, 3-bits, and so on. Hit collision is the parameter that gives the hit between the two keys, and it should be as low as possible to sustain a brute force attack. Hit collision analysis is carried out in Fig. 4 as follows:

1. Key2 differs Key1 only by single bit.
2. New keys (Key3 and Key4) are generated using Neural key exchange protocol.
3. Key3 and key4 are compared which gives number of hits (i.e., number of bits matched between Key3 and Key4).
4. Hit collision is calculated as **Number of hits/ Image Length**.

Results tabulated in Table 7 suggest that proposed method is more resilient to attacks as hit collision is small as compared to traditional AES.

Algorithm's sensitivity analysis to the original image

The algorithm's sensitivity analysis is carried out by finding the number of pixel changing rates (NPCR) and the unified average changing intensity (UACI). C1 and

Table 6 Key sensitivity analysis by comparing encrypted images

Cameraman		Encrypted Image1	Encrypted Image2 (1bit change)
Coins		Encrypted Image1	Encrypted Image2 (1bit change)
Pepper		Encrypted Image1	Encrypted Image2 (1bit change)

Note Encrypted image1 is of the original image and encrypted image 2 is obtained after a 1-bit change in key

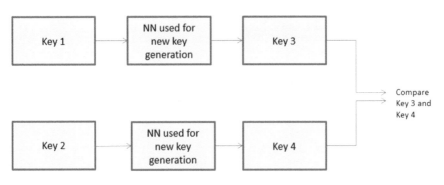

Key2 is generated by changing the bits in Key1

Fig. 4 Hit collision analysis (Key 2 differs Key 1 by a single bit)

Table 7 Hit collision analysis

Hit Collision	Pepper	Camera man	Coins
AES	0.776281	0.195423	0.5687938
Proposed algorithm	0.326954	0.109082	0.415958

Table 8 NPCR values

NPCR	Pepper	Camera man	Coins
Ref. [1]	99.6414	99.5697	NA
Proposed algorithm	99.8016	99.6243	99.4673

Table 9 UACI values

UACI	Pepper	Camera man	Coins
Ref. [1]	33.5864	33.4767	NA
Proposed algorithm	29.1362	33.4238	33.4901

C2 are the two encrypted images whose original images differ only by one bit. $D(i, j)$ is calculated as:

$$D(i, j) = \begin{cases} 1 & C_1(i, j) \neq C_2(i, j) \\ 0 & C_1(i, j) = C_2(i, j) \end{cases} \tag{9}$$

Pixel value in the (i, j) coordinate is denoted by $c1(i, j)$ and $c2(i, j)$.
Using (10), NPCR value is calculated as below:

$$\text{NPCR} = \frac{1}{x * y} \sum_{i,j} [D(i, j)] * 100 \tag{10}$$

x and y: The dimensions of the original image.
Equation (11) is used to find the UACI value:

$$\text{UACI} = \frac{1}{x * y} \sum_{i,j} (E_1(i, j) - E_2(i, j))/(2^1 - 1) \tag{11}$$

NPCR and UACI values are tabulated in Tables 8 and 9 and compared with the reference paper [1] values and with traditional AES. The NPCR and UACI values of the proposed algorithm are closer to the ideal value suggesting the better security.

4 Conclusion

Image encryption is a fast-growing research area due to the rapid growth in multimedia communication. Two key objectives of image encryption are encryption efficacy and robustness. A new image encryption technique is proposed in this paper by combining a TPM-based key generation approach and an AES algorithm. A new key is generated by TPM-based neural key exchange protocol which is different from the key shared between the two communicating parties. The value of the key changes dynamically when a user gives a different secret key as input to the TPM. This makes it impossible for the intruder to get the actual key that is used for image encryption. The experimental results demonstrate that the above-mentioned technique has several benefits: good key sensitivity, better correlation coefficient, uniform distribution of pixels, entropy proximity closer to ideal value, and sizable key.

The study conducted also suggests that the key generated by the neural key exchange protocol has the optimum behavior between Entropy, Correlation coefficient, Hit collision, and No. of bits changed in encrypted image after a single bit change in key.

References

1. Arab, A., Rostami, M. J., & Ghavami, B. (2019). An image encryption method based on chaos system and AES algorithm. *Journal of Supercomputing, 75*(10), 6663–6682. https://doi.org/10.1007/s11227-019-02878-7
2. Chakraborty, S., Dalal, J., Sarkar, B., & Mukherjee, D. (2015). *Neural synchronization based secret key exchange over public channels: A Survey*. arXiv preprint arXiv:1502.05153
3. Chourasia, S., Bharadwaj, H. C., Das, Q., Agarwal, K., & Lavanya, K. (2019). Vectorized neural key exchange using tree parity machine. *Compusoft, 8*(5), 3140–3145.
4. Gupta, M., Gupta, M., & Deshmukh, M. (2020). Single secret image sharing scheme using neural cryptography. In *Multimedia tools and applications* (pp. 1–22). https://doi.org/10.1007/s11042-019-08454-8
5. Jeong, S., Park, C., Hong, D., Seo, C., & Jho, N. (2021). Neural cryptography based on generalized tree parity machine for real-life systems. In: *Security and communication networks*. https://doi.org/10.1155/2021/6680782
6. Jogdand, R. M., & Bisalapur, S. S. (2011) Design of an efficient neural key generation. *International Journal of Artificial Intelligence & Applications (IJAIA), 2*(1), 60–69. https://doi.org/10.5121/ijaia.2011.2105
7. Kanter, I., Kinzel, W., & Kanter, E. (2002). Secure exchange of information by synchronization of neural networks. *EPL (Europhysics Letters), 57*(1), 141. https://doi.org/10.1209/epl/i2002-00552-9
8. Munir, R. (2012). Security analysis of selective image encryption algorithm based on chaos and CBC-like mode. In *2012 7th international conference on Telecommunication Systems, Services, and Applications (TSSA)* (pp. 142–146). IEEE. https://doi.org/10.1109/TSSA.2012.6366039
9. Pal, S. K., & Mishra, S. (2019). An TPM based approach for generation of secret key. *International Journal of Computer Network & Information Security, 11*(10). https://doi.org/10.5815/ijcnis.2019.10.06
10. Rosen-Zvi, M., Klein, E., Kanter, I., & Kinzel, W. (2002). Mutual learning in a tree parity machine and its application to cryptography. *Physical Review E, 66*(6), 066135. https://doi.org/10.1103/PhysRevE.66.066135
11. Salguero, D., Édgar, W. F., & Lascano, E. (2019). On the development of an optimal structure of tree parity machine for the establishment of a cryptographic key. In *Security and communication networks*. https://doi.org/10.1155/2019/8214681
12. Sarkar, A., & Mandal, J. K. (2015). Comparative analysis of tree parity machine and double hidden layer perceptron based session key exchange in wireless communication. In *Emerging ICT for bridging the future-proceedings of the 49th annual convention of the Computer Society of India (CSI)* (Vol. 1, pp. 53–61). Springer.https://doi.org/10.1007/978-3-319-13728-5_6

13. Stypiński, M., & Niemiec, M. (2011). *Synchronization of tree parity machines using non-binary input vectors.* arXiv preprint arXiv:2104.11105
14. Zeghid, M., Machhout, M., Khriji, L., Baganne, A., & Tourki, R. (2007). A modified AES based algorithm for image encryption. *International Journal of Computer Science and Engineering, 1*(1), 70–75.

Application Prototypes for Human to Computer Interactions

N. Soujanya, G. Sourabha, C. B. Jeevitha, and M. R. Pooja

Abstract In this growing era of human–computer interaction (HCI), the interaction between human and computer has a higher scope in automated food service industry. This research study attempts to design and develop a desktop application prototype for a vending machine that utilizes an individual's basic human features like age and gender. The proposed model is built by using the technologies such as python, computer vision, and Convolution Neural Network (CNN) to greet a customer who uses the vending machine. The primary objective of the proposed research study is to welcome the customers to the automated food vending machine. The proposed model will assist in enhancing the mental health of the customers. The proposed model can also be preferred for specially challenged customers, who would like to be at the receiving end of an automated food service industry.

Keywords Convolution neural network · Deep learning · Food vending machine · Image processing · Automation · Human · Computer interaction

1 Introduction

Human–computer interaction bridges the communication gap between customers and food servicing industries by improving the user experience. Increasing the user experience would lead to increase in the company's economy. Since there is a significant decrease in conversational exchange, this idea will improve the user's mental health by establishing a pleasant environment for the user.

This paper discusses about the technologies like Convolution Neural Networks (CNN), deep learning, and python. The ideologies with which this work will be built are based on the importance given to mental health of the customers and for the specially challenged customers to obtain a more easily understandable translation of the products that remain as a part of the industry.

N. Soujanya (✉) · G. Sourabha · C. B. Jeevitha · M. R. Pooja
Department of Computer Science and Engineering, Vidyavardhaka College of Engineering, Mysuru, India
e-mail: soujanya2016@gmail.com

© The Author(s), under exclusive license to Springer Nature Singapore Pte Ltd. 2023 27
S. Shakya et al. (eds.), *Sentiment Analysis and Deep Learning*,
Advances in Intelligent Systems and Computing 1432,
https://doi.org/10.1007/978-981-19-5443-6_3

Age and gender classification is an inherently challenging problem, and this remains as the main reason for this discrepancy in the complex nature of the data that is needed to train.

Here, images are used for training the datasets with age and gender labels. For such images, they need personal information of the subject present in the images so this research work incorporated the CNN technique. The main goal of the proposed system is to classify the age and gender based on the facial images.

2 Related Work

2.1 Methodology

The proposed system deals with the recognition of basic features of an individual like age and gender. This system performs facial recognition of the individual by using computer vision. The proposed system is a tool that is presently a desktop application that takes individual's facial image as an input, which is then processed by the system to give an output in the form of speech delivered via speakers present in the proposed system. This will be implemented on food vending machines to greet the customers.

The team working with the proposed system will be training the machine to understand and differentiate the facial images that allow the proposed system to recognize the individual, who is present at the situation. We will use data to train the system to learn the societal differences that an individual yields to.

This proposed system uses CNN or Convolutional Neural Networks. CNN is used to convert images into a form, which is obtained by reducing the image and is easier to compute without any loss of image characteristics, which are important for obtaining a good indicator and then the image itself can be a very complex process. The importance of this technology is that it can be used with a huge amount of datasets and is also acceptable when it comes to learning indicators or features.

The random samples of individuals of different age groups with different facial structures and gender are used. This will provide a vast area of data that is usually in the form of an image and later converted to a form, which is easy to compute without the loss of the image characteristics that is crucial for determining an individual's age and gender range bracket.

Once the age and gender of an individual are recognized by matching the criteria that were used to train the proposed system, then it will be made sure that the recognition happens as accurately as possible and we accept the system to be performing as expected; if in case there happens to be a lot of difference in the expected output and the output that is obtained, we will train the system with more vast data. Once we accept our system performance, we will implement it in a desktop as a prototype and test it with real-time samples. The output that is obtained will be in the form of speech. The output expected from our system happens to be recognizing a customer and greeting them accordingly.

Our team has taken privacy also into consideration. Our proposed system does not store images that are used to analyze customers because we believe in their privacy. We also want to increase human–computer interaction as in today's world we have lost a basic touch of communication of greeting one another face to face. This proposed system aims to keep the face-to-face greeting amidst the changing world.

2.1.1 CNN

A neural network is a system of interconnected artificial neurons that exchange messages between each other. The connections have numeric weights that will come during the training process, so that a properly trained network will help to respond correctly when presented with an image or pattern to recognize. CNN is used in variety of fields including image and pattern recognition, speech recognition, natural language processing, and video analysis. Why convolutional neural networks are important means among the traditional models for model identification, feature extractors are hand crafted. CNN provides better performance for applications with input local correlation (e.g., image and speech). In prototype and image recognition applications, the best possible identification rates are achieved by using CNN.

2.2 Literature Review

The most important part of this proposed system is to recognize a person's age and gender and greet them accordingly. Smart vending machines have been an evolving technology that is being worked on by famous companies like Coca-Cola and so on. The technology that this company used to compete in this industry is AI, where they used stored information to give discounts on what a customer purchase [1]. Age and gender recognition two systems are tested by using the common voice and Korean speech recognition. We can predict that recognizing the age gender speech signal is the best way to perform human-computer interaction, as shown in Fig. 1 [2]. Replacing the machine in the place of human and when the human comes in front of the machine, the machine understands their needs by their voice [3]. The gender is detected by using a voice to find a gender by a voice signal. There are some techniques that were implemented to detect gender from voice signals, many voice-recognition applications on the information contained in human speech, the important aspect of a voice's detection is the gender voice, and a collection of techniques has been used to find gender from a voice signal. The gender (male or female) of a vocal signal can be determined using this approach. The contributions are of three parts: (a) a well-known dataset to analyze information about well-known voice signal characteristics, (b) educating different machine learning models from

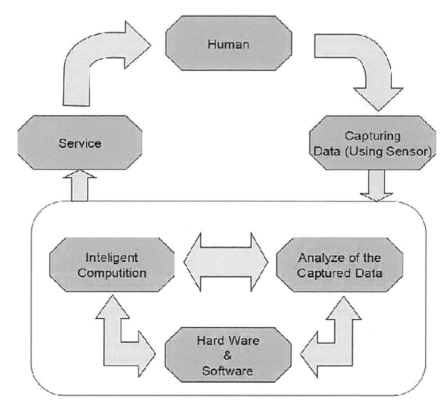

Fig. 1 Post-desktop human–computer interaction

various theoretical samples to classify voice gender, and (c) making use of three prominent feature selection algorithms [4]. The estimation of age algorithm works on the principle of multi-labeled sorting technique by using facial images, and it changes the method of multi-valued differentiation. Solution of a tedious step to detect age is found. By multi-label sorting learning, they find many fields have found success, it has the ability to communicate not just the complicated semantic content of learning objects, this provides a method for estimating the age of facial photographs, and this algorithm of estimation replaces the traditional multi-valued classification method in the absence of a facial age dataset, simplifies the problem of tedious steps to estimate age, and reduces model training time. The method has produced extremely good results in evaluating parameters such as mean absolute error, cumulative score, and convergence rate, according to a chain of experiments conducted on datasets of two age. The efficiency of this algorithm is superior to some traditional age estimation algorithms [5]. Two neural architectures are used to find gender and speaker recognized by Mel-frequency campestral coefficients, and MLP and CNN these two architectures are used to find the gender and speaker. To evaluate neural architectures of two for detection of gender and speaker identity tasks, one of

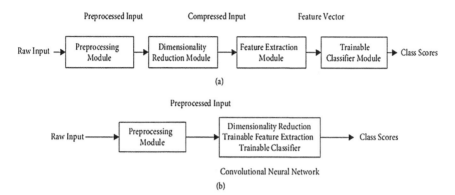

Fig. 2 Pattern recognition approaches: **a** conventional, **b** CNN-based [10]

the objectives is to differentiate neural architectures such as multi-layers and CNN, for both tasks in contexts to automatically learned gender and speaker properties. The results show that models that employ z-score and Grampian transformation of matrix produce finer outcomes than models that solely utilize MFCC's maximum-minimum normalization. MLP takes a larger understanding period to converge than CNN in terms of training time. In terms of generalization mistakes, further experimental data suggest that MLPs outperform CNNs in both tasks [6]. Deep CNN technique detects the gender of human beings from their facial images. In this case, VGGNet face network has been used for gender recognition, Online social networking website and social media, automatic gender detection is increasingly being used in a broader range of software and hardware, and the performance of the existing system is being used.

Physical world face photographs due to the proliferation of social networking websites online and social media, automatic gender detection is increasingly being used in a wider range of software and hardware. However, the performance of the existing system using physical world face photographs and images is not particularly good, especially when compared to the job outcome. Face recognition is linked. We have discovered in this research that learning and categorizing are two different things in terms of method and that Deep Convolutional Neural Networks (DCN) are used [7]. The human gender is similar to facial recognition, and by using CNN technique, the human gender is detected from images as in Fig. 2. Detection of human gender is a part of facial recognition that has gained a lot of attention; gender detection has previously been performed using the same characteristics such as eyebrows, hand shape, and face. Because of its various applications, human gender detection, which is a part of facial recognition, has gained a lot of attention. Gender detection has previously been performed using static bodily features such as the face, body shape, eyebrows, hand shape, fingernail, and so on. We have done a lot of research in this project and demonstrated human gender classification from human face photos using Convolution Neural Network (CNN) as CNN has been recognized as in the realm of picture classification, and this is the best algorithm. To put our method into action, we used a pre-processing procedure on every image. Processing of image is used. The image is pre-processed and then fed through the Convolution [8]. By using the speech signals, the speaker gender has identified; to identify the gender, the task has been used; it is automatic gender recognition. The task of detecting a speaker's gender from a voice signal is known as Automatic Gender Recognition, and this approach trains a binary classifier using parameters like frequency and campestral data extracted from the speech input. Here, the system listens to unprocessed audio as an input signal from beginning to conclusion. We present a gender classifier based on convolutional neural networks (CNNs), which includes: (i) convolution layers, which can be understood as feature learning and (ii) a multilayer perceptron (MLP). The system listens to unprocessed audio as an input signal [9]. Gender classification using a CNN is for real-time gender classification based on face images. Back propagation algorithm has been used, real-time gender classification is based on facial images, CNN technique was proposed and when we compare to other CNN methods which were used in pattern recognition, they combine the convolutional and subsampling layers. Neural networks are used to train the network, and this reduces the computational effort. The process of identifying patterns using a machine learning algorithm is called pattern recognition. Pattern recognition is the use of computer algorithms to identify data sequences and patterns. This type of detection can be done on a variety of input types such as biometric recognition, image recognition, and facial recognition. It has been applied in various fields such as image analysis, computer vision, health care, and seismic analysis. It has great accuracy in detecting patterns and to detect unknown objects. The sample identification system can detect partially hidden patterns. Pattern recognition is achieved by using the concept of learning. The pattern recognition model allows identification systems to become adaptable to train and provide more accurate results. Pattern recognition is used in digital image analysis to study images to gather needed information from images.

A pattern recognition model is used in the field of health care to improve health services, and data is stored and used by medical professionals for future analysis. Model recognition has the potential to transform into a more intelligent process that supports various digital technologies. An important advantage of CNN over conventional approaches in pattern identification is that ability to simultaneously capture features, reduce data size, and classify in one network structure [10]. Human gender identification systems using speech signals can easily identify the gender of persons by their speech [11]. Speaker recognition system implemented by using deep neural networks and CNN and determining the age of a speaker and age can be inferred from speech signal. The conversational speech recognition systems have with the growth such as Alexa and SIKI and with natural languages communication systems such as chat bot and voice-recognition systems are at an all-time high to find speaker age and inputs. It helps in speaker age classification using in-depth practice methods. A new technique is mentioned with the implementation model, deep neural network, and convolutional neural network [12]. Gender and age are detected from images by using convolutional neural networks. To categorize the age and gender, facial images are used. In this research, the challenge is to find gender and age classification for images which has built efficiently to improve results by expanding their tactics and

Fig. 3 A woman at a smart vending machine in Shanghai scans her palm (Photo by Daisuke Hiroshima)

also results in system performance. The challenge of gender and age classification for images is the subject of this research. The first major topic of research in this research is changing some of the most effective gender and age classification structures that have previously been published. By reducing the number of parameters in the network depth and that will change the dropout amount. As a result of these changes, system performance has suffered or has worsened. In comparison to the basic architecture I started with, at best, stay the same this confirmed my fears, age-related duties [13]. Combined dataset that consists of TIMIT, RAVDESS, and BGC datasets is used to find the gender and region of an individual that along with the combined datasets use the multi-output-based 1D CNN [14]. Images from a webcam are used to detect the age and gender with the help of technologies like PCA, LBP, SVM, VIOLA-JONES, and HOG that help in finding the emotions along hand with age and gender of an individual in various circumstances [15]. To detect the gender information of a singer. The recent popular convolutional neural networks (CNNs) with the support of spectrograms to obtain better accuracy over the traditional feature vector [16]. Hand-written Digit Recognition using Deep learning methods has been implemented also to implement a classification algorithm to recognize the written digits [17]. The design of a vending machine as shown in Fig. 3 uses fingerprint sensor for security with a rich UI that displays the product list. This type of machine is used for customization according to the user (institution) [18]. Convolutional neural network (CNN) model for classification of gender is based on images of fingerprint. The model determines the gender of the images from the fingerprint dataset [19]. The reduced speech bandwidth in this transmission system has an effect on the accuracy of speech emotion recognition. The impact of restricted speech bandwidth and transmission systems is low and compounding procedures on the accuracy of speech emotion recognition are investigated in the research. Speech (EMO-DB) data contain seven categorical emotions. The baseline technique achieved an average accuracy of 82%. The SER accuracy was reduced by roughly 3.3% when the sampling frequency was reduced from the baseline 16–8 kHz (i.e., bandwidth was reduced from 8 to 4 kHz, respectively). The companding process reduced average accuracy by 3.8% on its own, and the combined effect of companding while band reduction dropped average accuracy by 3.8% [20]. Fine-grained age estimation happens from a single face image in the human–computer interaction it conducts on audience and age regression performed by deep exception algorithm. Age estimate (calculation) of face photos in the wild has a low accuracy and which ignoring fine-grained features of age, this offers a unique method for fine-grained age estimate in the wild to the long-term memory network and this is motivated by the fine-grained categories [21]. Deep learning algorithms are used in recognition tasks and gender recognition by images when they proceed by protection filters. Deep learning-based algorithms have improved their performance in recognition and detection tasks. When trained with datasets, recent achievements have led to speculation that can identify and recognize object, and deep learning-based algorithms have improved their performance in recognition and detection tasks, particularly when trained on huge datasets. Such recent achievement has led to speculation that deep learning approaches can identify and recognize objects as well as, if not better than, the human visual system as well as their characteristics. This study concentrates on the task of gender recognition in photographs that have been altered [22]. General working of automated vending machines uses the coin as user input, and based on the amount (value of coin), the food product is given to the user(customer). This uses a microcontroller which does the basic computation of money to product mapping and gives out the product using mechanical components [23].

3   Conclusion

This paper has successfully provided a deep knowledge on the different real-time application prototypes of HCI by reviewing the existing research literature. The gained knowledge is further utilized to create a new system for the customers of the food servicing industry. The publications cited and considered illustrate that the concept of human–computer interaction begins with the consumer, which has been successfully evaluated in the proposed study effort. The proposed system is intended for the computer to start the conversation with a customer by greeting them as it is the best way to start any conversation in any servicing industry. The proposed system also works on another important aspect as it will allow specially challenged customers to enjoy the experience of a normal customer by making it a better environment for both the customers and the service industry.

References

1. De Bogotá. (2020). Cámara de Comercio. "Observatorio de Innovación–Estrategias y tácticas de marketing.
2. Ishaq, M., Son, G., & Kwon, S. *Utterance-level speech emotion recognition using parallel convolutional neural network with self-attention module.*
3. Sato, K., et al. (2001). Voice control of a tractor using vowel pitch characteristics. *Journal of the Japanese Society of Agricultural Machinery, 63*(1), 35–40.
4. Alkhawaldeh, R. S. (2019). DGR: gender recognition of human speech using one-dimensional conventional neural networks. In *Scientific Programming 2019.*
5. Zhu, Z., et al. (2018). Age estimation algorithm of facial images based on multi-label sorting. *EURASIP Journal on Image and Video Processing, 1*, 1–10.
6. Mamyrbayev, O., et al. (2020). Neural architectures for gender detection and speaker identification. *Cogent Engineering, 7*(1), 1727168.
7. Dhomne, A., Kumar, R., & Bhan, V. (2018). Gender recognition through face using deep learning. *Procedia Computer Science, 132*, 2–10.
8. Sumi, T. A., et al. (2021) Human gender detection from facial images using convolution neural networks. In *International conference on applied intelligence and informatics.* Springer.
9. Kabil, S. H., Muckenhirn, H., & Magimai-Doss, M. (2018). *On learning to identify genders from raw speech signal using CNNs.* Interspeech.
10. Liew, S. S., et al. (2016). Gender classification: A convolutional neural network approach. *Turkish Journal of Electrical Engineering & Computer Sciences, 24*(3), 1248–1264.
11. Djemili, R., Bourouba, H., & Korba, M. C. A. (2012) A speech signal based gender identification system using four classifiers. In *2012 international conference on multimedia computing and systems*, IEEE.
12. Kuppusamy, K., & Eswaran, C. (2020). *Speaker recognition system based on age-related features using convolutional and deep neural networks.*
13. Ekmekji, A. (2016). *Convolutional neural networks for age and gender classification.* Stanford University.
14. Uddin, M. A., et al. (2021). Gender and region detection from human voice using the three-layer feature extraction method with 1D CNN. *Journal of Information and Telecommunication*, 1–16.
15. Sharma, J., Dibya, S. D., & Bora, D. J. (2020). *REGA: Real-time emotion, gender, age detection using CNN—A review.*
16. Murthy, Y. V. S., Koolagudi, S. G., & Jeshventh Raja, T. K. (2021). Singer identification for Indian singers using convolutional neural networks. *International Journal of Speech Technology*, 1–16.
17. Alwzwazy, H. A., et al. (2016). Handwritten digit recognition using convolutional neural networks. *International Journal of Innovative Research in Computer and Communication Engineering, 4*(2), 1101–1106.
18. Sibanda, V., et al. (2020). Design of a high-tech vending machine. *Procedia CIRP, 91*, 678–683.
19. Jayakala, G. (2021). Gender classification based on fingerprint analysis. *Turkish Journal of Computer and Mathematics Education (TURCOMAT), 12*(10), 1249–1256.
20. Lech, M., et al. (2020). Real-time speech emotion recognition using a pre-trained image classification network: Effects of bandwidth reduction and companding. *Frontiers in Computer Science, 2*, 14.
21. Zhang, K., et al. (2019). Fine-grained age estimation in the wild with attention to LSTM networks. *IEEE Transactions on Circuits and Systems for Video Technology, 30*(9), 3140–3152.

22. Ruchaud, N., et al. (2015). The impact of privacy protection filters on gender recognition. In *Applications of digital image processing XXXVIII* (Vol. 9599). International Society for Optics and Photonics.
23. Zhang, W., & Zhang, X. L. (2010). Design and implementation of automatic vending machines based on the short massage payment. In *2010 6th international conference on Wireless Communications Networking and Mobile Computing (WiCOM)*. IEEE.

Feature Selection-Based Spam Detection System in SMS and Email Domain

S Aditya Chaturvedi and **Lalit Purohit**

Abstract Spam exists in several domains including SMS and Emails which are usually targeted by spammers to steal personal information, money, data, etc. There are several models exist for SMS and Email spam detection out of which supervised learning-based model is mostly efficient. However, the comprehensive study for spam detection with consideration to multiple domains simultaneously, is missing. In this paper an experimental evaluation on the effect of using feature selection techniques in the domain of SMS and Email spam detection is conducted. Parameters such as ROC and Train/Test time along with common parameters are used to evaluate performance of spam detection models. The experimentation results shows that the choice of feature selection technique has profound effect on the performance of the spam detection model and can be seen in the result generated using different evaluation measures out of which some are not used in the both domains previously.

Keywords Machine learning · Spam SMS detection · Spam Email detection · Feature selection

1 Introduction

Spam refers to an information (either SMS or Email) generated to result in loss of the receiver either as monetary/identity loss or data loss. Both Short Service Message (SMS) and Email domains are the major target of spammers. Every year plenty of user's loose their identity/money/data due to the spam SMS messages or spam Emails. SMS is the most common means for exchange of information without Internet [1]. It is a medium used for exchanging information for individual personal and busi-

S. A. Chaturvedi (✉) · L. Purohit
Department of Information Technology, Shri G S Institute of Technology and Science, Indore, Mathya Pradesh, India
e-mail: sheriadi00@gmail.com

L. Purohit
e-mail: purohit.lalit@gmail.com

© The Author(s), under exclusive license to Springer Nature Singapore Pte Ltd. 2023
S. Shakya et al. (eds.), *Sentiment Analysis and Deep Learning*,
Advances in Intelligent Systems and Computing 1432,
https://doi.org/10.1007/978-981-19-5443-6_4

ness purposes. One-time passwords, registration details, important information, etc. is exchanged using SMS by many public and private organizations including banks [2]. As text of SMS is limited to 160 characters, spammers send fake messages pretending to be the genuine organizations containing phishing links and payload along with the text which can result in identification /monetary /data loss [3]Application-to-Person (A2P) bulk messaging is one of the main reason of spamming in SMS domain as spammers use A2P bulk messaging majorly to spam the users [4]. Email as another most common and widely used medium for information exchange used by various organizations using Internet. Around 4147 million of users of around the globe use Email [5]. Since last decades it is also being used for spam. Spammers send fake Emails containing false information,phishing links and files which can result in loss of time, money and data too. With increasing threats due to spam SMS and Emails, many solutions are available to detect spam. In [1], supervised machine learning model-based algorithms are used for detecting spam in SMS domain. In this proposed work, NLP is used for text preprocessing and classification is done using the Naive Bayes (NB), Random Forest (RF) and Logistic Regression (LR) algorithms. Similarly, in [2] a comparative analysis of several machine learning algorithms to identify the most accurate model for SMS spam detection is done. A similar approach of using NLP for text preprocessing along with classification algorithm to build a model for SMS Spam detection is presented [3]. In [6], Natural Language Processing is used for preprocessing and out four machine learning techniques, best one is determined basis on evaluation measures. In [7], a comprehensive review for SMS spam detection based on deep learning methods is presented. The review work focuses on the application of Convolutional Neural Network(CNN) and Recurrent Neural Network. A hybrid approach using Particle Swarm Optimization (PSO) and Primal Estimated Sub-Gradient Solver for SVM (PEAGSOS) for spam detection on Email domain is available [5]. A Email filtering model based on integrated approach using Gray List filtration (GL) and NB for Email spam detection is also available [8]. A spam detection model for Email using NLP and NB and optimized using PSO is also effective approach [9]. A Hybrid model using Logistic Regression and PSO is presented in [10] for Email spam detection. Author presented result and compare them using accuracy evaluation measure with and without feature selection. A comprehensive review is presented in [11], where various research work available in Email spam detection using machine learning is reviewed.

The available solutions in the existing state-of-the arts for SMS and Email spam detection have some limitations. The feature selection/extraction is among the important step for spam detection either for SMS or Email. Also, the existing work does not focus on choosing the appropriate feature selection/extraction technique. Few comparison works are available [1, 3] and they are majorly using only three evaluation measures - accuracy, recall and precision. Time is most important evaluation measures which is not used for comparison of the results. Also, a comparative analysis consisting of more than one domain (such as email, SMS) is required to understand the effect of preprocessing steps on spam detection model.

Looking at the importance of spam detection in SMS and Email domains along with limitations of existing state-of-the-art, a work is carried out on spam detection for SMS and Email domains. Following are the important objectives of the work presented in this paper:

- Comparative analysis of effect of feature selection/extraction techniques on the overall performance of model.
- The performance evaluation is to be done by using Time (Build and Test) and Receiver Operating Characteristic Curve (ROC), Accuracy, precision, recall and F1 measure parameters.
- Improve an existing supervised learning model-based classification algorithm with the above 2 objectives and to show that algorithm performance varies from one domain to another.

The further road map of this paper is as follows—Sect. 2 describes the literature review available for spam detection in the domain of SMS and Email. Section 3, presents proposed model for spam detection in Email and SMS domain. In Sect. 4, carious results drawn from experimentation are discussed followed by various conclusions drawn and future scope are discussed in Sect. 5.

2 Literature Review

In this section, a literature review of existing state-of-the-art on spam detection in the domain of SMS and Email is presented. From the existing state-of-the-art on the spam detection in the domain of SMS and Email, it is observed that there are mainly three phases of processing [1, 3]. Phase 1 includes data acquisition and text preprocessing, Phase 2 focuses on feature selection/extraction and Phase 3 includes model used for spam detection and result evaluation. These three phases are found common in literature's and are used to identify SMS and Email as ham or spam. Table 1 shows summary of literature review conducted on problem addressed by the paper, algorithm used for spam SMS and Email detection and the limitations of the proposed approach. In the following, the literature review of the existing works is done on the the three phases identified above.

2.1 Phase 1

In phase 1, data is acquitted from several available sources for detecting spam on SMS and Email domains. Kaggle is one of the widely used platform for finding experiment data-sets for SMS and Email domains [1]. UCI repository is also popular and several data-sets for SMS spam detection are used in the past for conducting experimentation [2, 5, 12]. Data-sets from other source like Spam assign, Ling spam is used for Email spam detection [9]. After the data-set selection, text preprocessing is done. As part of text preprocessing, unwanted data is removed from the data-set.

Natural Language processing (NLP) is preferred for text preprocessing for both SMS and Email spam detection [1, 3, 8]. In NLP, firstly, segmentation is done to break the paragraphs into sentences. Tokenization results in further division of sentences in to words (tokens)[1, 3]. Stop words and are remove afterward [5, 8]. These three sub-steps are majorly used in NLP for spam detection in the domain of SMS and Email.

2.2 Phase 2

After the phase 1 is executed, some useful features are selected from the available feature set. Sometime, number of features are very large in number and involve unnecessary processing of data. This leads to degraded performance of subsequent processing involved in spam detection and sometimes features are generated from the available data-sets. After feature selection, feature extraction is done convert the data-sets into real time vector matrix. Several techniques like- TFIDF [2], Info Gain Martirx(IGM) [1], and Document Term Matrix [3] are used.

2.3 Phase 3

In the final phase, data-set is divided in two parts- test and train data [13, 14] and classification is performed afterward. Classification of ham or spam instances of Emails/SMS is done using -Support Vector Machine(SVM) [2, 3], LR [1, 15], Decision Tree [2], NB [1–3, 8, 9], RF [1, 2]is used majorly to analyze the results. Accuracy, Precision Recall are three major evaluation measures used for spam detection for SMS and Email [] spam detection. Comparison is drawn with other existing works or with the various machine learning algorithms to find the best performer for spam SMS and Email spam detection.

In phase 3 firstly, classification is done and then result is prepared based upon the different evaluation measures. For classification, is used for both SMS and Email domains. Accuracy, Precision Recall are three evaluation measures used in the past for spam detection in SMS and Email domain [2, 5, 8] . From Table 1 it can be observed that, NB is mostly used technique for both spam Email and spam SMS detection [1–3]. Also, [8, 9] have used NB as a hybrid model for spam Email detection. Feature selection/extraction can improve the model performance [1] and different feature selection/extraction techniques make direct impact on performance[2]. Further it can be observed from Table 1 that the existing evaluation measure does not include testing/training time [1, 2, 5, 8, 9]. Also from the study of state-of-the-art, it is observed that the study on the effect of feature selection/extraction technique on available spam detection techniques is missing. Further, performance comparison of various spam detection techniques using train/test time parameter is not performed,

Table 1 Literature review

Ref No.	Year	Problem addressed	Algorithm used	Data-set used	Limitation
[1]	2017	Increased model performance using Information Gain Matrix for SMS spam detection	NB, LR, RF	Kaggle	Proper analysis on performance evaluation is not done
[2]	2021	Using different feature extraction techniques and different algorithms to detect spam on SMS domain.	SVM,DT, KNN, LR, RF, NB, LSTM	Kaggle	Discussion and evaluation measures is not drawn for the most accurate algorithm.
[3]	2018	Proposed a simple SMS spam filter and identified the most accurate one among 3 algorithms	SVM, NB, MAXENT	Spam SMS Collection	Feature selection/ extraction is not done appropriately
[5]	2019	Hybrid approach is proposed for Spam Email detection and result is compared with work done using the same data-set under same evaluation measures	PSO, PEGA-SOS	UCI	Discussion on time taken by proposed integrated approach as compared to traditional algorithms is not done
[8]	2018	Optimizing the main algorithm using evolutionary computing for Email Spam detection	NB, PSO	Spambase	Detailed analysis of evaluation measure of integrated approach is not done
[9]	2017	Tested integrated approach for spam Email detection on 3different data-sets	GL, NB	Google, Yahoo mail	Analysis on time evaluation measure is not present and comparison is not done with other hybrid approaches

which is important in the current perspective. Therefore, in this work an experimentation evaluation is conducted to show the effect of feature selection technique and including time parameter for performance comparison.

3 Classification Methods Used for Spam Detection in SMS and Email Domains

In this section, SMS and Email spam detection system details are discussed. A phase diagram for SMS and Email spam detection is proposed as shown in Fig. 1 . The proposed system is consisting of three phases. Phase 1 includes data-set and preprocessing of data-set. Feature selection and extraction is done in Phase 2. In phase 3, classification algorithm is applied for detection of spam in the message (SMS/Email) followed by performance evaluation of the system. Details of each phase are as follows:

Fig. 1 Proposed phase diagram for SMS and Email spam detection

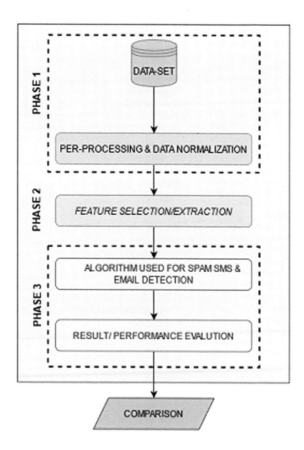

3.1 Phase 1

3.1.1 Data Acquisition

In the past, Email and SMS related data are obtained from various sources/repositories. In most of the work, the data-sets are chosen from Kaggle platform for both SMS and Email spam detection [1, 2]. One of the data-set from Kaggle for SMS domain, consists of 5572 total instances [16]. Out of which, 4825 are labeled as ham while 747 are labeled as spam. Data-set is divided into two attributes—Text and Class. Text attribute consist of SMS messages while class attribute consists of labels ham or spam for the text attribute. Another data-set for Email domain consist of 4601 instances and is divided into 58 attributes (after the preprocessing) [8, 14]. Class attribute shows the label of an instance. Out of 4601, 2788 are labeled as 0(ham) while 1813 are labeled as spam(1).Table 2 shows the example of data-set attributes used in SMS and Email domain.

3.1.2 Text Preprocessing

The unwanted data from the data-set is needed to be removed as it affects the performance of mode [2]. Hence text-preprocessing is performed over the data-set. The text preprocessing is performed using NLP on the data-set of Email and SMS spam detection [1, 5, 17]. Paragraphs from the data-sets are needed to be broken into various sentences [8, 14]. Segmentation is done for SMS data-set [3].After this, stop words are to be removed. This step is required for data-sets of SMS and Email domain. Multi-stop-words is used to filter out stop words [9, 16]. Once stop words

Table 2 Data-sets attributes example used for SMS and Email spam detection

Domain	Class	Text
SMS	Go until jurong point crazy...Available only in bugis n greatworld la e buffet...Cine there got amore wat...	Ham
SMS	Free entry in 2 a wkly comp to win FA Cup final tkts 21st May2005. Text FA to 87121 to receive entry question(std txt rate) T&Cs apply 08452810075over18s'	Spam
EMAIL	this deal is to book the teco pvr revenue. it is my understanding that teco just sends us a check, i have not received an answer as to whether there is a predermined price associated with this deal or if teco just lets us know what we are giving. I can continue to chase this deal down if you need	Ham
EMAIL	Photoshop, windows, office, cheap main trending abasements darer prudently fortuitous undergone lighthearted charm orinoco taster railroad affluent pornographic cuvier irvin park house blameworthy chlorophyll robed diagrammatic fogarty clears bayd inconveniencing managing represented smartness hashish academies shareholders unload badness danielson pure caffein spaniard chargeable levin	Spam

are filtered out, tokenization is to be done. N-gram-tokenizer is a popular technique used in SMS and Email domain [9]. All tokens are converted into lowercase and at last stemming is done using Lovin-Stemmer [18].

3.2 Phase 2

3.2.1 Feature Selection/Extraction

From the pre-processed data, few important features needs to be selected/extracted. The feature selection technique select subset of features from the original feature set. Feature extraction derives new features as a combination of features from original feature set. Both the methods ensure to reduce the model complexity and reduce the error introduced due to irrelevant features. Many feature selection/extraction techniques are under practice, however, appropriate feature selection and extraction methods needs to identified due to its profound impact on the overall performance of the model [2]. In [1, 3], for the SMS domain, features are selected using Correlation evaluator and searching is done using Ranker search. Similarly, in Email domain, Gain Ratio evaluator along with Ranker search is used for feature [8]. TFIDF is widely used for feature extraction in both SMS and Email domains[2, 8, 14].

3.3 Phase 3

In this phase, suitable classification algorithm is to be used for classifying Email/SMS as spam or Ham. A lot of machine learning models are available as a choice for spam detection. For experimentation purpose, we have used Random Forest and Multinomial-Naive Bayes. The details of both the learning models are discussed below.

3.3.1 Random Forest (RF)

RF is a tree-based classification algorithm consisting of several trees where each of the individual tree vote for the classification—ham or spam [1]. SMS and Emails after preprocessing and feature selection, are converted in to real time vector matrix from which samples and categories are drawn for the training purpose. As forest consists of several trees. For each tree, node value is calculated using threshold function and node is classify using split function and when all the nodes of a tree are gone through splitting process successfully then that tree is completed. Similarly, when all trees are completed, a forest is built [2]. After the building of forest, F-Values is used for final classification of SMS or Emails as spam or ham.

3.3.2 Multinomial-Naive Bayes(MNB)

This algorithm is based on NB and can be used for discrete and continuous data [2]. Algorithm try to classify each token of the SMS and Emails. Likelihood is calculated for each given token and output (ham or spam) is predicted. The token is classify as ham or spam based on largest posterior probability of class membership [8]. Large data-sets can easily be handled and classified using this algorithm hence it is used widely in spam detection [9].

4 Experiment and Result

4.1 Data-sets Used

Table 3 displays the data-sets used for SMS and Email domains for spam detection. For SMS domain, data-set is taken from Kaggle while for Email domain data-set is chosen from Spambase.

4.1.1 Tools Used

For experimentation and simulation, Java-based Weka tool is used and total of five evaluation measures are used. Weka is an open-source tool. Various machine learning algorithms are implemented in Weka tool to perform experiments. Simulation of ROC curves is done using Weka Tool only. The system used for experimentation and simulation consists of 12 GB RAM, Intel i3-6gen, 2.00 GHz CPU, 64 bit windows OS with 4 GB GPU is used. Results are prepared as graphs using Microsoft Excel.

4.1.2 Evaluation Measures

True Positive (TP) is defined as number of correctly identified spam SMS or Emails [1]. True Negative (TN) can be defined as number of correctly identified ham SMS or Emails [2]. Number of spam SMS or Emails classified as ham is referred to False

Table 3 Data-sets used in proposed model

Domain	Data-set name	Total	Ham	Spam
SMS	SMS Spam collection (Kaggle)	5572	4825	747
EMAIL	Spambase	4601	2788	1813

Positive (FP) [5] and False Negative (FN) [5] is defined as number of ham SMS or Emails classified as spam. Following are the evaluation measures used for both SMS and Email spam detection.

4.1.3 Recall

It is defined as probability of existing spam SMS or Emails [2]. Recall is also referred as negative predicted values can be formulated as shown in (1).

$$Recall = (TP)/(TP + FN) \tag{1}$$

4.1.4 Precision

It is defined as the probability of only true spam SMS or Emails [2]. Precision is also referred as true predicted values and formulated as defined in (2).

$$Precision = (TP)/(TP + FP) \tag{2}$$

4.1.5 Accuracy

Accuracy is the ratio of true values of spam SMS or Email detected with all true and negative values [5]. It can be defined using (3).

$$Accuracy = (TP + TN)/(TP + TN + FP + FN) \tag{3}$$

4.1.6 F Measure

It defines the overall performance of the algorithm [5]. It can be formulated as (4).

$$FMeasure = 2 * (Precision)/(Precision + Recall) \tag{4}$$

4.1.7 ROC

It is a plot which shows the relationship between true predicated rate with the negative predicated rate. Algorithm having higher accuracy has a curve near to left most corner and vice versa [13].

4.1.8 Time

Time taken to train a model is referred as training time while time taken to test the model is known as testing time. Both are generally measure in seconds [6, 17].

Using the data-sets defined in Phase 1 of Sect. 4 of SMS and Email domain, RF and MNB is applied. For first time all features are included and the experiment is repeated with feature selection/extraction technique. Table 4 shows the performance of RF and MNB (with and without using feature selection technique) for the SMS domain with 66% of data-set used for training and 33% used for testing the model. Similar experimentation is done for spam detection in Email domain and is presented in Table 5 with 66% of data-set used for training and 33% used for testing the model. It is evident from Table 4 that for SMS spam detection without feature extraction/selection, MNB algorithm results 97.8 % accuracy. The increase in the accuracy to 98.68% is observed when MNB algorithm is used with the proposed feature selection/extraction. Similarly, accuracy of RF algorithm for spam detection in SMS domain is 98.5% with feature selection/ extraction, as compared to the 96.4% accuracy without feature selection/extraction. Further, improvement in the precision, recall, F1 measure and time (train, test) is achieved for RF and MNB when feature selection is done.

Similar, for spam Email detection, RF and MNB has respective improved performance of 98.4%, 97.96% when feature selection technique is used as compared

Table 4 Performance of RF and MNB techniques for spam detection in SMS domain

	Without feature		With feature	
	RF	MNB	RF	MNB
Accuracy (%)	96.4	97.8	98.7	98.68
Precision (%)	100	97.9	100	98.7
Recall (%)	72.9	97.9	90.32	98.7
F1 (%)	98.01	97.9	99.9	98.7
Time train (66%) (s)	86.4	0.11	29	2.8
Time test (33%) (s)	1.1	0.02	0.9	0.6

Table 5 Performance of RF and MNB for spam detection in Email domain

	Without feature		With feature	
	RF	MNB	RF	MNB
Accuracy (%)	95.1	94.63	98.43	97.96
Precision (%)	96.6	94.21	97.3	97.63
Recall (%)	92.8	93.4	94.6	97.21
F1 (%)	96.01	94.5	97.1	97.23
Time train (66%) (s)	3.2	0.11	4	2.8
Time test (33%) (s)	0.31	0.02	0.1	0.6

to 95.1% and 94.63%, respectively, as shown in Table 5. Further, it can be observed from Table 5 that for spam detection in Email, 96.6% precision is achieved by RF when used without feature selection/extraction and it is improved to 97.3% when used with feature selection/extraction. Also, 97.63% precision is achieved when feature selection/extraction is used with MNB as compared to 94.21% precision without feature selection. Moreover, improvement in the Recall, F1 and time (train/test) is also observed when feature selection technique is used with RF and MNB in the Email domain for spam detection.

Train and testing time for SMS spam detection using RF and MNB with and without feature selection is shown in Fig. 2. It can be observed that training and testing time for RF is reduced and that of MNB is increased slightly. Similar observations are drawn for train and test time in SMS domain as shown in Fig 3. Figure 4 shows the accuracy comparison of RF-based proposed approach in SMS domain as compared to [1–3]. Also, proposed RF-based spam detection in Email domain is compared with the available techniques in [4, 6, 7]. It can be observed that the accuracy of proposed RF-based spam detection results in improved accuracy of 98.70% as compared to accuracy of 98.5% [1], 97.4% [2], 98.5% [3] in SMS domain.

Similarly, the proposed RF-based spam detection results in improved accuracy of 98.43% as compared to accuracy of 96.19% [4], 97.43% [6], 96.24% [7] in Email domain. Using proper feature selection/extraction techniques, proposed experiment

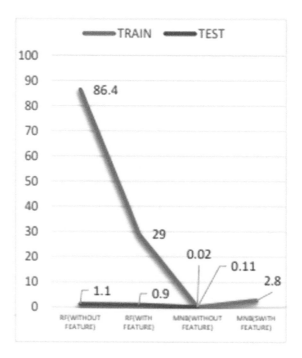

Fig. 2 Train/test time (in sec) for SMS spam detection

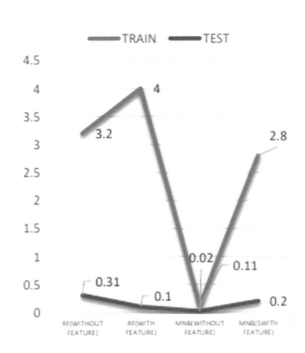

Fig. 3 Train/test time (in sec) for Email spam detection

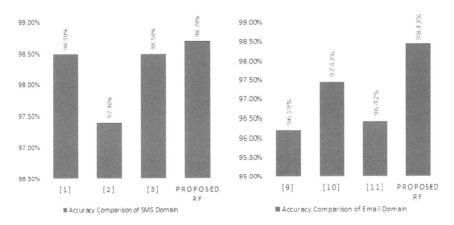

Fig. 4 Accuracy comparison for SMS and Email spam detection

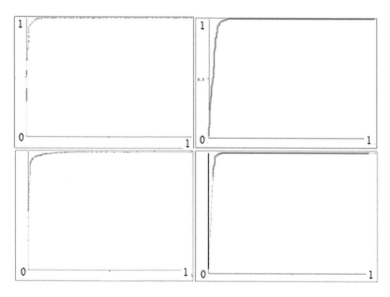

Fig. 5 ROC plots for SMS and Email spam detection

results even better than the hybrid models [9]. RF scores 0.99 ROC value in both model (with and without feature selection/extraction) for SMS spam detection. For MNB, improved 0.984 ROC value is achieved with feature selection/extraction than the 0.982 ROC value without feature selection/extraction for SMS spam detection. For Email spam detection, RF scores 0.98 ROC value for in both model (with and without feature selection/extraction). Also , MNB score 0.994 ROC value with feature selection/extraction as compared to 0.987 ROC value without feature selection/extraction. Further Fig. 5 represents the plot for ROC for spam detection in SMS and Email with feature selection/extraction.

5 Conclusions and Future Work

After experimentation on the selected state-of-the-art, various conclusions are drawn. It is concluded that by selecting appropriate feature selection/extraction techniques, the overall performance of the spam detection model can be improved. RF achieved 98.7% accuracy for spam detection in SMS domain and 98.43 for spam detection in Email domain. Also MNB with feature selection and extraction has accuracy of 97.8% and 97.96% for spam detection in SMS and Email domain, respectively. For spam detection in SMS domain, based on all evaluation measures it can be concluded that RF outperformed over other techniques and can be followed for spam detection. MNB perform better in SMS domain with less time to train and test the model. Thus, it can be used as spam detection technique in Email domain. Also, it can be concluded from Figs. 2 and 3 that time taken for training the model with and without feature selection/extraction varies for SMS and Email spam detection and can be seen in result of different evaluation measures used. As a future work, hybrid methods and evolutionary computing-based algorithms can be followed for spam detection for SMS and Email domains.

Acknowledgements Author's are thankful to Shri G. S. Institute of Technology and Science, Indore for providing all facilities and necessary support to carry out the research work.

References

1. Sethi, P., Bhandari, V., & Kohli, B. (2017). SMS spam detection and comparison of various machine learning algorithms. In *2017 International Conference on Computing and Communication Technologies for Smart Nation (IC3TSN)* (pp. 28–31). https://doi.org/10.1109/IC3TSN.2017.8284445
2. Gadde, S., Lakshmanarao, A., & Satyanarayana, S. (2021). SMS spam detection using machine learning and deep learning techniques. In *2021 7th International Conference on Advanced Computing and Communication Systems (ICACCS)* (p. 358362).
3. Navaney, P., Dubey, G., & Rana, A. (2018). SMS spam filtering using supervised machine learning algorithms. In *2018 8th international conference on cloud computing, data science & engineering (confluence)* (pp. 43–48). https://doi.org/10.1109/CONFLUENCE.2018.8442564
4. Tang, S., Mi, X., Li, Y., Wang, X., & Chen, K. (2022). *Clues in tweets: Twitter-guided discovery and analysis of SMS spam.* arXiv preprint arXiv:2204.01233. 2022 Apr 4.
5. ElBakrawy, L. M. (2019). *Hybrid particle swarm optimization and Pegasos algorithm for spam Email detection.* https://doi.org/10.25728/assa.2019.19.3.751
6. Jain, T., Garg, P., Chalil, N., Sinha, A., Verma, V. K., & Gupta, R. (2022). SMS spam classification using machine learning techniques. In *2022 12th international conference on cloud computing, data science & engineering (confluence)* (pp. 273–279). https://doi.org/10.1109/Confluence52989.2022.9734128
7. Annareddy, S., & Tammina, S. (2019). A comparative study of deep learning methods for spam detection. In *2019 third international conference on I-SMAC (IoT in Social, Mobile, Analytics and Cloud) (I-SMAC)* (pp. 66–72).
8. Agarwal, K., & Kumar, T. (2018). Email spam detection using integrated approach of Naïve Bayes and particle swarm optimization. *Second International Conference on Intelligent Computing and Control Systems (ICICCS), 2018,* 685–690. https://doi.org/10.1109/ICCONS.2018.8662957.
9. Iyengar, A., Kalpana, G., Kalyankumar, S., & GunaNandhini, S. (2017). Integrated SPAM detection for multilingual emails. *International Conference on Information Communication and Embedded Systems (ICICES), 2017,* 1–4. https://doi.org/10.1109/ICICES.2017.8070784.
10. Ponmalar, A., Rajkumar, K., Hariharan, U., Kalaiselvi, V. K. G., & Deeba, S. (2021). Analysis of spam detection using integration of logistic regression and PSO algorithm. In *2021 4th International Conference on Computing and Communications Technologies (ICCCT)* (pp. 396–402). https://doi.org/10.1109/ICCCT53315.2021.9711903
11. Raza, M., Jayasinghe, N. D., & Muslam, M. M. A. (2021). A comprehensive review on email spam classification using machine learning algorithms. *International Conference on Information Networking (ICOIN), 2021,* 327–332.
12. GuangJun, L., Nazir, S., Khan, H. U., & Ul Haq, A. (2020). Spam detection approach for secure mobile messgae communication using machine learning algorithms. In *Hindawi, security and communication netwroks* (Vol. 2020, Article id: 8873639)

13. Maloof, M. A. (2002). On machine learning, ROC analysis, and statistical tests of significance. In *2002 international conference on pattern recognition* (Vol. 2, pp. 204–207). https://doi.org/ 10.1109/ICPR.2002.1048273

14. Nandhini, S., & Marseline, K. S. (2020). Performance evaluation of machine learning algorithms for Email spam detection. In *2020 international conference on Emerging Trends in Information Technology and Engineering (ic-ETITE), Vellore, India* (pp. 1–4). https://doi.org/ 10.1109/icETITE47903.2020.312

15. GuangJun, L., Nazir, S., Khan, H. U., & Ul Haq, A. (2020). Spam detection approach for secure mobile messgae communication using machine learning algorithms. In *Hindawi, security and communication netwroks* (Vol. 2020, Article id: 8873639)

16. Sravya, G. S., Pradeepini, G., & Vaddeswaram. (2020). Mobile Sms spam filter techniques using machine learning techniques. *International Journal of Scientific & Technology Research, 9*(03).

17. Markova, E., Bajtoš, T., Sokol, P., & Mzéšová, T. (2019). Classification of malicious emails. In *2019 IEEE 15th international scientific conference on informatics, Poprad, Slovakia* (pp. 000279–000284). https://doi.org/10.1109/Informatics47936.2019.9119329

18. Ismailov, A., Jalil, M. M. A., Abdullah, Z., & Rahim, N. H. A. (2016). A comparative study of stemming algorithms for use with the Uzbek language. In *2016 3rd International Conference on Computer and Information Sciences (ICCOINS)* (pp. 7–12).https://doi.org/10.1109/ICCOINS. 2016.7783180

Discerning the Application of Virtual Laboratory in Curriculum Transaction of Software Engineering Lab Course from the Lens of Critical Pedagogy

Ashraf Alam and Atasi Mohanty

Abstract One of the critical features of computer science and engineering (CSE) education is learning by doing. The rapid upsurge in the use of Internet and complete shutdown of educational institutions during COVID-19 outbreak has drawn attention to the importance of online laboratory-based learning in CSE education. Online laboratory-based learning has emerged as a popular area of research among educational-technology researchers. In an online laboratory learning environment, the instructor has a significantly reduced role, and students take increased responsibility for their learning. This shows that in online lab-based learning, the involvement of students is higher. Traditional classroom-based laboratory learning has many limitations. To investigate these issues, we studied a Software Engineering Virtual Lab (SE VLab) (http://vlabs.iitkgp.ernet.in/se/). This laboratory exposes students to various fundamental topics related to software engineering. In a traditional laboratory environment, the evaluation process is inherently deficient, subjective, and subject to unfairness with possibilities of bias. In this research paper, we have evaluated the SE VLab in different pedagogical contexts. Results show that SE VLab is more effective than similar traditional SE labs in terms of learning gains.

Keywords Critical pedagogy · Software engineering · Curriculum · COVID-19 pandemic · Virtual laboratory · Computer science engineering education · Evaluation · Learning outcomes

1 Introduction

Information and Communication Technologies (ICTs) are key enablers in the growth of modern education [1]. In developing nations such as India, the engineering institutions are quite diverse in nature in terms of infrastructure, the quality of faculty, and

A. Alam (✉) · A. Mohanty
Rekhi Centre of Excellence for the Science of Happiness, Indian Institute of Technology Kharagpur, Kharagpur, India
e-mail: ashraf_alam@kgpian.iitkgp.ac.in

© The Author(s), under exclusive license to Springer Nature Singapore Pte Ltd. 2023　　53
S. Shakya et al. (eds.), *Sentiment Analysis and Deep Learning*,
Advances in Intelligent Systems and Computing 1432,
https://doi.org/10.1007/978-981-19-5443-6_5

the quality of education imparted by them. In many institutions, neither the structure of the curriculum fulfills the requirements of the international job market, nor it makes the students competent enough to face the challenges of the post-globalization economy. Consequently, the students lag behind in their understanding and knowledge gain due to the lack of appropriate hands-on sessions [2, 3]. In India, for example, a number of engineering institutions were established in the last three decades in order to meet the growing demands for engineering education [1–4]. However, very few institutions have the state-of-the-art real laboratory (lab) infrastructure to provide the necessary hands-on training to the students. Sometimes, exorbitant cost of lab equipment's makes it impossible to provide the much-needed facilities at all the locations. In other cases, repetition of experiments, multiple uses of tools and technologies, frequency of interaction with instructor, a skewed student–teacher ratio, and lack of competent teachers with competent expertise exist [5, 6]. This results in diminishing inclusion of labs in the course curricula.

To overcome the problems faced in traditional labs, the virtual labs were developed with funding from the Ministry of Human Resources and Development (MHRD), Government of India (GoI). Institutions in remotely located and educationally backward areas in India can be benefited by using such labs. A virtual lab provides great opportunities to the students due to the flexibility of access and study times it offers. It allows the students to repeatedly go through the presented concepts at their own pace, clarify their doubts, and prepare for the examinations at their own convenience. It enables having lab experience at anytime and from anywhere. The virtual labs also enhance students' enthusiasm for learning through interactivity. The major disadvantage of many other virtual labs is that they are distanced from the reality of physical labs. However, such feeling is reduced in the case of Software Engineering Virtual Lab (SE VLab).

2 Motivation for This Work

Software engineering (SE) is one of the most important subjects in engineering education [7–11]. In this research paper, the term "virtual laboratories" specifically refer to online laboratories, in which the laboratory resources can be shared among a wide range of community [12]. However, it should be clarified that the use of the term "virtual learning" is not limited to online learning only. It extends to immersive or augmented virtual platform on cloud. Therefore, online learning or e-learning should not be generalized as virtual learning in all contexts. The term "virtual learning" is indeed broad. In this research paper, we conceptualize from the perspective of online dissemination of courseware, with the possibility of synchronous interaction between the students and teacher. In this environment, various tools and technologies are available online.

The work reported in this research paper is important since it attempts to solve the problems occurring in actual abstract-laboratory-based technical education through the use of the virtual lab. However, it may be clarified that the lack of facilities and the lower student–teacher ratio are not the key issues in putting lessons online. Contemporary researches [12, 13] have demonstrated that the virtual environment could provide students with unique learning experience, which might not be obtained in traditional classroom or laboratory.

Though software engineering tends not to have physical laboratories, but in traditional software engineering laboratories, students often face difficulties concerning laboratory setup (e.g., space for lab infrastructure and cost of software) and skewed teacher-student ratio resulting in all students not receiving uniform attention/feedback [5, 6]. Further, learning is time-constrained in a traditional lab. Typically, a new experiment is covered every week, and there is often no scope for students to revisit them at any later point of time. This constrains both the students and the teacher in a course to cover experiments within the stipulated lab duration. Virtual laboratories overcome the said difficulties by enabling ease of use of lab course materials. To a large extent, students can use the SE VLab independently, without physically involving an instructor. A teacher or an instructor, however, can be involved in the process in order to get some specific feedback or to verify certain aspects of the design. A virtual laboratory on software engineering is ubiquitous. It can be accessed from anywhere over the Web. Reusability is needed for different batches of students. Moreover, a student can revisit a particular experiment at any point of time (e.g., before examination), and certain aspects are better illustrated using virtual labs rather than in class room scenarios (e.g., audio, visual presentation, group activities, and internal discussion among participation), which typically is not the case with the traditional labs. Many engineering students, irrespective of their educational background, are interested in software engineering jobs [7, 14–17]. In software industries, the fresh engineering graduate recruits face problems owing to design, coding, testing, and maintaining different software projects [17–19]. So, it is necessary to have prior undergraduate level knowledge in SE.

The SE VLab (http://vlabs.iitkgp.ernet.in/se/) was developed by IIT Kharagpur to address some problems characteristic of the traditional SE labs.

3 Objectives and Scope of This Research

Achieving practical experience is a very important aspect of engineering education [8, 20]. Concurrently, flexibility in time for laboratory classes is an important issue. Many students have extended commitment toward laboratory experiments. A strategic solution is to develop Web-based laboratories or virtual laboratories on an Internet platform, in which experiments can be conducted anytime and from anywhere. Research analysis manifests the clear trends of growing interest of engineering students toward virtual learning [12, 17, 21]. To keep up with this trend, as well as the pace of the rapid advancement in software technologies, the software engineering virtual lab was developed. The aim of this virtual lab is to present the state-of-the-art contents in software engineering in an interactive manner through the Web. The main focus is to enable the students to interact with the "virtual teacher" in an effective and efficient manner, compared to how they would do in a real lab on the subject.

The aim of this study was to review the online virtual laboratory learning environment for SE education for students of UG level and the evaluation of its use as a complementary tool to TLab-based instruction on the teaching of SE lab course. The evaluation was done in terms of learning evaluation in different pedagogical contexts and students' acceptance.

3.1 Specific Objectives of the Research

- Review of design and implementation of SE virtual lab course for UG engineering students, as per the commonality in the standard All India Council of Technical Education (AICTE) [22] and Indian Institute of Technology (IIT) syllabi.
- Evaluation of the effectiveness of IIT Kharagpur's SE VLab as compared to any other TLab.

3.2 Scope of the Research

The scope of this research is to review the design of SE VLab following a commonly accepted syllabus of SE lab and use it for evaluating the learning gains and outcomes, using the theory of critical pedagogy (CP).

4 Organization of This Research Paper

This paper is organized as follows:

Firstly, the authors review the related work on the broad research areas of issues relating to SE education and VLab learning environment in software engineering. Secondly, the paper presents our approach to designing and implementing the Software Engineering Virtual Laboratory (SE VLab) developed at IIT Kharagpur. Further, we discuss the architecture, database schema, related technologies, and user activities. Next, the paper presents the details of discussion on our research based on the findings. Finally, the paper concludes by highlighting the important contributions of our work. A few possible future extensions to our work are also outlined.

5   Virtual or Online Learning

The learning time-line is segmented into four eras: instructional, problem solving, mind-tool, and media wave eras [23].

- *Instructional era (1975–85)*: This was based on a normal theoretical basis of behavioral and early cognitive learning theory. There exist studies [23] that present that student learning using the instructional mechanism is not able to achieve the learning outcomes objectively.
- *Problem solving era (1980–90): solving era (1980–90)*: This is an intermediate era, where students were transferring from instructional era to mind-tool era. In this era, the focus was based on shifting to students toward the computer technology.
- *Mind-tool era (1985–95)*: In this period, there was increased focus on learning by using computer as a tool.
- *Media wave era (1995-present)*: Introduction of the media into digital form. This allows students to learn using Internet-connected computers. Students get significant inputs from the Web. This era motivated the concept of virtual or online learning.

In the early 1990s, computers became accepted as efficient devices for communication and information sharing. The use of computers enabled the capability to interact electronically, collaborative working for problem solving, and searching through databases of study materials. In this way, interactivity with computers became increasingly important to users.

6   Traditional Versus Virtual Laboratory

Laboratory practice plays a crucial role in engineering education [24]. Though both traditional and virtual lab learning processes have notably converged in the last decade, they can be still differentiated. Traditional lab learning imparts teacher-centered synchronous instruction of a group of students, constrained by lab availability of classrooms and technical infrastructure. Virtual lab-based learning, on the other hand, is student-centered, asynchronous, and available at anytime and from anywhere. The following are the motivational objectives for virtual lab-based learning:

- *On-demand availability*: A virtual lab can provide opportunity for the learners to decide at what time and place s/he wants to learn. Further, the students have the opportunity to revisit the contents for reinforcement of knowledge.
- *Self-pacing*: The lab can provide the options for the advanced learners to speed through or bypass instructions that will be redundant for them, whereas for novices to follow step-wise progress through content. In other words, the use of virtual lab can help a wide range of learning styles, preferences, and needs.

Table 1 Distinctions between traditional and virtual lab learning

Parameters	Traditional lab	Virtual lab
Content	Stable, durable	Dynamic, transitory
Distribution of materials	Hard copy	Electronic download
Flexibility	Standardized	Customized
Focus of content	Teacher-centered	Student-centered
Focus of course	Group	Individual
Form	Synchronous	Asynchronous
Instructor preparation	Some (transparencies)	Extensive preparation
Interaction	Spontaneous	Structured
Number of students	Space delimited	Without limits
Place	Classroom	Anywhere
Range of interactivity	Full interactivity	Limited interactivity
Time	Scheduled	Anytime

- *Interactivity*: Users can interact with the virtual lab in a relaxed mood, as per their preference, rather than in a fixed time-based traditional class room lab.
- *Confidence*: Learners can increase their confidence levels using feedbacks that would be generated by the virtual lab during interactions.
- *Flexibility*: Virtual labs provide the flexibility to choose specific learning styles that would suit the learners best and thus, enable adopting individual learning preferences.

The distinctions between traditional and virtual lab learning are given in Table 1. Table 2 shows the comparison of features of traditional and virtual laboratories. In Table 2, the "$\sqrt{}$" and " $\times$ " symbols indicate "available" and "not-available", respectively.

The main characteristics of SE VLab include on-demand availability, self-pacing, interactivity, and flexibility.

Table 2 Comparison of features of laboratories

Characteristics of labs	Hardware component-based lab	Software simulation-based lab	Virtual/remote lab
High realism	$\sqrt{}$	$\times$	$\sqrt{}$
Ease of use	$\times$	$\sqrt{}$	$\sqrt{}$
Low cost	$\times$	$\sqrt{}$	$\sqrt{}$
Ease of maintenance	$\times$	$\sqrt{}$	$\sqrt{}$

7 Critical Pedagogy Theory

Critical pedagogy [9, 13] is a theory and a philosophy of education. It was first described by Paulo Freire. Freire heavily endorses students' ability to think critically about their education situation; this way of thinking allows them to "recognize connections between their individual problems and experiences and the social contexts in which they are embedded". CP has since been developed by Henry Giroux [7, 21]. It is an "educational movement, guided by passion and principle, to help students develop consciousness of freedom, recognize authoritarian tendencies, and connect knowledge to power and the ability to take constructive action" [6, 19]. CP promotes relationships between teaching and learning. It is a continuous process of "unlearning", "learning", "relearning", "reflection", and "evaluation". In forthcoming sections, we will discuss different parameters based on CP theory.

8 Instructional Design Delivery (IDD)

After designing the VLab course, it is important for teachers to analyze the instructional design by seeking feedback from students. The output of the IDD analysis helps in getting insight on whether the course meets students' requirements or not [2, 19]. Research results indicate that when designing any VLab course, the designer should identify the requirements of the students with respect to attributes such as the course objectives, different hands-on activities, course reference materials, virtual presence of facilitator, self-assessment through quizzes or multiple questions, and presence of chart boxes for communication with virtual tutor through e-mail [7, 14]. A virtual tutor should consider the educational background and knowledge toward the subject. These are the general aspects governing the CP theory [6, 14]. Once the virtual tutor/designer clearly implements the instructional design, it can help the students to interact with their friends and virtual tutor. The major challenges toward IDD are to evaluate students' expected input for the course and measure their learning outcomes.

It is reported in the literature that rather than using the TLab environment, the students are allowed to complete their assignments or project work individually, and when required, there should be a provision to communicate with their friends or virtual teacher [20].

9 Students' Learning Outcomes (LO)

In the VLab environment, as in a TLab environment, it is important that tutors' expectations must be clearly defined [8, 19]. For instance, LO should be presented before the experiments start, e.g., "at the end of the course, students will be able to do ...".

Bloom [9, 17, 21] clearly identified different LO such as knowledge, comprehension, application, analysis, synthesis, and evaluation. In this case, the virtual tutor should not confuse between the learning objectives and learning outcomes [9, 17]. So, LO is defined with respect to how students will demonstrate lab experiments as per their understanding through virtual learning [7, 11].

In this context, virtual teacher should clearly define the LO prior to the start of the lab experiments. According to the CP theory, LO must reflect the testing of students' critical thought, rather than testing of memorization. So, it is required that students should think critically about the lab experiments [2, 19].

10 Assessments (ASS)

Students' performance evaluation is one of the most important aspects of teaching [3, 22]. Evaluation may be processed through examinations, portfolios, physical observation in lab, project work, and assessment of lab record assignments [3, 7]. These can be measured in the form of grades, scoring of marks, and other evaluation methods. It is very difficult to measure these without any valid measuring instrument [4, 19]. On the other hand, a teacher will be unable to ensure that the students comply with the LO of the lab course [5, 22]. Often students also do not understand the teachers' expectations from them. So, the evaluation procedure should be clearly defined in the lab syllabus [3, 17].

The CP theory indicates that students need to display their lab experiments as part of the evaluation process [7, 13]. It is also important that students involve themselves in the self-evaluation process [8, 15].

11 Students' Empowerment (EM)

In order to create empowered students, teacher should create an atmosphere that promotes personal responsibility, autonomy, continuous learning, and ability to cope with change [9, 21, 25–27]. Students' empowerment ensures that the teacher gives opportunities to choose any assignment from a set of given assignments [3, 15]. Teachers may empower their students by permitting friends to interact and enabling additional reading responsibility [7, 18]. It is also important that students should be given an opportunity to discuss with their peers [7, 17]. In the VLab environment, students are expected to be more actively involved and be proactive in learning and gaining knowledge is comparison with a TLab environment [4, 23].

In the context of the CP theory, interaction between students and teacher is a process that makes the students capable of learning about themselves and realizing their own potential [7, 13].

12   Critical Thinking (CT)

CT promotes responsible, reflective, skillful, and reasonable thinking [1, 4, 18]. A SE teacher should always encourage problem solving and CT among his/her students. A teacher can measure a student's CT performance by asking various open-ended questions. Researchers maintain that CT skills can be taught and learned [4, 15, 28–31]. After acquiring the CT skills, the students can apply these skills appropriately and can improve their thinking skills. It is reported that currently many students join colleges with very poor CT skills. To overcome this problem and to increase the level of critical thinking in a virtual learning environment, more innovative questions toward the experiments which will increase the thinking skills as well as students' self-responsibility for their solutions should be posted in the VLab environments.

In this context, the CP theory emphasizes the importance of CT in a virtual learning environment. Students reflect their own learning skills and make decisions based on their reflections are under CT skills [2, 13, 32–36].

13   Social Presence (SP)

Social presence has implications on the feelings of community that a student's knowledge and experiences in a virtual environment only [7, 19]. It is reported [8, 13] that interaction between the peers is an issue. Interaction is very important in creating better learning environment, and it is also true that social presence motivates learners to learn [9, 13].

It is a difficult task to create social presence in virtual learning environment as compared to traditional learning environment [7, 21]. In virtual learning environment, students may feel bored or alienated [4, 15]. In order to avoid this, in a virtual lab environment, a virtual tutor permits the user to send e-mail to the student peers. After joining a virtual learning course, all students need to post their short introduction, so that the virtual teacher and peers can interact with one another through mail [3, 9].

In the context of the CP theory, the users of virtual lab can communicate their critical thinking feedback toward virtual lab with one another, and the students can exchange their ideas with each other and can feel the active presence of student peers, as in a TLab environment [2, 15].

14   Alignment (AL)

Alignment in both the VLab and the TLab environments is similar. It indicates that both teachers and students work together to achieve same objectives. So, a teacher should design a course in such a way that all the objectives and outcomes are aligned to reinforce each other [7, 13].

In a virtual learning environment, a teacher should ensure that students have some technical skills for attending the laboratory course [4, 15]. In other cases, a teacher should be aware of the laboratory requirements and activities and is in alignment with student skills [2, 9]. When designing a course, a teacher should take into consideration the different levels of students—below average, average, and above average (with respect to SE knowledge), who will have access to the VLab.

Existing research also suggests some principles of alignment, where the following parameters are taken into consideration for effective teaching in virtual learning environments [2, 8, 19]:

- Assessment procedures and methods of results reporting.
- Curriculum for teaching.
- Methods of teaching.
- Climate for interactions with students and instructions toward students.

In the context of CP, a virtual tutor needs to address the alignment concept. The main objectives are positive interactions, LO, personal freedom, change in society, and some other aspects on virtual learning environments [7, 11, 19].

15 Approaches to Virtual Laboratories and Software Engineering Virtual Laboratory

In view of the above discussion on the various dimensions of virtual learning environment, we reviewed the design of SE virtual labs that would enhance the teaching effectiveness of SE lab courses.

Existing literature indicates that multiple efforts have been undertaken on virtual labs in subjects of various disciplines, including those of computer science and engineering [13–15]. To the best of our knowledge, there exists no work specifically on the SE labs. Here, we describe some related works on simulation-based virtual labs, and how virtual technologies enhance learning.

Barros et al. developed a lab by combining both the remote as well as local Chemistry labs [7, 16]. Their work was based on Web-based interactions with real lab devices. They worked on both the real and simulated activities related to the Chemistry laboratory. The lab developed by them was not a pure virtual lab, rather it was a collaborative one.

Wang et al. presented the key performance indicator (KPI)-based approach in a Web-based learning system [7, 11, 16]. They developed a prototype of the KPI-oriented learning system by using the Java programming language. Their proposed lab is essentially useful for software testing professionals.

Choi et al. developed an online educational laboratory based on electrical and computer science and engineering subjects such as digital control and power electronics circuits [8, 17, 21]. In their work, they described the design and implementation of a set of power electronic circuits. They evaluated the students' level of confidence, satisfaction, and perception with respect to the analysis, design, and digital control.

Chris et al. developed a virtual lab for both the mobile computing and the embedded systems subjects for software engineering students [7, 19, 22]. The pedagogical goals listed in their work describe how to understand embedded system domain, how to decide when to use software engineering tools and techniques, and when to specialize, and provide hands-on experience.

Dawy et al. developed a lab course on wireless communication, based on cellular network planning [10, 13, 23]. In their work, they used MATLAB tools for simulation purpose. However, their work did not provide the details of the software architecture and database schema used in the development.

Munch et al. developed a virtual software engineering laboratory (VSEL) in support of trade-off software business analyzes [1, 8, 23]. In their work, authors establish a virtual laboratory for software business professionals. This laboratory is not targeted for the undergraduate and postgraduate engineering students.

Piccoli et al. described a framework and preliminary assessment of the effectiveness of Web-based virtual learning environment [2, 14, 21]. They studied the use of virtual technologies is enhancing students' learning. On a similar note, Lee et al. analyzed the perspective from both undergraduate and postgraduate levels that there is significant difference between traditional learning and virtual learning [3, 11, 19]. They summarized from the learners' perspective that virtual technologies enhance students' learning.

16 Synthesis of Related Work and Concluding Remarks

A survey of the existing literature reveals the rising trend of the use of virtual labs and other ICT mechanisms to improve education. However, often such mediums are not publicly accessible to the common people in general. Moreover, while merely a few works target the SE domain, often they are for a focused target audience (e.g., business professionals [1, 8, 17]). This highlights the lack of a virtual lab on SE, where the students can have hands-on experience.

17 Educational Implications

The present study has the following educational implications:

1. An effective teaching–learning tool for distance learning and mass-education.
2. Can boost quality education through personalized instruction and self-learning.
3. As self-learners, the students will develop skills to own-up their learning responsibilities.

4. Knowledge construction, dissemination, and management would be enhanced across the subject topics.
5. Students' motivation and creativity (through problem solving) be improved by promoting learning communities through both synchronous and asynchronous communication networks.

18   Limitations

The limitations that we come across during our study are listed as follow:

1. In this work, we did not consider those studies that involved responses from teachers. As our sample of literatures reviewed was not very large, the drop rate toward the VLab users was almost zero. However, as per the existing research literature, drop rates tend to be higher in technology-mediated learning systems [5, 12, 15].
2. In this work, review of the development of SE VLab is based on only text and animations, and it does not include video and audio facilities.
3. As there are no other publicly available online virtual labs on software engineering, all the results reported in this study are specific to the developed SE VLab available at IIT Kharagpur.
4. Cross-cultural and trans-national aspects of evaluations in responses have not been particularly performed in this study.

19   Conclusion and Way Forward

In the modern society, software is developed for use in both offices and homes. This introduces challenge of training large number of software engineers in the undergraduate and postgraduate programs. As the real-world problems are evolving rapidly, software engineers of the twenty-first century face new challenges. In order to cope with the situation, the introducing of laboratory-based learning in software engineering education is indispensable. However, large number of academic institutions in developing countries currently involved in engineering education do not have necessary faculty and infrastructure that can ensure imparting quality hands-on training necessary for SE education. In order to address this lacuna in developing countries, particularly in India, SE VLab was developed at IIT Kharagpur.

19.1   Design Software Engineering Virtual Lab

Architecture and database schema are developed for SE VLab. Ten fundamental experiments are available in the designed SE VLab. The experiments included in the

SE VLab are standard for undergraduate SE course in engineering colleges/institutes in India. Students have found SE VLab to be suitable and useful for their course. Students' overall acceptance toward SE VLab was that they preferred instruction design delivery and learning outcomes more compared to other parameters. This indicates that making the SE VLab a part of the regular SE curriculum might be a viable option.

19.2   CP-VLLM: Development of a Virtual Learning Measurement Tool

In an undergraduate course, practical activities are of different types, such as experimental investigation, control assignments, and project works. The control assignments constitute a major part of the laboratory activities. Often it is very difficult to assess lab activities, as there is no such standard measurement tool/instrument available online/offline. In view of this, IIT Kharagpur has developed a learning measurement tool, named CP-VLLM, based on the theory of critical pedagogy (CP). Reliability and validity of the developed tool were statistically tested. This tool can be used for measuring students' learning outcomes and performance in online/virtual learning environments.

19.3   Evaluation of Students' Learning Performance in a Virtual Laboratory Learning Environment

As SE VLab is designed and implemented, it is important to evaluate this laboratory by considering students' feedback. In this context, IIT Kharagpur considered the CP-VLLM measurement tool, along with two SE subject knowledge test tools, and administered them for both pre-test and post-test data collection to come to its efficacy.

19.4   Directions for Future Research

There exist several unexplored points and open issues related to the researches reported here that warrants further investigation. We here briefly outline a few possible extensions to our work. Consideration of additional features in lab design: The current SE VLab provides some scope of improvement in future. Apart from the textual content provided here, audio and video materials can be added to enrich this lab further. Video-based tutorials and discussion groups to promote students' mutual cooperation and collaborative work can be included. Moreover, a tighter coupling with the social networking platforms (e.g., Facebook, Twitter, and Instagram) is envisioned to ensure a greater user participation, as well as a global audience. Such a feature would also enable group activities and interval discussion among the participants. More case-study, new experiments (as per different syllabus), and exercises can be added in future.

Consideration of alternative learning styles and cognitive theories for developing learning measurement tools: Here, only critical pedagogy theory is considered, and tool is developed following the said theory. In future, it is possible to develop other virtual learning measurement tools by considering Bloom's taxonomy [9, 12, 23], Kolb's model, Peter Honey, Alan Mumford's model, Anthony Gregorc's model, Neil Fleming's VAK/VARK model, and other cognitive approach to learning styles [5, 7, 13, 22].

Consideration of other procedures for evaluating students' performance: As this lab is available online on Web, students can use this lab and send their feedback online.

In future, it can be extended to change or add some new experiments as per the current syllabus. As mentioned earlier, currently, there is no publicly available SE VLab. In future, when other SE VLabs are developed, a comparison can be made between them. However, for extending the evaluation of this SE VLab in future, one can attempt to find similarities in user experiences between this VLab and virtual labs developed in other areas of computer science and engineering.

References

1. Triejunita, C. N., Putri, A., & Rosmansyah, Y. (2021, November). A Systematic Literature Review on Virtual Laboratory for Learning. In *2021 International Conference on Data and Software Engineering (ICoDSE)* (pp. 1–6). IEEE.
2. Alam, A. (2020). Challenges and possibilities in teaching and learning of calculus: A case study of India. *Journal for the Education of Gifted Young Scientists, 8*(1), 407–433. Retrieved from https://doi.org/10.17478/jegys.660201
3. Ramírez, J., Soto, D., López, S., Akroyd, J., Nurkowski, D., Botero, M. L., Bianco, N., Brownbridge, G., Kraft, M., & Molina, A. (2020). A virtual laboratory to support chemical reaction engineering courses using real-life problems and industrial software. *Education for Chemical Engineers, 33*, 36–44.
4. Tobarra, L., Robles-Gomez, A., Pastor, R., Hernandez, R., Duque, A., & Cano, J. (2020). Students' acceptance and tracking of a new container-based virtual laboratory. *Applied Sciences, 10*(3), 1091.
5. Alam, A. (2020c). Possibilities and challenges of compounding artificial intelligence in India's educational landscape. *International Journal of Advanced Science and Technology, 29*(5), 5077–5094. Retrieved from http://sersc.org/journals/index.php/IJAST/article/view/13910
6. Jamshidi, R., & Milanovic, I. (2022). Building virtual laboratory with simulations. *Computer Applications in Engineering Education, 30*(2), 483–489.
7. Alam, A. (2021, December). Should Robots Replace Teachers? Mobilisation of AI and Learning Analytics in Education. In *2021 International Conference on Advances in Computing, Communication, and Control (ICAC3)* (pp. 1–12). IEEE. Retrieved from https://doi.org/10.1109/ICAC353642.2021.9697300

8. Hao, C., Zheng, A., Wang, Y., & Jiang, B. (2021). Experiment information system based on an online virtual laboratory. *Future Internet, 13*(2), 27.

9. Alam, A. (2020d). Test of knowledge of elementary vectors concepts (TKEVC) among first-semester bachelor of engineering and technology students. *Periódico Tchê Química, 17*(35), 477–494. Retrieved from http://www.doi.org/10.52571/PTQ.v17.n35.2020.41_ALAM_pgs_477_494.pdf

10. Reeves, S. M., & Crippen, K. J. (2021). Virtual laboratories in undergraduate science and engineering courses: a systematic review, 2009–2019. *Journal of Science Education and Technology, 30*(1), 16–30.

11. Alam, A. (2020b). Pedagogy of calculus in India: An empirical investigation. *Periódico Tchê Química, 17*(34), 164–180. Retrieved from http://www.doi.org/10.52571/PTQ.v17.n34.2020.181_P34_pgs_164_180.pdf

12. Alam, A. (2021, November). Possibilities and apprehensions in the landscape of artificial intelligence in education. In *2021 International Conference on Computational Intelligence and Computing Applications (ICCICA)* (pp. 1–8). IEEE. Retrieved from https://doi.org/10.1109/ICCICA52458.2021.9697272

13. Lin, M., San, L., & Ding, Y. (2020, March). Construction of robotic virtual laboratory system based on Unity3D. In *IOP Conference Series: Materials Science and Engineering* (Vol. 768, No. 7, p. 072084). IOP Publishing.

14. Seifan, M., Robertson, N., & Berenjian, A. (2020). Use of virtual learning to increase key laboratory skills and essential non-cognitive characteristics. *Education for Chemical Engineers, 33,* 66–75.

15. Diwakar, A. S. (2020, February). Improvement in quality of virtual laboratory experiment designs by using the online SDVIcE Tool. In *International Conference on Remote Engineering and Virtual Instrumentation* (pp. 393–410). Springer, Cham.

16. Tejado, I., Gonzalez, I., Pérez, E., & Merchán, P. (2021). Introducing systems theory with virtual laboratories at the university of Extremadura: How to improve learning in the lab in engineering degrees. *The International Journal of Electrical Engineering and Education, 58*(4), 874–899.

17. Alam, A. (2022). Psychological, sociocultural, and biological elucidations for gender Gap in STEM education: A call for translation of research into evidence-based interventions. In: *Proceedings of the 2nd International Conference on Sustainability and Equity (ICSE-2021). Atlantis Highlights in Social Sciences, Education and Humanities.* ISSN: 2667-128X. Retrieved from https://doi.org/10.2991/ahsseh.k.220105.012

18. El Kharki, K., Berrada, K., & Burgos, D. (2021). Design and implementation of a virtual laboratory for physics subjects in moroccan universities. *Sustainability, 13*(7), 3711.

19. Gubsky, D. S., Daineko, Y. A., Ipalakova, M. T., Kleschenkov, A. B., Aitmagambetov, A. Z., & Vyatkina, S. A. (2021, November). Spectrum analyzer model for a virtual laboratory. In *2021 Photonics & Electromagnetics Research Symposium (PIERS)* (pp. 373–376). IEEE.

20. Deepika, N. M., Bala, M. M., & Kumar, R. (2021). Design and implementation of intelligent virtual laboratory using RASA framework. Materials Today: Proceedings.

21. Chatterjee, S. (2021). Revolutionizing science education through virtual laboratories. *Advances in Science Education, 118.*

22. Robles-Gómez, A., Tobarra, L., Pastor-Vargas, R., Hernández, R., & Cano, J. (2020). Emulating and evaluating virtual remote laboratories for cybersecurity. *Sensors, 20*(11), 3011.

23. Alam, A. (2021, December). Designing XR into Higher Education using Immersive Learning Environments (ILEs) and Hybrid Education for Innovation in HEIs to attract UN's Education for Sustainable Development (ESD) Initiative. In *2021 International Conference on Advances in Computing, Communication, and Control (ICAC3)* (pp. 1–9). IEEE. Retrieved from https://doi.org/10.1109/ICAC353642.2021.9697130

24. Solak, S., Yakut, Ö., & Dogru Bolat, E. (2020). Design and Implementation of Web-Based Virtual Mobile Robot Laboratory for Engineering Education. *Symmetry, 12*(6), 906.

25. Alam, A. (2022, April). Educational robotics and computer programming in early childhood education: A conceptual framework for assessing elementary school students' computational thinking for designing powerful educational scenarios. In *2022 International Conference on*

Smart Technologies and Systems for Next Generation Computing (ICSTSN) (pp. 1–7). IEEE. Retrieved from https://doi.org/10.1109/ICSTSN53084.2022.9761354

26. Schnieder, M., Williams, S., & Ghosh, S. (2022). Comparison of in-person and virtual labs/tutorials for engineering students using blended learning principles. *Education Sciences, 12*(3), 153.

27. Giray, G. (2021). An assessment of student satisfaction with e-learning: an empirical study with computer and software engineering undergraduate students in Turkey under pandemic conditions. *Education and Information Technologies, 26*(6), 6651–6673.

28. Kanij, T., & Grundy, J. (2020, November). Adapting teaching of a software engineering service course due to COVID-19. In *2020 IEEE 32nd Conference on Software Engineering Education and Training (CSEE&T)* (pp. 1–6). IEEE

29. Singh, G., Mantri, A., Sharma, O., & Kaur, R. (2021). Virtual reality learning environment for enhancing electronics engineering laboratory experience. *Computer Applications in Engineering Education, 29*(1), 229–243.

30. Glassey, J., & Magalhães, F. D. (2020). Virtual labs–love them or hate them, they are likely to be used more in the future. *Education for Chemical Engineers, 33*, 76.

31. Alam, A. (2022, April). A digital game based learning approach for effective curriculum transaction for teaching-learning of artificial intelligence and machine learning. In *2022 International Conference on Sustainable Computing and Data Communication Systems (ICSCDS)* (pp. 69–74). IEEE. Retrieved from https://doi.org/10.1109/ICSCDS53736.2022.9760932

32. Alam, A. (2022). Investigating sustainable education and positive psychology interventions in schools towards achievement of sustainable happiness and wellbeing for 21st century pedagogy and curriculum. *ECS Transactions, 107*(1), 19481. Retrieved from https://doi.org/10.1149/10701.19481ecst

33. Alam, A. (2022). Social robots in education for long-term human-robot interaction: Socially supportive behaviour of robotic tutor for creating robo-tangible learning environment in a guided discovery learning interaction. *ECS Transactions, 107*(1), 12389. Retrieved from https://doi.org/10.1149/10701.12389ecst

34. Alam, A. (2022). Positive psychology goes to school: conceptualizing students' happiness in 21st century schools while 'minding the mind!' are we there yet? evidence-backed, school-based positive psychology interventions. *ECS Transactions, 107*(1), 11199. Retrieved from https://doi.org/10.1149/10701.11199ecst

35. Alam, A. (2022). Mapping a sustainable future through conceptualization of transformative learning framework, education for sustainable development, critical reflection, and responsible citizenship: an exploration of pedagogies for twenty-first century learning. *ECS Transactions, 107*(1), 9827. Retrieved from https://doi.org/10.1149/10701.9827ecst

36. Crichigno, J., Kfoury, E., Caudle, K., Crump, P. (2021, July). A distributed academic cloud and virtual laboratories for information technology education and research. In *2021 44th International Conference on Telecommunications and Signal Processing (TSP)* (pp. 195–198). IEEE.

Chrome Extension for Text Sentiment Analysis

Tirumalasetty Satya Prabhasa, Sairam Maganti, Gelam Sai Sriram, Katakam Jayadeep Reddy, and Jayashree Nair

Abstract Sentiment analysis has been one of the trending research subjects in the field of natural language processing over the last decade. The main reason behind the interest in sentiment analysis is the wide range of its applications and the amount of data or the information available at the fingertips for processing these applications. This motivated us to develop a new sentiment analysis application with a new perspective of sentiment analysis utility. We built a sentiment analysis application for chrome, which analyzes the text used for Google search. Thus, our target is to extract sentiment from every text used in search engines. In contemporary times, the core theme is that every person uses search engines in daily life, and they turned out to be a necessity. People search for news, ask questions, and learn new things and techniques, but they are not always on the right track. So, based on observation of how high the usage of search engines is, we propose an application that can detect potentially harmful, self-harming or anomalous search texts. Hence, this chrome extension can potentially act as parental monitoring of an individual's digital web life, which helps in the detection of the person's extreme psychological condition in some suicidal, abusive cases. With the advantage of prior identification, we can use it to notify or warn people about this. Machine learning algorithms and JavaScript were used in developing this application to create models and the chrome extension, respectively.

T. S. Prabhasa (✉) · S. Maganti · G. S. Sriram · K. J. Reddy
Department of Computer Science and Engineering, Amrita Vishwa Vidyapeetham, Amritapuri, Kerala, India
e-mail: tsatyaprabhasa@am.students.amrita.edu

S. Maganti
e-mail: sairammaganti@am.students.amrita.edu

G. S. Sriram
e-mail: gsaisriram@am.students.amrita.edu

K. J. Reddy
e-mail: kjreddy@am.students.amrita.edu

J. Nair
Department of Computer Science and Applications, Amrita Vishwa Vidyapeetham, Amritapuri, Kerala, India
e-mail: jayashree@am.amrita.edu

© The Author(s), under exclusive license to Springer Nature Singapore Pte Ltd. 2023
S. Shakya et al. (eds.), *Sentiment Analysis and Deep Learning*,
Advances in Intelligent Systems and Computing 1432,
https://doi.org/10.1007/978-981-19-5443-6_6

69

Keywords Sentiment analysis · Search text · Natural language processing ·
Machine learning · Chrome extension · JavaScript

1 Introduction

The usage of Internet is on an immense scale; it is used to find people, communicate, learn, teach, discover, and so on. Undoubtedly, it has become an inevitable part of our lives. Search engines like Google, Bing, and Yahoo, are one of the key reasons for the rapid growth of internet users. Exclusively Google search blended into our daily lives so well that it comes to play whenever we have any queries, are left puzzled, or need to research. Over four billion people actively use Google and its products as of May, 2021 [1]. The figures of number of active users of Google shows how much people are interested in the Internet and social media platforms; therefore, the amount of data accessible to everyone is tremendous. So this allowed us to grab a chance, and present a novel application using this data. There will be various texts with different contexts in this data as search engines are used in distinct situations. Hence, these texts will be helpful in finding an individual's extreme state of mind by extracting the sentiments of the search texts used by them.

This paper focuses on identifying the sentiment of the text used for google search. Sentiment analysis is a technique of extracting, converting, and interpreting opinions from a text and categorizing them into different sentiments using Natural Language Processing (NLP) [2]. The majority of previous studies applied sentiment analysis to a product review [3] or movie review [4] to better understand their customer and make the necessary decisions to improve their product or services. Some of the studies are also based on tweet analysis through which a sentence is classified as positive, negative, or neutral [5]. All these studies and models explain the utility or application of sentiment analysis similarly; we put forward a new application or approach for the use of sentiment analysis.

Our application analyzes the text used in the google search engine to detects the sentiment of the query, whether it is sad, joy, love, anger, fear, and surprise. So, to brief when someone intends to do a text based google search, our model identifies the sentiment of that sentence. This can be useful in identifying potentially dangerous or abnormal behaviors that may lead to adverse effects. We understand that no one wants to know the polarity or sentiment of their statements, but it can be used by monitors or guardians who need to check on their clients or wards. The central theme of our model is that as every teen is using the Internet, this chrome extension can act as parental monitoring of their digital web life. It may also be helpful in the detection of the person's psychological condition in some suicidal, abusive cases. To achieve this successfully, we used machine learning techniques to train the sentiment predicting model and python programming language libraries as a coding framework. It provides many libraries, which are easy to use in collecting, processing, and visualizing the data. Javascript is used to build and collaborate the chrome extension with the sentiment analysis model. Further in section ii, and section iii, which includes literature survey, and proposed model respectively, every element, and function were explained thoroughly.

2 Related Works

There have been a lot of studies on sentiment analysis in the last few years. In all the recent studies, sentiment analysis is mostly used in e-commerce sites and online review sites to extract generalized user opinions on different aspects of products. These reviews by the users are segregated into different categories as per the requirement; here, we can understand that the objective is to classify text based on the polarity of the meaning. For instance, Minal T [6] created four language based models for sentiment analysis of covid tweets in Nepal pandemic out break. This sentiment analysis of tweets is used to evaluate, and analyze these tweets, to look into the requirements, and adoptions essential at the movement. This would be an added advantage to the government, or firms working in that sector. And Sangeetha lal et al. [7] in their study created a model to distinguish between crime tweets from normal tweets on the micro-blogging platform Twitter. To extract text from Twitter, a text-mining [8] based approach is used, and four classifiers are used for text classification; Random Forest, Naive Bayesian, ZeroR, and J48. The study results show that RF classifiers have the highest accuracy of 98.1%, and the ZeroR classifier performs the lowest, with a 61.5% accuracy. The authors strongly believe that automated detection or classification of the crime tweets can be effective in better managing crimes reported on Twitter. This paper has some similarity to our idea in a few aspects; the difference is that our aim is to find different sentiment like; sadness, happiness, surprise, love, etc. in the text, and here the authors [7] aim is only to find whether the text is related to crime, or not.

Similarly, the research done by Badjatiya et al. [9] is concerned with detecting hate speech in tweets and determining whether, or not a specific tweet is racist, sexist, or both. This paper includes the study of the applications of deep learning methods and explored various tweet semantic analysis methods like; Bag of Words Vectors (BoW) over Global Vectors for Word Representation (GloVe), char n-grams, word Term Frequency Inverse Document Frequency (TF-IDF) values, and task-specific embeddings learned using FastText, LSTMs, and CNNs. LSTMs captured long-range dependencies in tweets, which played a vital role in hate speech detection. This research concludes that the semantic identification of the text, when learned from deep neural network models, combined with gradient boosted decision trees resulted in the best accuracy values. Though the process is different, the notion of this paper is similar to our main goal. However, in our model, we defined different emotions rather than sexist or racist, such as happiness, sadness, surprise, and so on. A few more similar studies in this zone, wherein Anam et al. in their work [10], described about using machine learning models to distinguish between happy and unhappy texts. Suicidal tweets are identified with the help of machine learning algorithms and semantic analysis in a study by Marouane et al. [11]. Yarkareddy et al. in their research [12], sentiment analysis is performed on amazon food reviews to analyze whether the given review is positive or negative for a better understanding of the product.

The work by Jack [13], provided us with a way out to gain a prerequisite understanding of the chrome extension. In this case, a chrome extension is built to automate a simple task in the Reddit application. An option is given to open a new tab consisting of the subreddit's Top Posts of All Time, whenever that subreddit is highlighted or selected. Here, two files background.js and a manifest.json were used to build this simple model. This work provided us with the spine to construct our groundwork while developing the chrome extension for our model.

3 Proposed Model

This section explains the proposed model; every detail is elaborated and used in the process of building our algorithm, integrating the modules, and skills and techniques used in our model are also briefed.

3.1 Sentiment Analysis

Sentiment analysis is done using machine learning techniques. For creating a distinct sentiment detection model, there is a need for the data in categorized sentiment labels. So, we used Kaggle to find datasets with various types of texts that represent distinct emotions. As there is very limited text data present in each dataset, we merged a few datasets in order to increase the number of texts available for training our model and to increase the range of emotions that can be segregated, which also can result in better output prediction. So, the total statements used for the training sentiment analysis model are 56,824, which belong to different categories, such as sadness, happy, surprise, fear, anger, love, and normal. And these texts are undergone a long pre-processing using NLP techniques [14, 15] to get into a fine shape before being used for training as shown in Fig. 1.

As mentioned earlier, the raw text will get preprocessed in below eight steps.

1. Tokenize: Breaking the text into small words, called tokens using word-tokenize method from nltk natural language proccessing library.
2. Normalize: With the aid of LancasterStemmer function all those tokens are converted into it's canonical or base form.
3. Remove Non-ASCII: Removing special characters, non-alphanumeric characters with the assistance of python replace method.
4. To lowercase: Using python lower method all characters are converted into lowercase for better synchronization.

Fig. 1 Text preprocessing

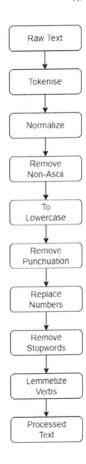

5. Remove punctuation: If any punctuation is left, they are cleaned with sub function from Regular expressions (re) module in python.
6. Replace numbers: All numeric characters are dismissed by checking each character with isdigit() inbuilt function from python.
7. Remove stopwords: All English stop words are downloaded, and removed so that only keywords are left out.
8. Lemmetize verbs: Every word is returned to its dictionary form using WordNet lemmetizer method.

The processed text, as shown in Fig. 2, is trained against multiple machine learning algorithms; SVM Classifier, Logistic Regression, Random Forest, and Naïve Bayes, from which the best algorithm with the highest accuracy or minimum error is chosen are our final sentiment analyzing the model. Hence, After a thorough comparative study, Random Forest Classifier has resulted in the returning desired output with high accuracy, which made us choose this as the appropriate algorithm, best suitable for our objective.

Fig. 2 Processed text

Before Text Preprocessing

	text
0	RT @SchudioTv: Want to know more about #autism...
1	We blame ourselves and feel worse. Start with ...
2	RT @PsychiatristCNS: 130,000 patient years and...
3	RT @SkypeTherapist: See a therapist online ove...
4	RT @PsychiatristCNS: 130,000 patient years and...

After Text Preprocessing

	text
0	rt schudiotv want know autism anxiety watch sh...
1	blame feel worse start selfcompassion deserve ...
2	rt psychiatristcns one hundred and thirty thou...
3	rt skypetherapist see therapist online skype l...
4	rt psychiatristcns one hundred and thirty thou...

3.2 Chrome Extension

The chrome extension built has two important modules; a background.js file and a manifest.json, which are used for successful execution. The manifest.json file tells Chrome about the required crucial information about the extension out layer, its parts, and how to handle each one. And the background script (background.js file) listens for key events or actions in the browser and reacts with specific code. To brief, the javascript file is injected, which detects the search bar changes and sends that to a service worker. After receiving these changes, it is further sent to a web socket server and is processed via a machine learning model. The final output predicted from our sentiment analysis model is sent to the foreground and a simple Html pop-up is displayed with the predicted result (sentiment of the input text). We can simply set up the chrome extension on our local machine by switching to developer mode on the extensions page and uploading our source code folder to apply our desired functionalities (Fig. 3).

To summarize, the text entered in the search engine is taken as input to the Random Forest model via chrome extension using service worker and WebSocket server. The trained Random Forest model and its weights are saved using a library called joblib and are further used in our python script to establish a connection between our pre-trained model and extension. The predicted output from the model will be displayed on the web page immediately, or as per the requirement.

Fig. 3 Flow chart of
proposed model

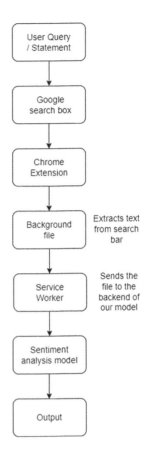

4 Experimental Analysis and Results

Succeeding with the implementation of the model, in the testing phase several experiments were done to obtain a report on the working of the model. Through the process, several datasets were used, and predictions were made, which were briefed in the later paragraphs respectively.

4.1 Dataset Description

There are a lot of datasets scattered over the internet; we managed to retrieve datasets with our requisites, like different types of texts with different emotions from the Kaggle website. We choose three different datasets, Tweet Sentiment and Emotion Analysis [17], and Emotion Dataset, respectively, for training the Emotion detection

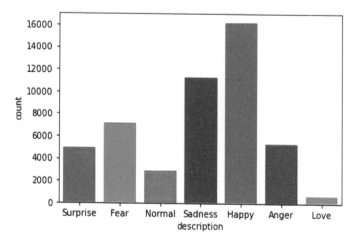

Fig. 4 Count of each emotion in final dataset

model. And all the datasets are combined and synchronized to execute coordinately (Fig. 4).

Tweet Sentiment and Emotion Analysis dataset consists of four columns [16]; sentiment, text, user, label. Since all these are tweets, the user column represents the username, whereas text represents tweets, on-the-other-hand sentiment represents polarity; positive, negative, and neutral. The tweet's emotion is represented by the label column. This data set has a total number of 6034 tweets and their corresponding emotion. We choose the text and label columns from this dataset. The second dataset [17] has 34792 rows, with three columns, namely 'emotion', 'text', and 'clean text', where emotion is the sentiment of the text and clean text is the preprocessed text. And from the dataset, we extracted emotion and clean text columns to combine into the final dataset. And the final dataset, "Emotion Dataset for Emotion Recognition Tasks" [18], has two columns in both sub-files, namely text and label, where the text contains statements or sentences, and the label is represented with numerical values which are further assorted with suitable emotions. All of the datasets were combined and formed into a final dataset with a total of 56824 sample sentences and respective emotions, as they were all in the same formatted into the same syntax.

4.2 Performance Evaluation

In this section, we evaluate the performance of the various models used and investigate every model. Each model is compared against other models based on their accuracies and other evaluation metrics so that we can find a suitable method, with less margin of error, for our proposed model.

Fig. 5 Confusion matrix for logistic regression testing

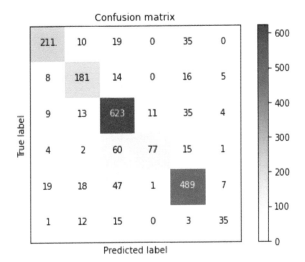

4.2.1 Logistic Regression

Logistic regression, one of the most used classification models in machine learning, was earliest used as logistic regression as a broad statistical model [19]. The field of statistics is the ground for many machine learning algorithms [20]. Logistic regression is also a technique borrowed from statistical practices. This model is known for predicting categorical variables, where the value predicted either must be categorical, discrete, or binary [21], which depends upon the data previously used. This model is also referred as the probabilistic classifier, as move forward with the probabilities obtained on each model (Fig. 5).

4.2.2 Naive Bayes

Naive Bayes [22] is the supervised classification model in machine learning, which is considered the fastest and most simple. Basically, it performs the classification of task-based upon probabilities obtained for each class. Bayes' Theorem [23] is the core idea of this classifier. And there are different types of Naive Bayes classifiers; Gaussian, Bernoulli, Multi-nominal Naive Bayes. All these classifiers work considering the value of the predictor feature is independent of other features, given the class variable [24]. Recommendation systems, Sentiment analysis, and spam filtering are the areas where this classifier is used most (Fig. 6).

Fig. 6 Confusion matrix for
Navie Bayes testing

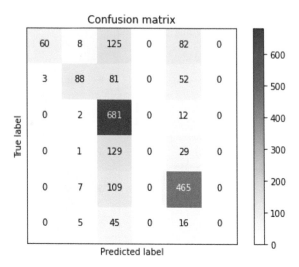

4.2.3   Random Forest Classifier

This classifier uses multiple decision tree classifiers to encounter a classification or regression model. It is adopted from the concept of ensemble learning, where complex decisions and predictions were made using multiple classifiers combined together. In sub-sections, different decision trees were given the freedom to predict the output, the majority voting technique is used for classification, and the mean value was chosen in case of regression to improve the accuracy of the prediction [25]. As different irrelevant models are working together, the decision of one tree might get correct if not the other, which made this model more outstanding than other individually working models (Fig. 7).

4.2.4   Support Vector Machine (SVM)

Support vector machine is also a learning algorithm, which can be used to predict discretely, and limited continuous values as a logistic regression model. Based on statistical learning frameworks, SVMs are titled one of the most accurate prediction algorithms [26]. In support vector machines, the model is trained over the data, plotted in an n-dimensional to create a hyperplane, or an optimal limiting boundary, which helps in categorizing the data points into classes. The extreme coordinates of the individual data points scattered over the plane, which assist in segregating classes within the face of the hyperplane, are called support vectors. And the unit controlling these support vectors to get organized for the best outputs is support vector machine [27] (Figs. 8 and 9).

    As shown in Table 1, Naive Bayes has an accuracy of 64.70, followed by Logistic Regression and SVM classifier with accuracies of 80.20 and 80.80, respectively, and

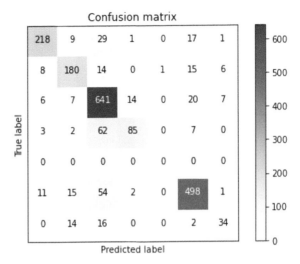

Fig. 7 Confusion matrix for random forest testing

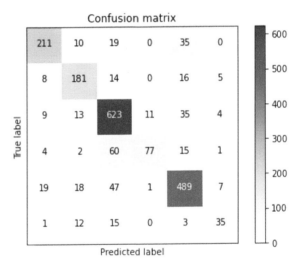

Fig. 8 Confusion matrix for SVM testing

furthermore, Random Forest Classifier has the highest accuracy of 82.80. So, we used Random Forest Classifier for our proposed model and connected it with the chrome extension with joblib. Integration is performed on the chrome extension with ML code. When we type in something on the google chrome search engine, the output is predicted, and can be stored in any desired structure based further usage of the data. (Note: that only output can be stored not the input, the query statement). In this case, the output is shown on the top right of the chrome tab in the real time, as shown in Fig. 10.

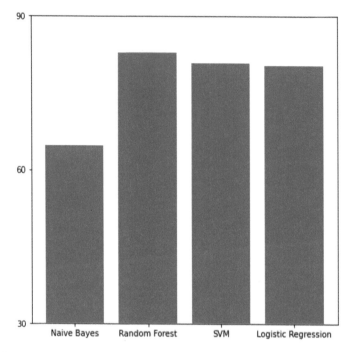

Fig. 9 Comparative analysis of ML algorithms

Table 1 Performance comparison for all models used

Model	Accuracy
Logistic regression	0.8020
Naive Bayes	0.6470
Random forest classifier	0.8280
SVM classifier	0.8080

5 Conclusion

In this paper, we developed a chrome extension to collect the texts used in the chrome search box and extract sentiment from those texts. As the function works in real-time, it does not exploit data privacy. No data repositories are used in our model, which can conclude that this extension is risk-free in terms of data security. Our main idea is to introduce a new application in the sentiment analysis research ground. Nowadays, everyone is using computers or smartphones; this extension can be implemented to know the nature of their search thoughts. This can help identify any self-harming or dangerous nature as early as possible without compromising your data privacy. Our model has achieved an acceptable level of accuracy to be implemented in real life. Furthermore, this model can be extended, where a notification or warning is sent to

Fig. 10 Sample of final result

the registered member if any self-harming search is done in the Google search engine. Thus, this can act as a parenting control system to detect sadness or depression in their children.

References

1. Walsh, S. (2021, May 26). [web log]. Retrieved from https://www.semrush.com/blog/google-search-statistics/.
2. Khurana, D., Koli, A., Khatter, K., & Singh, S. (2017). Natural language processing: State of The Art, Current Trends and Challenges.
3. Panthati, J., Bhaskar, J., Ranga, T. K., & Challa, M. R. (2018). Sentiment Analysis of Product Reviews using Deep Learning. In *2018 International Conference on Advances in Computing, Communications and Informatics (ICACCI)* (pp. 2408–2414). https://doi.org/10.1109/ICACCI.2018.8554551
4. Vanaja, S., & Belwal, M. (2018). Aspect-level sentiment analysis on E-Commerce data. *International Conference on Inventive Research in Computing Applications (ICIRCA), 2018*, 1275–1279. https://doi.org/10.1109/ICIRCA.2018.8597286
5. Bhaskar, J., Sruthi, K., & Nedungadi, P. (2014). Enhanced sentiment analysis of informal textual communication in social media by considering objective words and intensifiers. In *International Conference on Recent Advances and Innovations in Engineering (ICRAIE-2014)* (pp. 1–6). https://doi.org/10.1109/ICRAIE.2014.6909220
6. Tripathi, M. (2021). Sentiment analysis of Nepali COVID19 tweets using NB, SVM AND LSTM. *Journal of Artificial Intelligence, 3*(03), 151–168.
7. Lal, S., Tiwari, L., Ranjan, R., Verma, A., Sardana, N., & Mourya, R. (2020). Analysis and classification of crime tweets. *Procedia Computer Science, 167*, 1911–1919. https://doi.org/10.1016/j.procs.2020.03.211
8. Marti, H. (2003, October 17). [web log]. Retrieved April 26, 2022, from https://people.ischool.berkeley.edu/hearst/text-mining.html:text=Text

9. Badjatiya, P., Gupta, S., Gupta, M., & Varma, V. (2017). Deep learning for hate speech detection in Tweets. In *Proceedings of the 26th International Conference on World Wide Web Companion—WWW '17 Companion*. https://doi.org/10.1145/3041021.3054223

10. Yousaf, A., Umer, M., Sadiq, S., Ullah, S., Mirjalili, S., Rupapara, V., & Nappi, M. (2021). Emotion recognition by textual tweets classification using voting classifier (LR-SGD). *IEEE Access*. https://doi.org/10.1109/access.2020.3047831

11. Marouane Birjali, A.B.-H.M.E. (2017). Machine learning and semantic sentiment analysis based algorithms for suicide sentiment prediction in social networks. *Procedia Computer Science, 113*(1877–0509), 65–72.

12. Yarkareddy, S., Sasikala, T., & Santhanalakshmi, S. (2022). Sentiment analysis of amazon fine food reviews. In *2022 4th International Conference on Smart Systems and Inventive Technology (ICSSIT)*, pp. 1242–1247. https://doi.org/10.1109/ICSSIT53264.2022.9716410

13. Joseph, J. (2020, October 13). [web log]. Retrieved from https://levelup.gitconnected.com/make-your-first-chrome-extension-with-javascript-7aa383db2b03

14. Nair, J., Nithya, R., & Vinod Jincy, M. K. (2020). Design of a morphological generator for an English to Indian languages in a declension rule-based machine translation system. *Lecture Notes in Electrical Engineering*, 247–258. https://doi.org/10.1007/978-981-15-5558-9-24

15. Nair, J., Nair, S. S., & Abhishek, U. (2021). Sanskrit stemmer design: A literature perspective. *Advances in Intelligent Systems and Computing*, 117–128. https://doi.org/10.1007/978-981-16-3071-2-11

16. Pramudya, W. B. N., & Galande, K. (2021). Tweet sentiment and emotion analysis, Version 1. https://www.kaggle.com/datasets/subhajournal/tweet-sentiment-and-emotion-analysis?resource=download

17. Jcharis. (2021). end2end-nlp-project/data/. https://github.com/Jcharis/end2end-nlp-project/tree/main/data

18. Pandey, P. (2021). Emotion dataset for emotion recognition tasks, Version 1. https://www.kaggle.com/datasets/parulpandey/emotion-dataset

19. Wikipedia contributors. (2022). Logistic regression. In Wikipedia, The Free Encyclopedia. Retrieved 05:50, April 29, 2022, from https://en.wikipedia.org/w/index.php?title=Logistic-regression&oldid=1082953998

20. Brownlee, J. (2018). [web log]. Retrieved from https://machinelearningmastery.com/logistic-regression-for-machine-learning/

21. authors, java t point. (n.d.). Logistic regression in machine learning - javatpoint. www.javatpoint.com. Retrieved April 29, 2022, from https://www.javatpoint.com/logistic-regression-in-machine-learning:text=Logistic

22. Authors, J. T. P. (n.d.). Naive Bayes classifier in machine learning - javatpoint. www.javatpoint.com Retrieved April 29, 2022, from https://www.javatpoint.com/machine-learning-naive-bayes-classifier:text=Na

23. Authors, G. (2022). Naive Bayes classifiers. GeeksforGeeks. Retrieved April 29, 2022, from https://www.geeksforgeeks.org/naive-bayes-classifiers/

24. Wikipedia contributors. (2022). Naive Bayes classifier. In Wikipedia, The Free Encyclopedia. Retrieved 06:02, April 29, 2022, from https://en.wikipedia.org/w/index.php?title=Naive-Bayes-classifier&oldid=1075188874

25. Ajesh, A., Nair, J., & Jijin, P. S. (2016). [IEEE 2016 International Conference on Advances in Computing, Communications and Informatics (ICACCI)—Jaipur, India (2016.9.21-2016.9.24)] 2016 International Conference on Advances in Computing, Communications and Informatics (ICACCI)—A random forest approach for rating-based recommender system., 1293–1297. https://doi.org/10.1109/ICACCI.2016.7732225

26. Wikipedia contributors. (2022). Support-vector machine. In Wikipedia, The Free Encyclopedia. Retrieved 06:04, April 29, 2022, from https://en.wikipedia.org/w/index.php?title=Support-vector-machine&oldid=1079167701

27. Authors, J. (n.d.). Support Vector Machine (SVM) algorithm - javatpoint. www.javatpoint.com. Retrieved April 29, 2022, from https://www.javatpoint.com/machine-learning-support-vector-machine-algorithm

Performance of RSA Algorithm Using Game Theory for Aadhaar Card

R. Felista Sugirtha Lizy and V. Joseph Raj

Abstract Data security is ensured by the use of cryptography. Data security refers to the protection of data and privacy to prevent hackers from gaining unauthorized access to applications, computers, and data servers. Cryptography is a process of encrypting data and storing it in databases in a manner that anyone who gains access to it by accident is rendered useless. To encrypt and decrypt data, the RSA algorithm, the ECC algorithm, and other encryption methods are routinely employed. Game Theory—Rivest–Shamir–Adleman (GT-RSA) is a new hybrid algorithm that combines Game Theory and RSA to improve the efficiency of the RSA algorithm by altering the function. By improving the speed, throughput, avalanche effect, and power consumption performance of the GT-RSA algorithm, it is proposed. The performance of the GT-RSA algorithm has been enhanced, and experimental results have been shown.

Keywords Cryptography · Game theory · Mixed strategy · RSA · Throughput

1 Introduction

Cryptography encrypts data and communications so that only those with access to it may read and process it. The approaches of cryptography are based on mathematical principles and algorithms, which are a a collection of calculations based on rules that modify communications in such a way that they are difficult to decipher. Data privacy, Internet browsing, and secret transactions such as credit and debit card transactions are all protected by these algorithms [1]. They are employed in the creation of cryptographic keys, digital signatures, and verification.

Every resident Indian has an Aadhaar number, a 12-digit unique identification that comprises all of a person's information, including demographic and biometric data. Aadhaar is a significant amount of data that must be properly stored and maintained.

R. Felista Sugirtha Lizy (✉) · V. Joseph Raj
Department of Computer Science, Kamaraj College, Manonmaniam Sundaranar University, Thoothukudi, Tamil Nadu 628003, India
e-mail: 21felistaa@gmail.com

To process such massive amounts of sensitive data, several processing algorithms and privacy safeguards have been created [2].

Symmetric encryptions, such as Advanced Encryption Standard (AES), Blowfish, Data Encryption Standard (DES), Rivest Cipher 2 (RC2), Rivest Cipher 4 (RC4), Rivest Cipher 5 (RC5), and Rivest Cipher 6 (RC6), are the most basic and oldest types of encryption since they only require one secret key to encrypt and decrypt data. The key is shared by the sender and recipient, which is a severe problem because it lets an attacker listen in on the key exchange channel and decrypt the contents. To compute the secret key, the sender and receiver must communicate over a secure channel. Asymmetric encryption, on the other hand, encrypts plain text with two keys—public and private—and is used by digital signature algorithm (DSA), RSA, and elliptic curve cryptography (ECC). Anyone with the public key can send a message, but the private key is kept hidden and is only used to decrypt the message, ensuring greater security. They all include turning plaintext to ciphertext and vice versa [3].

2   Related Work

The interconnection of today's information networks has created new security concerns. In terms of safeguarding the confidentiality, integrity, availability, and authenticity of well-defined goals, Traditional security has progressed significantly. Existing techniques of prevention may become prohibitively expensive as assaults get more sophisticated and the system becomes more complicated. To perceive security from a strategic and decision-making stance, a fresh perspective and theoretical grounding are required. The aggressive and defensive exchanges between an attacker and a defender can be shown using Game Theory. It includes a quantitative security assessment, security outcome prediction and a mechanical design tool that can enable security by design and put the attacker on the defensive [4, 5].

As a result of the massive increase of information accessible on social media platforms, privacy issues have developed. Although differential privacy provides theoretical foundations for privacy protection, the privacy-data utility trade-off still has to be improved. Most previous studies, on the other hand, do not account for the adversary's quantitative influence when evaluating data utility [6].

Panda [7] looked at two RSA and ECC public key methods. Compared to the RSA approach, the ECC algorithm has various advantages, including faster computation and data transmission. The RSA algorithm is a way for creating a public key encryption system whose security is based on huge prime number factoring complexity. Data can be encrypted and digital signatures can be created using this technology. The elliptic curve algebraic structure and elliptic curve size challenges are connected to the ECC algorithm. The primary benefit of this technique is that it uses smaller keys, which decreases storage and transmission costs.

In [8] the RSA approach was used with Round-Robin priority scheduling to improve security and limit infiltration effectiveness. Low overhead, increased throughput, and confidentiality are just a few of the advantages. The user uses the RSA technology to encrypt messages, which are then prioritized and sent. The receiver decrypts messages in order of priority using the RSA technique.

Elliptic curve cryptography is powerful cryptography with unique features, such as a short key size in comparison to other key public infrastructures. The National Institute of Standards and Technology (NIST) elliptic curve random generator is used in [9] to generate a sequence of arbitrary numbers based on curves. The generation phase uses a common key and a G generator point, which is a curve generator, to generate a random sequence.

3 RSA Algorithm

Ron Rivest, Adi Shamir, and Leonard Adleman developed the RSA algorithm in 1977. This is a cryptographic algorithm for public key encryption (Hoffstein et al. 2008). If a person makes their public key available, other users can encrypt and send communications to them. Each user, on the other hand, has their own decryption key. The decryption key is significant since deriving it from the public key is extremely tough. Because the public key is created by multiplying two huge prime numbers, this is the case.

Although RSA is now considered obsolete, it is still used to encrypt messages by some organizations and the government.

The weaknesses of symmetric algorithms, such as authenticity and confidentiality, have been overcome by RSA. So that, using RSA algorithm in this research. The RSA algorithm is fairly simple to implement. For transferring secret data, the RSA algorithm is safe and secure. Because it requires sophisticated mathematics, cracking the RSA algorithm is extremely difficult. It is simple to distribute a user's public key.

Key Creation

- Choose your s and t secret primes.
- Calculate $N = s\,t$
- Decide on an encryption factor, ex, and make sure $\gcd(ex, (s-1)(t-1)) = 1$
- Make N and ex

Encryption

- Choose a plaintext message to encrypt, pm.
- Use the receiver's public key (N, ex) to compute $co \equiv pm^e \pmod{N}$
- Send the ciphertext message, co, to the receiver

Decryption

- Compute de: $de \equiv 1(\bmod\,(s-1)\,(t-1))$
- Calculate $pm' \equiv co^{de} \pmod{N}$
- Then, $pm' = $ plaintext pm

4   Proposed GT-RSA Algorithm and Analysis

Figure 1 shows the proposed GT-RSA technique, which is based on the RSA and Game Theory's Mixed Strategy approaches [10]. The GT-RSA algorithm is described in greater depth below.

4.1   Game Theory

A game is a broad term that encompasses a wide range of competitive circumstances. When two or more competent opponents are engaged, each attempting to optimize his choice at the expense of the others, Game Theory is a set of tools and approaches for making difficult decisions. An opponent is referred to as a player in Game Theory. Each player can choose from a number of strategies, which can be finite or endless. The game's results, or payoffs, are determined by the various strategies employed by each player.

1. The number of players can be two or more. A player might be a single person or a group of people who are all working toward the same goal.
2. Timing—at the same time, the opposing parties reach the same conclusion.
3. Competing goals—Each side is hell-bent on achieving its own objectives at the expense of the others.
4. Repetition—The bulk of the time, solutions are duplicated.
5. Payoff—All players are aware of the payoffs for each combination of decisions.
6. Access to information—All stakeholders have access to all critical information. Each player is aware of all of the options available to the adversary, as well as the anticipated payoffs.

Fig. 1 Block diagram of GT-RSA algorithm

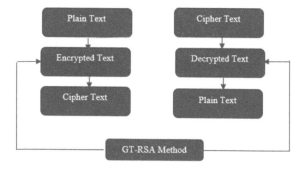

4.2 Games Are Classified in Several Ways

1. Zero-sum games are those in which the winner(s) receive the entire payment amount provided the loser's perspective (strictly competitive).
2. Nonzero sum games: In these games, one player's earnings do not equal the other's losses. Other environmental parties may benefit or suffer as a result of the benefits or losses (not strictly competitive).

4.3 Pure Strategy Characteristics of a Two-Person, Zero-Sum Game

1. There must be two participants, each with their own set of strategies.
2. In a zero-sum game, one player's losses are exactly equal to the other's gain.
3. A preset approach in which each participant is given only one option at a time is known as pure strategy.
4. Bargaining is prohibited. There is no way to strike an agreement that is beneficial to both parties.

Take a look at the game matrix below from the standpoint of player C.

	D_1	D_2	...	D_n
C_1	W_{11}	W_{12}	...	W_{1n}
C_2	W_{21}	W_{22}	...	W_{2n}
⋮	⋮	⋮	⋮	⋮
C_m	W_{m1}	W_{m2}	...	W_{mn}

A labour contract between a corporation and a union, for example, is about to expire. It is necessary to negotiate a new contract. The group (C0) agrees that the following are feasible goals for the organization to pursue after reviewing recent experience:

S1 = all-out attack; strong, forceful negotiating.
S2 = a rational and logical attitude.
A legalistic approach is denoted by the letter S3.
S4 = a conciliatory and agreeable approach.

Assume that the union is evaluating the following options:

T1 denotes a frontal attack; aggressive, harsh negotiating.
T2 is a method that is built on logic and reasoning.
The letter T3 stands for a legalistic attitude.
T4 is an approach that is nice and conciliatory.

We create the following game matrix with the help of an outside mediator:

Union gains on a conditional basis				
Union strategies	Company strategies			
	S_1	S_2	S_3	S_4
T_1	2.0	1.5	1.2	3.5
T_2	2.5	1.4	0.8	1.0
T_3	4.0	0.2	1.0	0.5
T_4	0.5	0.4	1.1	0.0

The above table, commonly known as a game matrix, has numerous interpretations. If the employer agrees S1 and the union chooses T1, the final agreement will include a P2.0 wage rise (hence, a P2.0 loss to the company). According to the table above, if the company choose T3, the union will select T1. If the union opts for U3, the firm opts for T2.

4.4 Pure Strategy Problems

Change in strategy—Due to the repetitious nature of games, both players may alter their methods. However, there is no motivation to change in pure strategy games. A player's reward is almost always lowered if the players does not follow the recommended approach [11]. Numerous ideal methods—In some games, you can utilize multiple ideal tactics at once.

The dominating row will have entries that are greater than or equal to the dominated row's comparable items (with at least one entry larger than). The entries in the dominating column will be smaller and/or equal to those in the dominated column (with at least one entry smaller than). The prominent rows and columns of the table can be eliminated.

4.5 GT-RSA

The F function [12] has been improved to the point that GT-RSA now outperforms the RSA approach in terms of speed and security. The updated F function can encrypt and decode the entire contents of a file. To reduce the amount of time required on each decryption, these procedures are carried out in three steps.

The specified texts of the full file can be encrypted and decrypted using this modified F function.

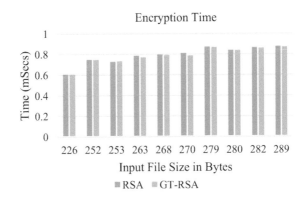

Fig. 2 Performance of GT-RSA algorithm for encryption time (in milliseconds)

5 Experimental Results

A laptop with an Intel(R) Celeron(R) CPU3865U@1.80 GHz processor. The text files in this study range in size from 226 to 289 bytes. The average of the ten values is used to compute the file size (ten times). Encryption time, decryption time, execution speed, encryption throughput, decryption throughput, and the avalanche effect were the overall performance metrics. MATLAB is used to implement the GT-RSA method.

The experimental results for a variety of overall performance measures for the GT-RSA algorithm are listed below. The implications are detailed below.

5.1 Encryption Time

Figure 2 depicts the encryption time, which is defined as the amount of time it takes to encrypt a plaintext message into ciphertext [13]. For various input sizes, the average encryption time is computed. When compared to the RSA method, the average encryption time for the GT-RSA algorithm takes the shortest time in the bar chart. The results are detailed in Table 1.

5.2 Decryption Time

One of the overall performance metrics is decryption time, which is defined as the time it takes to convert a ciphertext message into plain text at the time of decryption (see Fig. 3). Clearly, the GT-RSA algorithm takes the least amount of decryption time when compared to the RSA algorithm from the bar chart. Table 2 summarizes the findings.

R. Felista Sugirtha Lizy and V. Joseph Raj

Table 1 Performancce of GT-RSA algorithm for encryption time (in milliseconds)

Input size in bytes	RSA	GT-RSA
226	0.6023	0.6022
252	0.744	0.7434
253	0.7238	0.7309
263	0.7821	0.7673
268	0.7915	0.7876
270	0.8053	0.7828
279	0.8669	0.8633
280	0.8342	0.8341
282	0.8621	0.8571
289	0.8715	0.8666
Average time (milliseconds)	0.7884	0.7835

Fig. 3 Performance of GT-RSA algorithm for decryption time (in milliseconds)

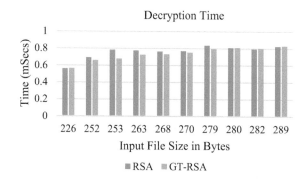

Table 2 Performancce of GT-RSA algorithm for decryption time (in milliseconds)

Input file size in bytes	RSA	GT-RSA
226	0.5603	0.5636
252	0.6932	0.6601
253	0.7824	0.6792
263	0.7758	0.7274
268	0.7643	0.7321
270	0.7694	0.7517
279	0.8349	0.7948
280	0.8095	0.8109
282	0.7943	0.8032
289	0.8249	0.8317
Average time (milliseconds)	0.7609	0.7355

Fig. 4 Performance of GT-RSA algorithm for execution time (in milliseconds)

Table 3 Performancce of GT-RSA algorithm for execution time (in milliseconds)

Input file size in bytes	RSA	GT-RSA
226	1.1626	1.3664
252	1.4373	1.3645
253	1.5061	1.4101
263	1.5579	1.4947
268	1.5558	1.5097
270	1.5747	1.5345
279	1.7018	1.6581
280	1.6439	1.6512
282	1.6564	1.6303
289	1.6965	1.6983
Average time (milliseconds)	1.5493	1.5318

5.3 Execution Time

As shown in Fig. 4, the execution time is the time it takes to construct a ciphertext from plain text and plain text from the ciphertext. The GT-RSA method has the shortest execution time when compared to the RSA algorithm, as seen in the bar chart. Table 3 shows the results.

5.4 Encryption Throughput

Figure 5 depicts the GT-RSA technique's encryption throughput with various input files. The GT-RSA method has the highest encryption throughput when compared to the RSA algorithm, as shown in the bar chart.

The encryption throughput is calculated by dividing the full plaintext in Megabytes by the encryption time in seconds.

Fig. 5 Performance of
GT-RSA algorithm for
encryption throughput
(KB/s)

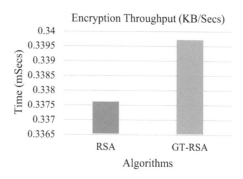

Table 4 Performancce of
GT-RSA algorithm for
encryption throughput

Algorithms	RSA	GT-RSA
Throughput (KB/s)	0.3376	0.3397

Total plaintext in megabytes/encryption time = Throughput.
Table 4 summarizes the findings in great detail.

5.5 *Decryption Throughput*

Figure 6 depicts the GT-RSA technique's decryption throughput with various input
files. When comparing the GT-RSA algorithm to the RSA algorithm, the bar chart
clearly reveals that the GT-RSA algorithm has the highest decryption throughput.

Decryption throughput is calculated by multiplying the whole plaintext in
Megabytes by the decryption time in seconds.

Total plaintext in megabytes/decryption time = Throughput.

Table 5 presents a summary of the findings.

Fig. 6 Performance of
GT-RSA algorithm for
decryption throughput
(KB/s)

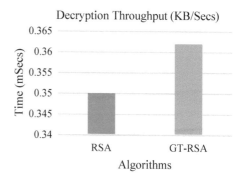

Table 5 Performancce of GT-RSA algorithm for decryption throughput	Algorithms	RSA	GT-RSA
	Throughput (KB/s)	0.3499	0.3619

Fig. 7 Performance of GT-RSA algorithm for execution throughput (KB/s)

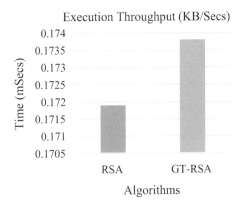

Table 6 Performancce of GT-RSA algorithm for execution throughput	Algorithms	RSA	GT-RSA
	Throughput (KB/s)	0.1719	0.1738

5.6 Execution Throughput

Figure 7 shows the GT-RSA algorithm's execution throughput with various input files. When compared to the RSA algorithm, the GT-RSA algorithm offers the highest throughput.

The execution throughput is calculated by dividing the total plaintext in megabytes by the execution time in seconds.

Total plaintext in megabytes divided by execution time equals throughput. The findings are summarized in Table 6.

5.7 Power Consumption

The aforementioned statistics clearly show that the GT-RSA method, which has the highest execution throughput when compared to the RSA algorithm, will consume the least amount of power [14].

Fig. 8 Avalanche effect

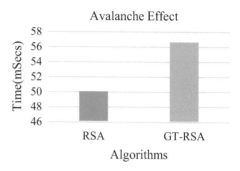

Table 7 Avalanche effect

Encryption technique	Avalanche effect
RSA	50
GT-RSA	56.6

5.8 Avalanche Effect

The avalanche effect happens when a single bit of the original text or one bit of the key scheduling techniques is modified, causing changes in several bits of the ciphertext. As a result, the higher the avalanche value, the greater the security.

Figure 8 depicts the avalanche effect of the RSA and GT-RSA algorithms with various input data files. When comparing the RSA and GT-RSA algorithms, the bar chart clearly reveals that the RSA algorithm has the lowest avalanche impact. Table 7 summarizes the findings.

6 Conclusion

Three factors must be established for an encryption system to be complete and secure: confidentiality, authenticity, and integrity. In sensor networks, confidentiality refers to the capacity to hide messages from an attacker. The capacity to detect unread or unrecognized messages is known as integrity. Authentication refers to the message's origin's trustworthiness. Cryptography is the study of concealing and verifying data, and it encompasses rules, methods, and secure procedures for preventing unauthorized access to sensitive data. Encryption is a method of ensuring the integrity of a connection's components. An encryption algorithm cannot supply all encryption principles on its own. A cryptographic system's security is ensured by a hybrid encryption technique that combines symmetric and asymmetric encryption techniques. The proposed algorithm is the most time, memory, and complexity efficient. The proposed the Game Theory—Rivest–Shamir–Adleman (GT-RSA) as a new hybrid algorithm that combines Game Theory and RSA to improve the efficiency of the RSA algorithm by altering the function. Then, the novelty in this research work is to improve the speed and security of Aadhaar card. The intricacies and fundamental concepts of the approach are given in such a way that the algorithm can be put into practice. The RSA and Game Theory hybrid algorithm has been proven secure by hackers.

References

1. Dhivya, N. (2020). Network security with cryptography and steganography. *IJERT Journal, 8,* 1–4. ISSN 2278-0181
2. Jayashree, R. (2019). Analysis of aadhaar card dataset using big data analytics. *Emerging Trends in Computing and Expert Technology, LNDECT, 35,* 1208–1225
3. Al-Shabi, M. A. (2019). A survey on symmetric and asymmetric cryptography algorithms in information security *IJSRP, 9,* 576–589. ISSN 2250–3153
4. Casey, W., Morales, J. A., Wright, E., Zhu, Q., & Mishra, B. (2016). Compliance signaling games: Toward modeling the deterrence of insider threats. *Computational and Mathematical Organization Theory, 3,* 318–349.
5. Chen, J., Touati, C., & Zhu, Q. (2017). A dynamic game analysis and design of infrastructure network protection and recovery. *ACM SIGMETRICS Performance Evaluation Review, 2,* 128.
6. Cui, L., et al. (2019). Improving data utility through game theory in personalized differential privacy. *Journal of Computer Science and Technology, 34,* 272–286.
7. Panda, M. (2014). Security in wireless sensor networks using cryptographic techniques. *American Journal of Engineering Research, 3,* 50–56.
8. Shankar, M., & Akshaya, P. (2014). Hybrid cryptographic technique using RSA algorithm and scheduling concepts. *International Journal of Network Security & Its Applications, 6,* 39–48.
9. Shahryar, T., Fathi, M. H., & Sekhavat, Y. A. (2017). An image encryption scheme based on elliptic curve pseudo-random and advanced encryption system. *Signal Processing, 141,* 217–227.
10. Joseph Raj, V., & Felista Sugirtha Lizy, R. (2020). Performance enhancement of RSA using runge-kutta technique. *Test Engineering and Management* 11850–11857.
11. Felista Sugirtha Lizy, R., & Joseph Raj, V. (2021). Image encryption using RK-RSA algorithm in aadhaar card. *Turkish Journal of Computer and Mathematics Education,* 4683–4693.
12. Felista Sugirtha Lizy, R., & Joseph Raj, V. (2021). Improvement of RSA algorithm using euclidean technique. *Turkish Journal of Computer and Mathematics Education,* 4694–4700.
13. Josephraj, V., & Shamina Ross, B. (2016). Security evaluation of blowfish and its modified version using GT's one-shot category of nash equilibrium. *International Journal of Control Theory and Applications,* pp. 4771–4777.
14. Josephraj, V., & Shamina Ross, B. (2015). Enhancing the performance of blowfish encryption algorithm in terms of speed and security by modifying its function. *International Journal of Applied Engineering Research, 10*(79), 621–624.

Drought Prediction Using Recurrent Neural Networks and Long Short-Term Memory Model

P. Shobha⬤, **Kabeer Adlakha**⬤, **Shivam Singh**⬤, **Yash Kumar**⬤, **Mukesh Goit**⬤, **and N. Nalini**⬤

Abstract Predicting drought is the process of forecasting and classifying the dryness of the weather. The availability of long term meteorological data in the form of time-series data allows for various multivariate time-series forecasting techniques to be applied on it. Datasets consist of various meteorological parameters such as 'Precipitation' and 'Surface Pressure', and a target variable which specifies the degree of dryness. Various machine learning techniques such as recurrent neural networks (RNN) and long short-term memory (LSTM) can be used to forecast the value of score for various lead times. RNN is used to forecast for shorter lead times of 1 to 2 weeks, whereas LSTM is used for forecasting for lead times of 8 to 12 weeks. The results of the prediction can prove to be helpful in many ways. Prior knowledge of weather conditions allows us to minimize the catastrophic effects of extreme weather conditions given appropriate actions are taken for mitigation and prevention. In order to make an action plan for mitigation, it becomes necessary to predict weather conditions as well as the severity of it. Mitigation strategies can include alternate crop recommendations to farmers and special aid to regions which are at most risk.

Keywords Recurrent neural network (RNN) · Long short-term memory (LSTM) · Time-series data

1 Introduction

Extreme weather conditions can have an adverse impact on various industries, the environment and society in general. They can lead to catastrophic events such as crop failures and forest fires. Tools which can provide insights about future weather conditions with some certainty can prove to be helpful and allow for some action plans to be developed for mitigation and prevention purposes. Thus, making weather forecasting an important and active area of research.

P. Shobha (✉) · K. Adlakha · S. Singh · Y. Kumar · M. Goit · N. Nalini
Department of Computer Science and Engineering, Nitte Meenakshi Institute of Technology, Yelahanka, Bangalore, Karnataka 560064, India
e-mail: shobha.p@nmit.ac.in

S. Shakya et al. (eds.), *Sentiment Analysis and Deep Learning*,
Advances in Intelligent Systems and Computing 1432,
https://doi.org/10.1007/978-981-19-5443-6_8

Machine learning has proven to be a desirable technique to carry out complex tasks such as forecasting and classification. With huge amounts of weather and meteorological data being generated and collected, it has become possible to create time-series datasets using this data and then apply various machine learning algorithms on it to generate fruitful results. The dataset is split into train, test and validation sets. Several techniques such as autoregressive integrated moving average (ARIMA), seasonal autoregressive integrated moving average (SARIMA), linear regression, vector auto regression (VAR), artificial neural networks (ANN), long short-term memories (LSTM) have been used on time series datasets to carry out the forecasting tasks.

1.1 Literature Review

Algorithms such as support vector machine(SVM) and artificial neural networks(ANN) are used for making predictions for a lead time of 24, 48, and 72 h [1, 2].

Long lead time drought forecasting using lagged climate variables and a stacked long short-term memory model: In this paper, the authors have put forth the idea that droughts are not preventable but they are predictable, if early warning systems are developed. In order to forecast drought, meteorological data of the past 30 years was used. Indices such as standardized precipitation evaporation index (SPEI) were used as drought indicators. Drought severity was categorized into 7 categories based on the SPEI value. Considering the overall complexity of the system and the fact that the dataset is in the form of time-series data, deep learning algorithms such as recurrent neural networks (RNN) and long short-term memory (LSTM) were used. LSTM were found capable of retaining information for a longer period of time due to its recurrent and gate architecture. Various algorithms such as Support vector machine (SVM), artificial neural networks (ANN), random forest algorithm (RF), and LSTM were used to make predictions for a lead time of 1 month. LSTM was found to outperform other algorithms and thus, was used to forecast drought for higher lead times. Authors were able to obtain a threat score of 0.93 and 0.91 for lead times of 1 and 3 months, respectively. The performance of the model dropped with increasing lead times [3–5].

Certain parameters like North Atlantic Oscillation Index, air pressure, etc., were found to be of much more importance than the rest. The authors also concluded that seasonality had an important role in determining the accuracy of prediction [6–8].

Nationwide Prediction of Drought Conditions in Iran Bases on Remote Sensing Data: Mahdi Jalili et al. in this study suggest getting information about intensity, duration, and spatial coverage of drought in order to minimize the impact of drought episodes. This information becomes crucial when we talk about a country like Iran which falls in the drier part of the world and is frequently struck by droughts. Due to unavailability of long-term meteorological data for various parts of the country authors resorted to using satellite based remote sensing data that give information about vegetation cover and land cover. An artificial neural network model was constructed to forecast drought. The authors were able to get an accuracy of up to 90% by applying the model to those regions for which precipitation index (PI) normalized in a period (e.g., 30 years) was available. The study also found TCI to be the best feature to be extracted from satellite images. Also, the authors concluded Multilayer Perceptron was found to be better than radial basis function and support vector machines for predicting drought episodes [9–11].

In "A Novel Approach for Early Prediction of Drought" the authors have proposed to make use of LSTM models to predict the values of standard precipitation index (SPI) and normalized difference vegetation index (NDVI). The dataset consists of attributes such as location, precipitation, cumulative precipitation, minimum temperature, average temperature, maximum temperature, surface soil moisture, percent soil moisture, SPI, and NDVI for five districts of India. Stacked LSTM with autoencoder was used to make predictions for a lead time of 30 days. Autoencoder LSTM acts as feature extractor whereas Stacked LSTM acts as forecaster. The authors were able to obtain the best result of root mean squared error (RMSE) as 0.238 and 0.053 for SPI and NDVI, respectively [12–14].

2 Methodology

A five-step methodology can be developed for drought prediction.

2.1 Read the Data

Dataset consisting of various meteorological parameters is read into a data frame. The dataset is split into test, train, and validation sets. Each set is then read into a separate data frame.

2.2 Data Cleaning

This step is the first step in data pre-processing. It involves cleaning the data and removing the NaN values.

Scaling time-series data: The features are standardized by scaling to unit variance. In addition to that time-series is realized by replacing the values from the date column by consecutive integers starting from 0. Then, data for a single region is extracted.

2.3 Data Preparation and Visualization

RNN and LSTM require data to be fed in three-dimensions—[batch-size, Time-Steps, Number of elements in a single time-step].

2.4 Applying RNN and LSTM

Trained RNN and LSTM models are used to make predictions for various lead times. Drought prediction with RNN and LSTM models is ideal because each hidden layer of the neural network feeds its output to the next layer and also feeds itself the output at time-step 't' while training on data at time-step 't + 1'. As a result, RNN/LSTM attempt to train on a sequence of data rather than individual data records, which is a necessity for predicting tasks on time-series datasets. The architectural design is shown in Fig. 1.

3 Results

3.1 Experimental Methodology

The experimental methodology consists of extracting the meteorological data for a single region, from the pre-processed data-frame. Then, a fixed number of rows from this data are fed as input to a trained RNN or LSTM model in the form of a 3D matrix and score value is predicted for some lead time.

Input: First 10 weeks of meteorological data (week 0 to week 9) for a single region (fips code = 1) is given as input to a RNN model. Prediction is made for a lead time of 1 week, that is, for the 10th week, as shown in Fig. 2.

3.2 Results

In this paper, mean squared error (MSE) metrics used as loss function and encountered a loss of 0.80 as shown in Fig. 3.

4 Conclusion

Occurrence of drought is dependent on a number of meteorological parameters. In order to get accurate results, proper preprocessing and scaling of data is necessary

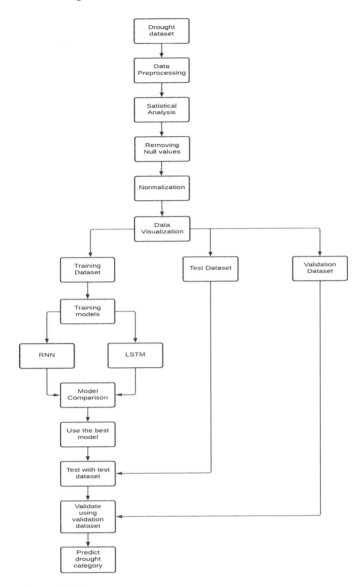

Fig. 1 Architectural design

before applying machine algorithms on the dataset. Most of the papers made use of ANN and LSTM to forecast drought. Some papers forecasted drought and categorized it in only 2 categories, namely, drought or no drought, whereas other papers have categorized it in multiple categories ranging from extremely dry to extremely wet. ANN and RNN were found to be useful in forecasting drought for shorter lead times, whereas LSTM was used for forecasting for longer lead times. The influence of

```
test_X[0:100]
```

	fips	date	PRECTOT	PS	QV2M	T2M	T2MDEW	T2MWET	T2M_MAX	T2M_MIN	T2M_RANGE	TS
0	1001	0	0.0133381	0.912798	0.434486	0.670724	0.78623	0.785817	0.60954	0.707917	0.20993	0.650301
1	1001	1	0.000296402	0.914408	0.39172	0.64986	0.761515	0.76111	0.643739	0.686586	0.35055	0.630462
2	1001	2	0	0.927824	0.177434	0.526281	0.583855	0.584309	0.506943	0.572542	0.273909	0.515696
3	1001	3	0.000177841	0.929702	0.21929	0.554634	0.629915	0.618322	0.545899	0.536305	0.463179	0.542943
4	1001	4	0.0343233	0.908774	0.16151	0.503812	0.562189	0.552382	0.480329	0.487625	0.411863	0.493973
...	...	...	...	...	...	...	...	...	...	...	...	...
95	1001	95	0.0343233	0.907164	0.644222	0.775578	0.884128	0.883042	0.763435	0.799809	0.384872	0.751256
96	1001	96	0	0.930775	0.175614	0.620837	0.581448	0.581261	0.640396	0.577465	0.607797	0.578729
97	1001	97	0.0075286	0.896163	0.652866	0.765414	0.886054	0.885128	0.735665	0.800766	0.310563	0.740457
98	1001	98	0	0.935605	0.178344	0.603852	0.585781	0.586395	0.616611	0.619308	0.444185	0.586765
99	1001	99	0	0.918165	0.246588	0.615487	0.656074	0.652976	0.642325	0.592643	0.575808	0.599196

100 rows × 20 columns

Fig. 2 Test input

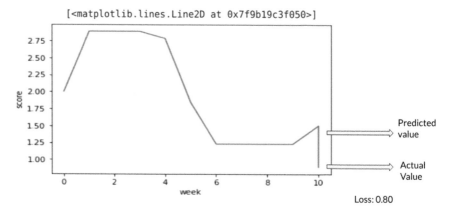

Fig. 3 Score versus week output plot

climatic variables greatly affected the forecasting results at higher lead times. Hybrid deep learning models such as CNN-LSTM need to be used to improve forecasting at higher lead times [15].

References

1. Dakshin, D., Rupesh, V. R., & Praveen Kumar, S. (2019). Water hazard prediction using machine learning. *International Journal of Innovative Technology and Exploring Engineering Regular Issue, 9*(1), 1451–1457. https://doi.org/10.35940/ijitee.a4245.119119
2. Elsafi, S. H. (2014). Artificial neural networks (ANNs) for flood forecasting at Dongola Station in the River Nile Sudan. *Alexandria Engineering Journal, 53*(3), 655–662. https://doi.org/10.1016/j.aej.2014.06.010

3. Dikshit, A., Pradhan, B., Alamri, A. M. (2021). Long lead time drought forecasting using lagged climate variables and a stacked long short-term memory model. *Science of The Total Environment 755*(Part 2), 142638. https://doi.org/10.1016/j.scitotenv.2020.142638

4. Abbot, J., & Marohasy, J. (2014). Input selection and optimisation for monthly rainfall forecasting in Queensland, Australia, using artificial neural networks. *Atmospheric Research, 138*, 166–178. https://doi.org/10.1016/j.atmosres.2013.11.002

5. Aghakouchak, A., Farahmand, A., Melton, F. S., Teixeira, J., Anderson, M. C., Wardlow, B. D., & Hain, C. R. (2015). Remote sensing of drought: Progress, challenges and opportunities. *Reviews of Geophysics, 53*(2), 452–480. https://doi.org/10.1002/2014rg000456

6. Felsche, E., & Ludwig, R. (2021). Applying machine learning for drought prediction using data from a large ensemble of climate simulations. https://doi.org/10.5194/nhess-2021-110

7. Bueh, C. & Nakamura, H. (2007). Scandinavian pattern and its climatic impact. *Quarterly Journal of the Royal Meteorological Society, 133*, 2117–2131. https://doi.org/10.1002/qj.173

8. Trenberth, K. (2011). Changes in precipitation with climate change. *Climate Research, 47*(1), 123–138. https://doi.org/10.3354/cr00953

9. Jalili, M., Gharibshah, J., Ghavami, S. M., Beheshtifar, M., & Farshi, R. (2014). Nationwide Prediction of Drought Conditions in Iran Based on Remote Sensing Data. *IEEE Transactions on Computers, 63*(1), 90–101. https://doi.org/10.1109/tc.2013.118

10. Heim, R. R. (2002). A review of twentieth-century drought indices used in the United States. *Bulletin of the American Meteorological Society, 83*(8), 1149–1166. https://doi.org/10.1175/1520-0477-83.8.1149

11. Silva, R. P., Dayawansa, N. D., & Ratnasiri, M. D. (2007). A comparison of methods used in estimating missing rainfall data. *Journal of Agricultural Sciences, 3*(2), 101. https://doi.org/10.4038/jas.v3i2.8107

12. Kansara, M., Maity, P., Malgaonkar, H., & Save, A. (2020). A Novel Approach for Early Prediction of Drought. In *2020 6th International Conference on Advanced Computing and Communication Systems (ICACCS)*. https://doi.org/10.1109/icaccs48705.2020.9074405

13. Aswathi, P. V., Nikam, B. R., Chouksey, A., & Aggarwal, S. P. (2018). Assessment and monitoring of agricultural droughts in Maharashtra using meteorological and remote sensing based indices. *ISPRS Annals of the Photogrammetry, Remote Sensing and Spatial Information Sciences, IV-5*, 253–264. https://doi.org/10.5194/isprs-annals-iv-5-253-2018

14. Dodamani, B. M., Anoop, R., & Mahajan, D. R. (2015). Agricultural drought modeling using remote sensing. *International Journal of Environmental Science and Development, 6*(4), 326–331. https://doi.org/10.7763/ijesd.2015.v6.612

15. Dikshit, A., Soleimanimatin, S., & Abdollahi, A. (2020, December 12). Using deep learning for Meteorological Drought Forecasting for New South Wales, Australia. Retrieved from https://www.youtube.com/watch?v=fiREt30Jj7g LNCS Homepage, http://www.springer.com/lncs. Last accessed 21 Nov 2016.

Recogn-Eye: A Smart Medical Assistant for Elderly

U. Sai Sreekar, V. Vishnu Vardhan, and Linda Joseph

Abstract In recent days, individuals work on PCs for quite a long time. This leads to lack the opportunity and willpower to deal with themselves. The chaotic timetables and utilization of unhealthy food influences the soundness of individuals, mostly heart. So, this research study executes a coronary illness expectation framework by utilizing the information mining strategy called naive Bayes and k-implies grouping calculation. It is a combination of both the calculations. This paper gives an outline for something very similar in nature. It helps in anticipating the coronary illness by utilizing the different properties to predict the result as in the expectation structure. It uses k-implies calculation for gathering distinct qualities and naive Bayes calculation for prediction. Nowadays, many elderly people fail to take their medications for a variety of reasons, the most common of which is mental stress. As a result, it may extend the time required to recover from illnesses. When elderly individuals gulp medication, the dose quantity becomes incorrect, resulting in a significant problem. As a result, the patient must take the required drugs at the precise dose and timing. Hence, this research work has developed a novel smart medication dispenser (SMD) technique to overcome these difficulties. The ATmega328p and ESP8266 models are used in the proposed system. The proposed system also includes microcontroller integrated circuit, LCD display, and Real-time clock (RTC) microcontroller, LCD and real-time Clock (RTC), and an Android application for checking the status.

Keywords RGB images · Image compression · k-mean clustering · Mean square error (MSE) · Structural similarity index measure (SSIM) · Peak signal-to-noise ratio (PSNR) · Malls · Clustering · Social media · Websites

U. Sai Sreekar (✉) · V. Vishnu Vardhan · L. Joseph
Department of Computer Science and Engineering, Hindustan Institute of Technology and Science, Chennai, India
e-mail: 18113139@student.hindutauniv.ac.in

V. Vishnu Vardhan
e-mail: 18113140@student.hindustanuniv.ac.in

L. Joseph
e-mail: Lindaj@hindustanuniv.ac.in

© The Author(s), under exclusive license to Springer Nature Singapore Pte Ltd. 2023
S. Shakya et al. (eds.), *Sentiment Analysis and Deep Learning*,
Advances in Intelligent Systems and Computing 1432,
https://doi.org/10.1007/978-981-19-5443-6_9

1 Introduction

The recent medical advancements have guaranteed a longer average lifespan of human beings. The availability of modern medical drugs can successfully treat even the most severe ailments. In most cases, patients must take medication at least once or twice a day, and they often forget to do so. Most medications must be taken on a regular basis or at a specific time of a day in order to be effective. Patients especially the elderly population may forget to take their medication, making it difficult to get the desired result. The Internet of Things (IoT) in healthcare offers a glimmer of hope since it has the potential to make medical facilities more efficient and improve patient care. Throughout the case of a medical emergency, real-time tracking could save lives, and this technology could indeed enhance the standard as well as the well-being of patients. According to a research, remote patient monitoring has resulted in a 50% decrease in 30-day readmission rates for heart failure patients.

2 Literature Survey

Nowadays, there are different ventures are centered on IOT-based wellbeing checking frameworks using remote sensor techniques. At first, the scientists built a patient health monitoring system using an Atmega-8 micro controller and a sensing network. The sensors used in this project include a temperature sensor, an ECG sensor, and a heartbeat sensor. Arduino Uno is used in the proposed setup for monitoring patient health indicators. In the past several years, Arduino has grown tremendously in popularity, making it possible for users to share data. Sensor data is gathered, stored, and analyzed in this area of the system. Several sensors may be connected or wireless, depending on the application at hand. Additionally, there are certain drawbacks, such as the necessity for high-resolution, precise identification, which necessitates high-quality data and the ability to forecast in arbitrary settings [1–3]. Embedded sensors on the human body allow doctors to monitor their patients' health without disrupting their treatment plans, and the precise noticing system includes sensors, micro controllers, displays, and GSM modems to transmit or receive achievement or any hazard occasions from the qualified professional. A similar GSM modem is thus utilized at a doctor's office. It takes some time and effort to inform patients but also their family and friends regarding the actual results, but it will be performed in a simple manner with no interference or mix-ups [4–7]. GSM modules were integrated into pulse-checking devices in a few cases. Essentially, this type of device evaluates the information received by sensors and conveys it to small regulator, where the microcontroller delivers it over the air by using the GSM technology. The transmitter distributes data via SMS to the appropriate persons by resulting in a functional reaction for perceptual structures on every structural as well as part level. GSM is used to transmit the results of boundary checks. This building has an office, where the patients may be screened for circulatory strain. There's no doubt that the GSM module specialist is responsible for the prosperity border.

Stating few

It is possible to create an electric medicine dispensing unit using a pic microcontroller, keyboard, and LCD display, allowing for human management of a patient's medicine regimen. The patient is alerted by an alarm sound when he receives the medication. Additionally, if the medication is not taken, a SMS is delivered to career's phone number.

1. Infrared sensors and Arduino micro controllers with alarms are used to construct pill dispensers [8]. Patients may benefit from the system if they take their medication at the appropriate time. The pop-up window was used to set up the alarm system. Notifications on a smartphone
2. Yet another Arduino-based pill box to avoid overdosing, the micro controller only administers one pill at a time [9]. When pill is poised to be consumed, it sends an SMS message to the user. He was also linked to an android app that is utilized by personnel to adjust date and time of medication dispensing.
3. The SMD system suggested in this paper is the first of its kind since it contains unique characteristics that are not found in some other vending machine [10–13]. It's only patients and caregivers with the proper documentation that may access ONE accounts on behalf of their patients. On top of that, a few stats about oral pills and their alarm systems are also included. Project design was also supported by the project's online database of consumers, medicines, and alarms. Android applications may be used to adjust and generate alarms from a smartphone.

3   Problem Statement

The advancements in medicine have increased human longevity. There are currently medications that can successfully cure even the most intractable of ailments. In most circumstances, patients must take medication at least once or twice a day, and they often forget to do so. Most medications must be taken on a regular basis or at specific schedule of day in order to be effective. Patients in this sector may forget to take their medication, making it difficult to attain the desired effect.

4   Proposed Method

Module 1:

Sensing the patients' physiological parameters in this module includes detecting the patients' temperature and pulse as shown in Fig. 1. The DHT11 sensor has three pins for power, ground, and output. The Arduino pulse sensor is an excellent plug-and-play heart rate sensor. There is a 24-in cable with color-coded (male) header connectors. The skin of the people is in direct contact with the heart logo. On the rear part of the sensor, there is a small LED hole. These two sensors may be connected by using a breadboard or an Arduino. A serial monitor/serial plotter included into the Arduino serves as a visual representation of what the sensors are learning. Patients' physiological information may be altered remotely by using the Octabricx ESP8266 chip, which is often used to analyze Arduino data.

Module 2:

Measurement data is retrieved through sensors and stored into cloud server. The goal of such module is displaying patient's physiological constraints using an esp8266 WIFI module. And using machine learning algorithm predicts heart disease of the penitent.

Module 3:

Every string value into array are continually contrasted with the output values of time by RTC module and only when one of them is equal, the final two digits of that specific string value are verified to precisely administer the appropriate medication at the exact time required. Because of this, an Arduino-generated signal is sent to centrifugal pump, which in turn sends appropriate quantity of liquid medication, depending on the final two numbers. Servo motor has configured so that just one axis may be used.

Block diagram:

Fig. 1 Block diagram

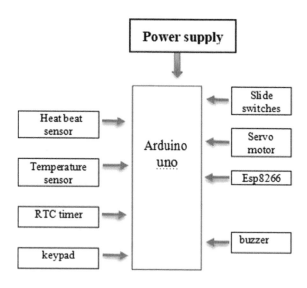

5 Project Demonstration

Dataset:

For training the dataset, the existing patient records are considered with the following attributes. We have listed the attributes and min–max ranges as in Figs. 2 and 3.

System Execution flow

Fig. 2 Uploading data set

Fig. 3 Display of uploaded data set in the webpage

Fig. 4 K-Means clustering for classification of data

6 Results and Discussions

This paper proposed results shown in Figs. 4, 5 and 6 is IoT module collaborating with machine learning [ML] concepts to monitor, mobilize health issues and on-time medicine indication for users. By taking advantage of machine learning concepts and advanced medical embedded components, the prediction results will be applied in heart diseases. This paper has also proposed the monetization of coma patient (MCP) using advanced medical embedded components and web technologies. The proposed model has conducted several successful testing cases and finally the expected results are obtained.

7 Conclusion

The developed prototype will be able to predict the heart attack, which can be used at a particular time based on the parameters of the patient at that time, as well as inform and alert the patient for taking a suitable dosage of pills at correct time. This research work utilizes a heart rate sensor, temperature and humidity sensor to collect the readings of the patient and the Arduino micro controller will send these readings to the server with the help of the ESP8266 module. Using the RTC timer, servo motor, and buzzer, the pill remains as per timing. Based on the dataset available and with the previous readings, the results are obtained with the help of two algorithms namely KNN and naïve Bayes. This research study proposes an IoT-based application collaborating with machine learning (ML) concepts to monitor and mobilize the health issues for users. By taking this as an advantage of machine learning concepts

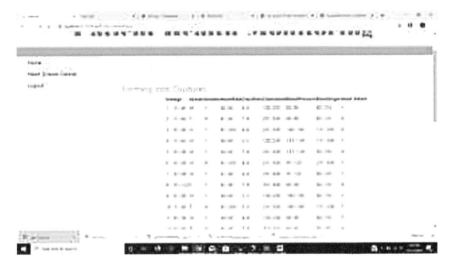

Fig. 5 Forming into clusters

Fig. 6 Frequency tables

and advanced medical embedded components, the proposed model can apply the prediction results in heart diseases. This research study has also proposed the Monetization of Coma Patient (MCP) using advanced medical embedded components and web technologies. The proposed model has conducted several successful testing cases and obtained the results.

References

1. Paris, France 22–23 June 2018
2. Senthamilarasi, C., Jansi Rani, J., Vidhya, B., & Aritha, H. (2018). A smart patient health monitoring system using IoT. *International Journal of Pure and Applied Mathematics, 119*(16)
3. Akhila, V., Vasavi, Y., Nissie, K., & Venkat Rao, P. (2017). An IoT based patient health monitoring system using Arduino uno. *International Journal of Research in Information Technology, 1*(1).
4. Raja Gopal, S., & Patan, S. A. (2019). Design and analysis of heterogeneous hybrid topology for VLAN configuration. *International Journal of Emerging Trends in Engineering Research, 7*(11), 487–491.
5. Ciocan, C. F., Fizeúan, R., & PalaghiĞă, N. (2015). Weekly electronic pills dispenser with circular containers. In *2015 IEEE 21st International Symposium for Design and Technology in Electronic Packaging (SIITME)*, pp. 125–129.
6. Othman, N. B., & Ek, O. P. (2016). Pill dispenser with alarm via smart phone notification. In *2016 IEEE 5th Global Conference on Consumer Electronics*, pp. 1–2.
7. Chawla, S. (2016). The autonomous pill dispenser: Mechanizing the delivery of tablet medication. In: *2016 IEEE 7th Annual Ubiquitous Computing, Electronics & Mobile Communication Conference (UEMCON)*, pp. 1–4.
8. Machine learning models for IoT-based heart disease prediction and diagnosis in healthcare.
9. IoT Based Smart Health Care Medical Box for Elderly People.
10. An Intelligent Medical Monitoring System Based on Sensors and Wireless Sensors Network.
11. An IoT Based Health Care System For ElderlyPeople.
12. Intelligent Medicine Box For Medication Management Using IoT.
13. Krishnan, S. R., Gupta, S. C., & Choudhury, T. (2018). An IoT based patient health monitoring system. In *2018 International Conference on Advances in Computing and Communication Engineering (ICACCE-2018)*.

Lossless Image Compression Using Machine Learning

Eli Veda Sai Rochishna, Vutukuri V N S M P Ganesh Hari Prasad Rao, Gandhavarapu Bhargav, and S. Sathyalakshmi

Abstract As communication networks evolve, so does the technology for exchanging and transmitting information. With millions of photos uploaded on social media every day, wireless sensor networks are rapidly being utilized to take images in a variety of applications such as traffic lights, roadways, social media, assembly, websites, and malls. As a result, it is required to reduce the size of these photos while retaining a suitable degree of quality. To compress RGB photos in this study, we used the K-mean clustering algorithm. Image quality is evaluated using after image compression, PSNR, MSE, and SSIM. The higher the K in the SSIM index, the more comparable the two photos are. This is a good sign that the image size has been greatly reduced. The proposed compression approach, which is based on the KMean clustering algorithm, is used to compress photos to minimize their size, server burden, and data transmission speed.

Keywords RGB images · Image compression · k-Mean clustering · Mean square error (MSE) · Structural similarity index measure (SSIM) · Peak signal-to-noise ratio (PSNR) · Malls · Clustering · Social media · Websites

E. Veda Sai Rochishna · V. V N S M P Ganesh Hari Prasad Rao · G. Bhargav (✉) ·
S. Sathyalakshmi
Computer Science and Engineering, Hindustan Institute of Technology and Science, Chennai, India
e-mail: 18113120@student.hindustanuniv.ac.in

E. Veda Sai Rochishna
e-mail: 18113090@student.hindustanuniv.ac.in

V. V N S M P Ganesh Hari Prasad Rao
e-mail: 18113119@student.hindustanuniv.ac.in

S. Sathyalakshmi
e-mail: slakshmi@hindustanuniv.ac.in

1 Introduction

According to Cisco's Annual Internet Report, digital media account for around 80 per of global traffic on the Internet. This figure is only anticipated to rise over time, as the demand for digital media continues to rise. Image compression is more important now than it has ever been due to the large number of images being created, stored, and transferred.

Picture compression seeks to decrease duplicate image data so that it can be stored or transmitted more efficiently. To put it another way, it aids in the reduction of a digital image's size (in bytes) without compromising image quality. This is now achievable thanks to a property found in almost all images: neighboring pixels are connected and hence convey redundant information.

Traditional image compression methods, such as JPEG2000 (Joint Photographic Experts Group), PNG (Portable Network Graphics), JPEG [Joint Photographic Experts Group]. Used fixed transforms, such as the To eliminate spatial redundancies, used the Discrete Cosine Transform (DCT) and Discrete Wavelet Transform (DWT) (to first convert the picture from its image pixels representation to a frequency domain representation), as well as quantization and the entropy encoder. Several new techniques have demonstrated promising results in replacing the classic DCT/DWT algorithm. Recent breakthroughs in the field of machine learning, which were projected to significantly enhance photo compressing in past, have proven to be perfection, allowing for new ways of using improved, more efficient procedures. We shall compare some of these strategies in this study.

Instead of being manually engineered, the structure is automatically found using machine learning techniques. In machine learning-based image compression, there are two basic steps: Encoding is done by selecting the most representative pixels, and decoding is done through colorization. The first phase is primarily concerned with active learning, whereas the next phase is concerned with semi-supervised learning [1]. We will use the K-Means clustering methodology for picture compression, which is a sort of compression transformation method. We'll quantize the colors in the image using K-Means clustering, which will help compress the image even further. The remaining of this document is structured as follows. Section 1 delivers a literature review on image compression using machine learning. Section 3 is all about how digital images are represented with colors. Section 4 is our problem statement. Section 5 is about clustering techniques. Section 6 is about Image compression Techniques. Section 7 is our project methodology. Section 8 is about the difference between lossless and loss image compression techniques. Section 9 is the implementation part. Section 10 is experimental results. Section 11 is about future work, and the final one is for references.

2 Related Work

[1] When K is increased, the compression ratio increases; when K is decreased, the compression ratio decreases. Compressing by the K-Means algorithm is known as loss compression. The increasing number of 3 iterations when the image's pixels are big will result in greater compression.

[2] Image compression is obtained by using clustering methods like K-Means, Fuzzy C-Means, and DBSCAN (Density-Based Spatial Clustering of Applications with Noise). The clustering techniques were used for the segmentation of images for effective compression. The segmentation results are encoded using Bit plane slicing, Huffman coding, and LZW (Lempel–Ziv-Welch) coding for efficient transmission and storage. By observing the decompressed images, the quality of images obtained by the Huffman encoding technique is high then the other two techniques for all three clustering techniques.

[3] An algorithm with an improved objective function such that supervised learning is used to train regression models to reduce overall prediction errors. and thus avoid the computation burden of the color seeds selection. In the end, the luma information and parameters of the regression model are transmitted to the decoder for color prediction and then the color image is reconstructed.

[4] Low-power embedded devices may be used to compress images using K-means clustering for image-based WSNs (Wireless Sensor Networks). Similar pixel colors have been utilized for grouping pixels to reduce source pictures specifically. While K-means learning on wireless sensor nodes might exceed the benefits of data compressing, outsourcing learning steps and only conducting compression might save significant amounts of energy.

[5] The K must be chosen before starting the algorithm. Always K should be smaller compared to the image pixels number, i.e. ($K ¡ N$). If the K value is smaller, the compression rate in both algorithms is higher. The distance Formula used here is Euclidean distance. Euclidean distance is the absolute value of the difference between the data point and centroid.

[6] In this paper the authors discussed, The JPEG picture coding and decoding process are explained in this algorithm. The image's encoding section can use JEPG to convert a BMP (Bitmap) file to a binary file for real-time storage. The related decoding application can decompress the picture.

[7] By assessing numerous features or specifications like Peak Signal to Noise Ratio, Mean Squared Prediction Error, Bits per Pixel, also Compression Ratio, Huffman was primarily used to remove unneeded bits 4 from data. Various input photographs of various sizes, as well as a new method of separating the input images into the same rows and columns. The aggregate of all individual compressed photos delivers the best outcome and information content in the final phase.

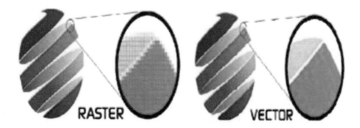

Fig. 1 Raster images versus vector images

3 Image Representation

Images are represented in a variety of ways in computer science, depending on how they are saved and what color data they contain. Raster (bitmapped) pictures and vector images [2, 3] are the two types of digital images that can be encoded. The distinction between raster and vector images is illustrated in Fig. 1.

Represent this type of image in vector graphics, mathematical equations are utilized to create points, lines, forms, and polygons. These graphics are mostly made by computer programs like computer-aided design (CAD), and they don't use resolution, thus they may be scaled down without losing their fundamental appearance. Graphs are frequently used to display these types of images. Scalable vector graphics (SVG), encapsulated postscript (EPS), portable document format (PDF), and AI [1] are some of the vector graphics file types. Graphics pixels are a set of dots that make up raster graphic graphics. These points are organized as a matrix with several columns and rows, each with its color. This method may be used to represent any form of picture or color. JPEG, PNG, and graphics interchange format (GIF) are all raster graphics file formats. Colors can be represented using a variety of color models [2]. The two-color models Cyan-Magenta-Yellow-blacK (CMYK) and RED GREEN BLUE (RGB) [4] are used to preserve the majority of raster graphics. Colors are shown in the RGB model by blending red, green, and blue primary colors, as seen in Fig. 2.

Three values for the red, blue, and green colors must be provided to represent each pixel color in the picture. Each fundamental color value may be represented by 8 bit = 1 byte, as illustrated in Fig. 3, therefore, each pixel requires 3 bytes of storage space [5].

4 Problem Statement

In the current world, every social platform consists of various types of media, which occupies terabytes of data, and cloud storage is becoming expensive every day websites handle a lot of images, which increases the load on the servers, and slow

Fig. 2 RGB color model

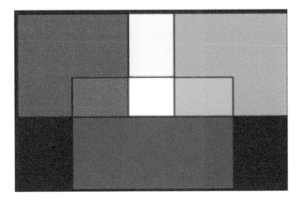

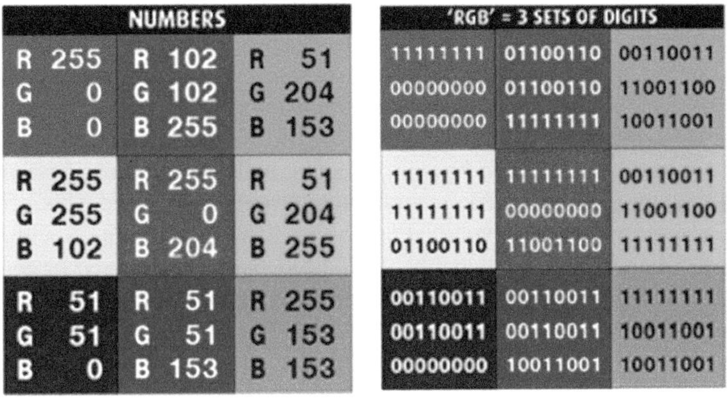

NUMBERS						'RGB' = 3 SETS OF DIGITS		
R	255	R	102	R	51	11111111	01100110	00110011
G	0	G	102	G	204	00000000	01100110	11001100
B	0	B	255	B	153	00000000	11111111	10011001
R	255	R	255	R	51	11111111	11111111	00110011
G	255	G	0	G	204	11111111	00000000	11001100
B	102	B	204	B	255	01100110	11001100	11111111
R	51	R	51	R	255	00110011	00110011	11111111
G	51	G	51	G	153	00110011	00110011	10011001
B	0	B	153	B	153	00000000	10011001	10011001

Fig. 3 RGB image has three sets of numbers per pixel

response of the website. We aim to develop a model using K-Means clustering that compresses the images with less loss of data.

5 Clustering Techniques

A descriptive method for assessing and grouping data is data clustering [4]. In the literature, scholars have tackled a variety of clustering techniques. There are numerous categories under which these algorithms could be placed. Hierarchical methods, partitioning methods, grid-based techniques, and density-based techniques are some of the most common categories for these methods.

A set of data objects is partitioned into non-overlapping groups is partitioned clustering. This approach may be used to divide the degree of color in photographs, allowing for easier compression. Also, techniques that use partitioned clustering may manage large amounts of data (pictures) better than other methods. Various algorithms fall within this category, the most common of which being k-Means. K-mean clustering is a straightforward unsupervised approach that works with any type of data. It is also simple to construct and performs well, as it is quicker than other clustering methods [5].

6 Image Compression Techniques

Based upon the route, it then creates a novel division by assigning each point to the nearest starting centroid. We then recalculate ate centroid of every collection. The approach is continued until convergence is achieved by alternating between such two steps. The eigenvalue of each picture block is determined in the proposed method, which divides images into 4×4 pixel blocks. Afterward, the algorithm divides the input picture pieces into blocks based on their eigenvalues values. A threshold is used to classify the picture blocks, which is the average eigenvalue value of all of them. All of the photo's main components are contained in blocks with higher eigenvalues.

Blocks having low eigenvalues contain fewer precise image elements which could be greatly compressed, while blocks containing higher eigenvalues have the most detailed picture component which could be significantly compressed. This is followed by clustering of each partition based on several high-eigenvalue units within every partition (K1 and K2). The global beginning codebook is composed of codebooks generated by higher Eigen as well as lower Eigen partitions. Lastly, K-means runs are done using a worldwide codebook used as a beginning codebook to build a global code.

6.1 Generative Adversarial Networks

Fabian Mentzer presented a method built upon the utilization of Generative Adversarial Networks (GANs), particularly Conditional GANs. Ian Goodfellow and his colleagues created the GAN family of machine learning frameworks in 2014. In this 'game,' two neural networks compete against one other. New data generated using this approach will have the same statistical properties as those generated from a training batch. In the realm of deep learning, GANs have been regarded as game-changer in recent years. A discriminator and a generator are the desired outcomes. The discriminator D's (training) goal is to improve the discriminator's acuity.

The objective of generator G (training) is to make its reconstructed image as authentic as feasible. GAN uses an alternating optimization strategy in this training process. Rather than creating a random selection with an undetermined sound distribution, Conditional GAN (CGAN) is a machine learning system that learns to generate a false sample for a certain condition or set of features (As an example, a label that is connected to a picture or a more specific tag). Mentzer et al. present an approach that uses a CGAN to supplement the neural image compression formulation.

When working by lower bpp value, the inclusion of GAN does have a better influence on performance enhancement since low bpp typically leads to insufficient bit allocation in non-important areas.

6.2 Gaussian Mixture Models

Gaussian mixture models are likely probability-based models which may be used to describe normally distributed subpopulations within a larger population. In general, mixture models do not require the user to know whichever subpopulation data piece represents, Subpopulations may be learned dynamically by the algorithm. This is an example of unsupervised learning since subpopulation allocation is unclear.

Learning image compression may be achieved using discretized Gaussian mixture probabilities and attention modules, as Cheng et al. demonstrated [8]. A neural attention mechanism, in this case, allows a neural network for focusing upon inputs of the subset.

To parameterize the distributions, discretized Gaussian mixture likelihoods are utilized, which reduces the residual redundancy to produce an accurate entropy model, resulting in less necessary encoding bits. Network designers have implemented a basic form of an attention module. A broad receptive field and improved rate-distortion performance are achieved by using residual blocks in the network design. To keep more information, the decoder uses subpixel convolution as upsampling units rather than transposed convolution. The model capacity is represented by N, which stands for the number of channels. We employ the Gaussian mixture model, which necessitates $3 \times N \times K$ channels for the auxiliary autoencoder's output.

7 Methodology

Bandwidth and storage are both expensive, and pictures often have a lot of redundant data. As a result, compressing photos saves storage space and bandwidth while not influencing image quality. Although image compression techniques such as PNG and Huffman coding exist, none of them are lossless. There is currently no compression technology that incorporates machine learning techniques on the market. There are some pre-built models, however, the accompanying libraries are not available for existing lossless models, resulting in just a 10 per success rate.

We intend to build a model from the ground up and test it with various activation functions, loss functions, and several centroids to achieve minimal loss when compressing photos. In a picture, we can only see a few hues, far less than the aforementioned amount. As a consequence, the K-Means clustering strategy adopts the utilization of the human eye's visual processing and uses a minimum amount of colors to describe a picture. Intense colors with different RGB values seem identical

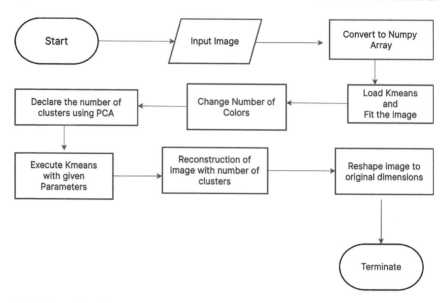

Fig. 4 System flow diagram

to the human eye. This is used by the K-Means algorithm, which creates groupings of colors that are visually similar.

The intensity of each color in an image may range from 0 to 255, but every pixel in a picture has three bytes (RGB). According to combinatorics, the total number of colors which might be depicted is $256 \times 256 \times 256$.

Compression in various image processes involves lowering the image's color components. This is what we're doing with k-means clustering. The value of k is pre-defined as the number of color components in the image that we want to save. The rest of the k-means algorithm is carried out in the same manner as the previous steps as shown in Fig. 4. While cluster numbers enhance, the image will develop as more similar to the original image, but at the cost of additional disc space for storing and a higher computational cost. We can play about the values of k to see what we can come up with. Because it gives us the variance of the cluster centroids, we can also calculate the within-cluster sum of the squared error to see if the clusters are well fitted and accurately assigned.

Note: It's advised to keep the value of k as a multiple (more preferably, a power) of 2, same as the conventional image formats, to get better results.

8 Lossless and Lossy

Image compression is a way to reduce image size and improve website/app/server load speed and overall performance. Both lossy and lossless compression decrease image file size, but methods and results are different. The main difference between lossy image compression and lossless image compression is that lossy compression completely deletes some image data, whereas lossless compression does not degrade image quality and only deletes unimportant data.

Lossy image compression is a method of reducing the size of your picture file by removing part of the data. The information in the file will be permanently erased as a result of this procedure, which cannot be reversed. fractal compression, transform encoding and discrete wavelet transform are some of the methods used for lossy compression. It is possible to minimize file sizes by using lossy compression technologies. Still, there is often a trade-off: image quality also degrades. JPEG format is a great example of lossy compression.

Lossy compression may be advantageous for e-commerce websites, blogs, or portfolio websites because it significantly reduces file size and improves site performance. This will improve the user experience and improve your search engine optimization (SEO) ranking. For example, to reduce load times, you must compress previews pictures as well as thumbnails on your e-commerce site utilizing lossy compression.

Lossless compression, compared with lossy compression, doesn't reduce the quality of a picture. This is because that lossless compression only removes excess data that was spontaneously enumerated by equipment utilized utilizing shot. Sinshotshere's no visible decrease in picture file size, there is a trade-off. As a consequence, it's unlikely that you'll save much space on your hard drive. Arithmetic coding, run-length coding, and Huffman coding are among the most used lossless compression methods. Lossless image compression is useful for a variety of image formats, including BMP, RAW, PNG, and GIF.

Photo websites, on the other hand, should use lossless compression. Such data compressing technique prefers for higher quality upon size reduction and allows you to view photos in a much more detailed manner. However, lossless compression should be used for product images to prevent quality degradation. Earrings and handicrafts are two examples of tiny and sophisticated objects that are particularly subject to this.

Large image files may slow the performance of the website and have a negative impact on SEO and user experience. Non-optimized images may cause the website to load slowly, become unresponsive, or become completely inaccessible. As a result, you must add the endpoints to your website. Change the image's resolution or complete the data for the Style sheet to accomplish this. In either case, the file size will be reduced and the website will load more quickly. According to a Google survey, 45 percent of visitors are unlikely to return to the site if their investment is unpleasant. Even Google agreed that page speed is an important ranking factor for SERPs (Search Engine Results Page). Slow page performance can also affect conversion rates. Faster page speeds increased mobile sales by around 45%, according to Dakine, an outdoor lifestyle business.

9 Implementation

Using K-means algorithms, we created a lossless image compression system. To implement, we used Python as a programming language and some machine learning components such as ELBOW METHOD, NumPy, and the Flatten Image technique to achieve image compression with minimal data loss. Lossless compression preserves image quality while reducing the size of the resulting image file. Increasing the performance of your website without sacrificing image quality is a win-win situation. The following are a few of the components we used during the implementation process.

[2] The Elbow Method is used to determine the optimal number of clusters for each image. We can find the best value to compress the image using this method. This method is based on calculating the sum of squared errors (WSS) within Cluster Sum for a number of different clusters (k) and selecting k where the change in WSS begins to decrease. The idea behind the curving method is that the explained transformation changes rapidly for a small number of clusters and then it slows down, resulting in a sharp bend of the curve. The knee point is the number of clusters we can use for our clustering algorithm.

Kumari et al. [3] when you use the Flatten Image technique, all of an image's layers are combined into a single layer with no alpha channels. After flattening, the image retains its original appearance. Because there is no transparency in the image content, everything is contained in a single layer. The primary reason for flattening or merging images is to compress the information so that the files can be made smaller and more portable.

Liu and Yang [1] Multidimensional arrays require more memory, while 1D arrays require less memory. Because this is the primary reason for smoothing the pictarraysrray before processing/feeding the model's information, Flattening may save time and memory when dealing with a huge amount of pictures, which is often the case while training a model.

Paek and Ko [4] arrays of varying sizes must be subjected to math and logical operations Because an image can be thought of as an array, NumPy can be used for a variety of image processing tasks. NumPy arrays can be used to perform complex arithmetic and other operations on large datasets. When compared to Python's built-in sequences, these operations are often executed more quickly and with less code [5] Arrays are at the core of NumPy. NumPy arrays have the benefit of requiring less memory and running more quickly than comparable python data structures (lists and tuples). Linear algebra is among the scientific functions supported by NumPy.

The steps involved in the Implementation Pare art are as follows.

Step 1: Which is nothing but importing required libraries into our development environment.
Step 2: Which is loading the image we want to compress.
Step 3: Here, We need to store the image dimensions into a variable.
Step 4: Now, transform the image into a NumPy array with help of the Numpy Library.

Step 5: Change the NumPy array shape into Flatten Image to Avoid reshape error after.

Step 6: Major Step, Implement K-means clustering to form k clusters to achieve image compression.

Step 7: Now, Find the Optimal number of clusters with a variance of 98% with help of the Elbow Method.

Step 8: Begin the K-means and Fit the image.

Step 9: With help of the k value, replace each pixel value in the image with a nearby k value (centroid value/centroid).

Step 10: Now, reshape the image after applying changes to its previous/original dimension.

Step 11: Return the final output to the end-user to save/ share/ use in his or her project.

By performing the above steps in our Development environment, we have achieved compressed images using machine learning. K means helps in reducing the colors in the image and reducing the image size.

10   Experimental Results

We tested existing algorithms as we mentioned in the section Image compression Techniques against our proposed method using the same image of size 1024 kb.

While our Method gave an output compressed image of 780 kb with minimal loss of data, existing online web services such as a short pixel gave an output of 980 kb. When Tested against other Machine Learning algorithms like Eigen Based Kmeans, Gans and, Gaussian the compression rate of our proposed method is far higher than existing ML methods as shown in Figs. 5 and 6. Individual Compression rates are:

The major drawback of lossless compression is larger files than if you were to use lossy compression but better quality when compared to lossy.

Drawbacks: We carefully selected the number of clusters using various methods like principal component analysis to get 98 percentage variance or to preserve the 98 percentage of the data. Hence, time taken to compile images is based on the size of the image, which can be optimized through further refactoring.

11   Conclusion

A method for compressing images with the machine is proposed in this paper. Lossless picture compression was achieved using the K-Mean clustering approach. With elbow method and the K-Mean clustering technique are introduced, followed by how to apply it to picture compression. The results indicate the advantages of picture compression by lowering the size of the photographs while maintaining acceptable

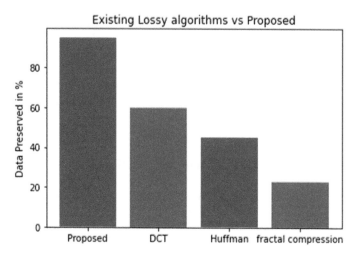

Fig. 5 Existing lossy algorithms versus proposed

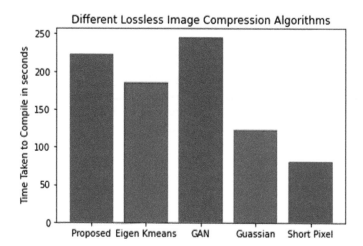

Fig. 6 Existing lossless versus proposed

image quality and data. We plan to broaden the media file range used to include evaluating the proposed algorithm in high resolution.

References

1. Liu, X., & Yang, J. (2018). Fast and high efficient color image compression using machine learning. In 2018 *2nd IEEE Advanced Information Management, Communicates, Electronic and Automation Control Conference (IMCEC).*

2. Wan, X. (2019). *IOP Conference Series: Materials Science and Engineering 563*, 052042.
3. Kumari, G. V., et al. (2020). Image compression using clustering techniques for biomedical applications.
4. Paek, J., & Ko, J. (2017). K-means clustering-based data compression scheme for wireless imaging sensor networks. *IEEE Systems Journal*.
5. Nandy, O., & Battu, J. V. (2020). A comparative analysis of machine learning methods for image compression. *International Journal of Engineering Research Technology (IJERT), 09*(08).
6. Tyagi, V., Kumar, V., Asthana, V., & Patel, R. (2016). A fast and improved image compression technique using Huffman coding. In *2016 International Conference on Wireless Communications, Signal Processing and Networking (WiSPNET)*.
7. Hong, A., Wan, N., Chen, X., & Xiao, W. (2020). A fast JPEG image compression algorithm based on DCT. In *2020 IEEE International Conference on Smart Cloud (SmartCloud)*.
8. Cisco Annual Internet Report (2018–2023). White Paper, March 2020.

An Energy-Efficient Approach to Transfer Data from WSN to Mobile Devices

A. Jaya Lakshmi and P. M. Joe Prathap

Abstract Wireless sensor network (WSN) has obtained noteworthy attention for investigation by researchers and industrialists. Nowadays, WSN has evolved as a prospering technology that supports a wide range of applications over the Internet. Now, with the propagation of internet-enabled mobile devices, mobile users can interact with sensor networks to collect environmental data without any boundaries of place and time using easily operated mobile applications. The usage of sensor networks has many issues in real time with certain obstacles in both theoretical and practical. Usually, sensor nodes include batteries with limited capacity. The wireless linkage between the sensor nodes is found to be vulnerable to many features in the surrounding. However, there are not many results about the network capability of a WSN with unpredictable wireless links. This work addresses an energy-efficient and light-weighted encoding technique termed lineage encoding that controls the obstacles (battery) to achieve the development of a large scale sustainable WSN. Our research focuses on the system of independent communication among the mobile data collected casually for mobile surroundings (environment). The lineage encoding scheme depicts the parent–child association of the XML nodes as an arrangement of bit-stream, called lineage code (V, H), and includes a productive twig pattern query processing at the clients (mobile). Experimentations have been carried out, to assure minimal time for processing the query and maximal lifetime of the network with lineage encoding (LE). The results from the simulation have proven the network range of large-scale WSN supports a boundless number of clients under diverse network circumstances.

Keywords Lineage encoding · G-node · Query processing time · Residual energy · Network lifetime maximisation

A. Jaya Lakshmi (✉)
Gojan School of Business and Technology, Chennai, India
e-mail: a.jayalakshmi92@gmail.com

P. M. Joe Prathap
RMD Engineering College, Chennai, India
e-mail: drjoeprathap@rmd.ac.in

1 Introduction

A WSN is a collection of sensor nodes and transmitting devices. Sensors are portable devices to measure environmental conditions and their changes and organise them to give data to a sink that collects the data from each sensor node over the wireless medium [1]. Internet of Things (IoT) is the evolved method of connecting devices with sensors. Generally, the sensors are deployed to capture the change in the respective environment. The sensor is a boon where humans cannot notice the change. For example, in nuclear power plant, sensors are used to will sense various factors like pressure, temperature, gas leaks, etc., which saves human life from an explosion or accident. Monitoring the physical events, such as climate, atmospheric conditions, pollution, or detection of any natural disasters is done by the sensor [2]. The database management of the sensor network involves relational models. To untangle data management [3] and extraction from WSN, stages like data collection, data aggregation [4, 5] and query processing. With the traditional model of data management in WSN, structured query language (SQL) seems more convenient for data storage and query processing [6]. But to broadcast the sensor values to the mobile devices, SQL is replaced with extensible markup language (XML) [7]. XML was designed to stow and send data in a way, that is easily readable by both humans and machines. As the routing tree [8] is more convenient for query processing in a sensor network, XML is found to reinforce the query processing more than SQL. The main two reasons for the replacement of the SQL are (i) SQL returns only tabular data; (ii) It is not convenient to traverse the hierarchical structure of the sensor network. To achieve compatibility in the mobile environment, the sensed data is broadcast in XML format rather than the conventional format (SQL). And so the sensed values available in the SQL format are converted into XML format, so as to make the query processing in the mobile environment easier.

Even if the sensor values are broadcasted to the mobile device in the XML format, there are so many drawbacks in case of energy loss in both client and server (Base station) [9]. In order to reduce energy loss in both client and server, and energy-efficient encoding technique called lineage encoding (LE) and a streaming unit called G-node (G-N) is introduced.

The rest of this paper is organised as follows: Sect. 2 discusses some associated work on WSN, XML and energy of sensor nodes, while Sect. 3 presents the proposed architecture model with LE. Section 4 explains the lineage encoding and G-Node. Section 5 describes query processing, and Sect. 6 deals with the performance evaluation. Section 7 concludes the paper with the points for subsequent research directions.

2 Related Work

With the speedy evolution of WSN and mobile computing, all available information in the sensor network is provided to the mobile user.

Park et al. [10] proposed an encoding technique called lineage encoding for energy-efficient XML dissemination in mobile computing. This approach is found to be both energy and latency efficient.

Considering the issues in sending the sensor values obtained from the WSN using traditional SQL, the use of XML for broadcasting to the mobile device was proposed by Elias [7]. In SQL databases, the sensor values are available in the form of tables. Broadcasting the values from these tables to the mobile environment [11] provides only the tabular form of content to the mobile user. But, this is not in the case of using XML data. The sensor values from the SQL database are being parsed by the XML parser for the ease of mobile users. But, there is a huge loss of energy for both client and server which should be focused on.

Conventional query processing is replaced with the energy-efficient Twig pattern query processing [12–14], which focuses on achieving child-to-parent traversal. This makes the query traversal as early as possible. For a long XML document, traversing always from parent to child is not convenient. Thus twig pattern query processing makes the traversal phase to be done also in the reverse order, i.e. from child to parent. Hence, query processing time shall be reduced which in turn reduces the battery power of the mobile client.

There are several approaches dealing with the broadcasting of sensor values to the mobile environment. But, the loss of energy is the major issue to be concentrated on both the server and client [15]. This paper proposes an encoding method (LE) that can support twig pattern query.

3 Proposed Architecture

The following are the major elements in this proposed architecture. (i) a relational database, (ii) XML streaming, and (iii) energy-efficient processing of query in the client (Mobile). The architectural flow is depicted in Fig. 1.

Sensor nodes deployed in various locations periodically send the sensor values to the SQL server with the intermediate gateway [16]. The available values in the server are to be broadcasted to the mobile device. So, initially, the XML parser retrieves the values from the server and the XML stream is generated in the wireless environment. When the mobile client issues a query, selective tuning is done so as to tune to the respective broadcast channel and to download (selectively) the XML stream for processing the query [17]. The XML broadcasting is completed fastly such that the server can reinforce the dynamic spreading of a G-N without any disturbance to broadcasting.

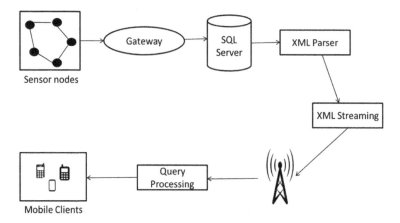

Fig. 1 Architecture diagram

The broadcasted data stream will be in the form of G–N with their corresponding calculated lineage code (V, H). When the mobile client issues a query, with the proposed architecture usual traversal from parent to child and also child-to-parent tree traversal (twig pattern) [13] can be done. This reduces the query processing time of the mobile client. Query processing from child to parent is made possible with lineage code. At the same time, this architecture will provide service to many mobile clients and thus achieves extensibility in terms of service. The wireless sensor network is an area that provides a massive number of data elements periodically to the server. Thus, broadcasting these large volumes of data periodically to the mobile client can be done without much consumption of energy which is the major issue for both server and client.

4 Wireless XML Stream

In this section, an advanced method of streaming XML data is proposed. Figure 2 depicts the XML document which is the running example in this research parsed by the XML parser. XML parser includes methodology and has properties to ingress and edit XML documents.

4.1 G-Node (G-N)

G-N is a series of node groups that involve the data stream (XML) for wireless communication. It is a streaming unit that eradicates conflicts on the nodes of XML documents and allows clients (mobile) to download only relevant data while

Fig. 2 Sample XML document

```
<Monitor>
        <Sensor id="01" name="S1">
                <location> p1 </location>
                <temp> 34 </temp>
                <humid> 12 </humid>
        </Sensor>
        <Sensor id="02" name="S2">
                <location> p2 </location>
                <temp> 35 </temp>
        </Sensor>
        <Sensor id="03" name="S3">
                <location> p3 </location>
                <temp> 26 </temp>
                <humid> 45 </humid>
        </Sensor>
</Monitor>
```

processing the query. Each G-N has three parts, namely (a) group descriptor (GD) (b) attribute value list (AVL) c) text list (TL).

GD is the collection of node name, location path, child index (CI), lineage code, attribute index (AI), and text index (TI). All these components of the GD are helpful in processing the queries (XML) by mobile clients. The index values of CI, AI and TI are used to choose and download the next G-N, attribute values and text.

4.2 Lineage Code

The proposed inventive encoding strategy called lineage code includes two types of code, i.e. vertical code and horizontal code [10]. The vertical code is indicated by the lineage code (V) and the horizontal code is indicated by the lineage code (H). The main advantage of Lineage code generation is to describe the parent–child relationship between the G-Nodes. The lineage codes for the available elements of the considered sample XML document is given in Fig. 3.

5 Query Processing

This section explains about (1) simple query processing and (2) twig pattern query processing.

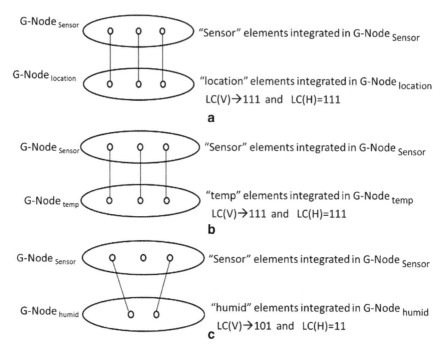

G-Node _Sensor_ — "Sensor" elements integrated in G-Node _Sensor_

G-Node _location_ — "location" elements integrated in G-Node _location_
LC(V)→111 and LC(H)=111

a

G-Node _Sensor_ — "Sensor" elements integrated in G-Node _Sensor_

G-Node _temp_ — "temp" elements integrated in G-Node _temp_
LC(V)→111 and LC(H)=111

b

G-Node _Sensor_ — "Sensor" elements integrated in G-Node _Sensor_

G-Node _humid_ — "humid" elements integrated in G-Node _humid_
LC(V)→101 and LC(H)=11

c

Fig. 3 Lineage codes of **a** G-Node location. **b** G-Node temp **c** G-Node humid

5.1 Simple Query Processing

In this type of query processing, first, the client (mobile) creates a query tree and determines the pertinent G-Nodes, and downloads the respective group descriptor. When the present node is found to be the leaf node, the mobile client uses the AI and TI to download the text and attributes. Figure 4 explains the parent to child traversal for the query,

Q1: "//sensor [name/text () = "S3"] /humid/".

The above query is generated for the sample XML document discussed in Sect. 4.

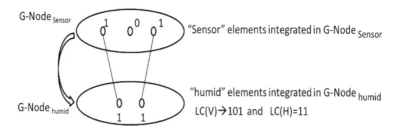

G-Node _Sensor_ — "Sensor" elements integrated in G-Node _Sensor_

G-Node _humid_ — "humid" elements integrated in G-Node _humid_
LC(V)→101 and LC(H)=11

Fig. 4 Example parent-to-child traversal

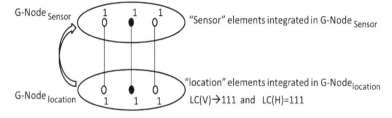

Fig. 5 Example child-to-parent traversal

5.2 Twig Pattern Query Processing

This type of query processing involves three phases, namely the tree traversal phase, sub paths traversal phase and main path traversal phase. The query in the form of the tree is traversed in depth-first manner and downloads the GD of the appropriate GN. The twig pattern query processing becomes easier as simple query processing with the lineage code is calculated. Figure 5 explains the child to parent traversal of the query,

Q2: "//sensor/location [p2]/".

This query is generated for the sample XML document discussed in Section IV.

Selective tuning is an active approach that makes the client's work easier by reducing the tuning time and the access time. And also, it effectively switches between the twig pattern query and simple query processing. Tuning is enhanced with the help of the XPath query pattern which holds the predicates.

6 Simulation Result

In the simulation environment, 100 sensor nodes are deployed in a random manner. The following results are obtained by experimenting with the NS2 simulator.

6.1 Query Processing Time with Lineage Encoding

The performance of the method lineage encoding (LE) is compared with the already available streaming methods like twig stack (TS) and extend chaining (EC) [18]. TS algorithm involves the processing of twig pattern queries. But TS will include processing of raw XML document. EC supports structure indexing which helps in resolving redundant XML nodes. But with LE, the generated codes are helpful in resolving both the issues of twig query and redundancy of XML nodes. The experiment is done considering the sample XML document in Sect. 4. Figure 6 shows the tuning time evaluation for the query Q1 generated from the client to know

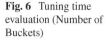
Fig. 6 Tuning time evaluation (Number of Buckets)

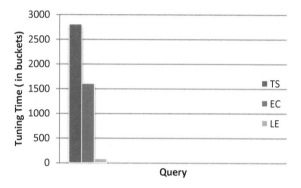

the sensor values from the XML document. Lineage encoding is found to be better when compared to other methods.

Q1 "//sensor [name/text () = "3"]/humid/".

Here, the tuning time is considered in terms of the bucket. The bucket is the unit to measure the data in a wireless medium [19]. The main reason for the reduced tuning time using lineage encoding is because of downloading only the relevant data values from the XML stream. Since we are considering G-Nodes downloading of irrelevant elements is avoided completely, thus reducing the tuning time, which in turn reduces the battery power of the client for obtaining the continuously sensed values from the wireless sensor networks (WSNs).

6.2 Maximum Residual Energy

Initially, the maximum energy of the sensor node is 2.5 J. In considering a sensor network with 100 sensor nodes the network will contain a maximum of 250 J in the initial stage. This graph shows the average residual energy remaining in the sensor network after a series of transactions with lineage encoding technique. The average residual energy in the network after performing 60 transactions will be 200 J as in Fig. 7.

6.3 Maximum Number of Sensor Nodes with Optimum Energy

The sensors in the sensor network will drain out of energy very soon on processing queries. But, with the application of lineage encoding in the WSN, the sensor nodes sustain their energy for a long time even after processing a large number of queries. The above graph shows the number of sensor nodes available after processing a certain percentage of queries with lineage encoding and without lineage encoding.

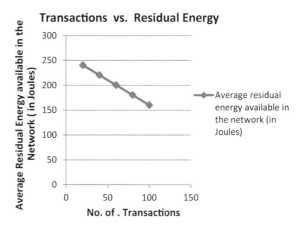

Fig. 7 Average residual energy remains in the network after transaction

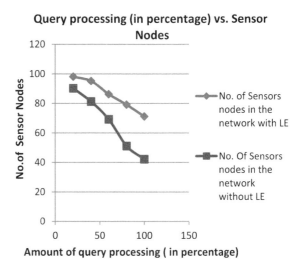

Fig. 8 Available sensor nodes with optimum energy after processing queries with lineage encoding and without lineage encoding

Figure 8 shows that after processing 60% of query processing 86 out of 100 sensor nodes will be available with optimum energy with the lineage encoding, whereas without this encoding technique only 69 sensor nodes will be available with optimum energy.

7 Conclusion and Future Work

The work concludes with the influence of energy-efficient light-weighted lineage encoding in allowing the sensor nodes to retain their energy for a longer time. This increases the residual energy of the WSN, which in turn increases the lifetime of the network. With this technique, the mobile client will also not drain out of energy in a short interval of time. Thus, this technique has shown good results in simulation to be efficient in the case of maintaining the energy (battery power) of both the sensor network and the mobile client.

References

1. Joe Prathap, P. M., & Praveen, K. V. (2021). Energy efficient congestion aware resource allocation and routing protocol for IoT network using hybrid optimization techniques. *Wireless Personal Communications, 117*(2), 1187–1207. Springer US.
2. ElHakim, R., Orfy, A., & ElHelw, M. (2010). Monitoring and visualization of large WSN deployments. *Sensors, IEEE*, 1377–1381.
3. Vajsar, P., & Rucka, L. (2011). Monitoring and management system for wireless sensor networks. In *34th International Conference on Telecommunications and Signal Processing (TSP 2011)*.
4. Joe Prathap, P. M., & Selvi, M. (2017). WSN data aggregation of dynamic routing by QoS analysis. *Journal of Advanced Research in Dynamical and Control Systems, 9*(18), 2900–2908.
5. Joe Prathap, P. M., & Selvi, M. (2019). Analysis and classification of secure data aggregation in wireless sensor networks. *International Journal of Engineering and Advanced Technology, 8*(4), 1404–1407.
6. Joe Prathap, P. M., & Menon, V. G. (2019). Moving from topology-dependent to opportunistic routing protocols in dynamic wireless ad hoc networks: Challenges and future directions. In *Algorithms, Methods, and Applications in Mobile Computing and Communications, IGI Global*, pp. 1–23.
7. Elias, A. G. F., Rodrigues, J. J. P. C., Oliveira, L. M. L., & Zarpelao, B. B. (2012). A Ubiquitous model for wireless sensor networks monitoring, innovative mobile and internet services in ubiquitous computing (IMIS). In *2012 Sixth International Conference*, pp. 835–839.
8. Joe Prathap, P. M., & Vasudevan, V. (2008). Revised variable length interval batch rekeying with balanced key tree management for secure multicast communications. *IJCSNS, 8*(4), 232.
9. Joe Prathap, P. M., & Alappatt, V. (2021). Trust-based energy efficient secure multipath routing in MANET using LFSSO and SH2E. *International Journal of Computer Networks and Applications, 8*(4), 400–411.
10. Park, J. P., Park, C.-S., & Chung, Y. D. (2013). Lineage encoding: An efficient wireless xml streaming supporting twig pattern queries. *Knowledge and Data Engineering, IEEE Transaction on, 25*(7), 1559–1573.
11. Joe Prathap, P. M., & Menon, V. G. (2017). Towards optimal data delivery in highly mobile wireless ad hoc networks. *International Journal of Computer Science and Engineering, 9*(1), 1–6.
12. Al-Khalifa, S., Jagadish, H.V., Koudas, N., Patel, J.M., Srivastava, D., & Wu, Y. (2002). Structural joins: A primitive for efficient XML query pattern matching. *IEEE International Conference on Data Engineering (ICDE)*, 141–152.
13. Jiang, H., Lu, H., & Wang, W. (2004). Efficient processing of XML twig queries with OR-Predicates. In *Proceedings ACM SIGMOD International Management of Data Conference*, pp. 59–70.
14. Altinel, M., & Franklin, M. (2000). Efficient filtering of XML documents for selective dissemination of information. In *Proceedings of International Conference Very Large Data Bases (VLDB)*, pp. 53–64.
15. Chung, Y. D., Yoo, S., & Kim, M. H. (2010). Energy- and latency-efficient processing of full-text searches on a wireless broadcast stream. *IEEE Transactions on Knowledge and Data Engineering, 22*(2), 207–218.

16. Joe Prathap, P. M., & Dhinakaran, D. (2022). *Lecture notes in networks and systems book series* (LNNS, Vol. 317), Springer.
17. Kumar, M., Panigrahi, M., & Verma, S. (2012) XQuery based query processing architecture in wireless sensor networks. *International Journal of Computer Applications (0975–8887), 43*(23).
18. Kaushik, R., Krishnamurthy, R., Naughton, J. F., & Ramakrishnan, R. (2004). On the integration of structure indexes and inverted lists. In *Proceedings of ACM SIGMOD International Management of Data Conference.*
19. Kim, D., An, B., & Kim, N. (2008). Architecture model of real-time monitoring service based on wireless sensor networks. In *10th International Conference on Advanced Communication Technology (ICACT 2008).*

Worker Safety Helmet

Sailasree Perumal, A. S. V. L. Neha, and R. Krishnaveni

Abstract Workers strive hard to satisfy their daily requirements, but sometimes the workplace environment is never on their side, making their working circumstances terrible. In particular, the mining employees face far worse working conditions. The main purpose of the proposed initiative is to save millions of lives by increasing the safety of the workplace. This research initiative intends to inform workers in advance by sending notifications based on previously determined risk factors, thereby preventing disasters before they occur. In the mining industries, people have been working in dangerous environments for a long time. Millions of lives are lost as a result of unsafe working conditions. However, with the advent of technology, there is a possibility of providing better working conditions for people and thereby saving millions of lives. The proposed model is a small step toward a brighter future. The model is designed with an RF-based tracking system and a smart gadget, which uses RF technology to transfer signals over an IoT network by using an ATmega microcontroller. This proposed model also includes an ML-based risk factor analysis system by employing support vector machine (SVM) algorithm, and the data obtained will be utilized to send signals to the prototype by utilizing a RF technology.

Keywords RF technology · SVM algorithm · Panic button · Arduino UNO

1 Introduction

Workers health and safety are the most sensitive topic. With the technological advancements, the lives of workers can be positioned in a better place. All these hazards are at the most serious level for mining workers. Technology will help them

S. Perumal (✉) · A. S. V. L. Neha · R. Krishnaveni
CSE Department, Hindustan Institute of Technology & Science, Chennai, India
e-mail: 18113241@student.hindustanuniv.ac.in

R. Krishnaveni
e-mail: krishnaveni@hindustanuniv.ac.in

to predict the risk before it occurs. This project concentrates on proposing a better place for the workers.

This project consists of a worker safety helmet, which is divided into two parts, namely control unit and prototype. Both the units are connected as one using RF technology. This technology uses radio signals to establish a communication. It also contains a SVM-based risk factor analysis system. A NodeMCU-based RF. The data sent by the helmet nodes is received by using a tracker circuit. The system uses this information to map the present location of workers as they travel about the mine in real time. A panic (emergency) button is also included in this helmet. If this button is pressed, an emergency sign will appear on the IoT Web interface. This interface will then inform the management to take the required precautions in order to ensure the safety of the personnel.

This project contains two domains, namely IoT: RF-based technology, ML and SVM algorithm. Many goods such as radios, mobile phones, Wi-Fi routers and other radio-based devices rely on radio technology and radio frequency design, commonly known as RF design. Numerous types of radio signals can be employed. Different techniques and technologies are used in radio frequency engineering, ranging from simple forms of modulation such as amplitude, frequency and phase modulation to complex waveforms by using techniques such as direct sequence spread spectrum and orthogonal frequency division multiplexing. The end project connects both the domains with a Web interface.

The main motto of the proposed project is to make betterment in the lives of mining workers to some extinct that they work in a better workplace where it suits them to sustain either higher temperatures or abnormal humidities. The physical panic button present in the main prototype helps one to face the natural disasters and a machine cannot predict.

The addition of RF technology alongside the GSM module helps one to stay connected to the user in more than one way when a certain way fails. In this project, the data can be viewed in three different forms such as the LCD display present on the control room, the messaging service which is provided through the GSM module and the last and most efficient way is the visibility of data on the interface. This mode is the most efficient way because it not only helps in observing the data at a particular time but also helps to look at all the data that happened before time.

2 Literature Review

Kumarvel [1] proposed IoT mining tracking and worker safety helmet. Safety helmet has been created which displays the readings on LCD display and send an alert message. But there is no risk factor analysis. The prototype was good and provides accurate results, but it performs pre-intimation of previous risks or incidents taken place in that area.

Pradeep Kumar [2] proposed the system of smart helmet for industrial system which used LoRaWAN as their technology. This technology helped in finding the results to a very certain area limiting its expectations of focusing on a wide range. When there are changes in values, it sends alert.

(B Yakub [3]). In this paper, the author is focused on to present a comprehensive survey on smart safety helmet surface prediction of danger which was done using sensors by the readings detect, the readings were displayed on base station and usage of NodeMCU made the sensors faital.

(Somya Sista Lakshmi [4]) In this paper, the authors have proposed a system based on RFID tracking system. Safety helmet using RF-based tracking system, displayed and sent alert about the hazard. There is also panic button which sends the signals for help to the base station. But there was no risk factor analysis.

(Dhanalakshmi [5]) It helped in alerting the central console in the case of critical conditions. GPS helps us to track the miner's location during abnormalities in the sensors information [6–9]. Usage of PIc16F877A microcontroller could not provide the acquired results [10–14].

3   Methodologies

NODE MCU

It is an open source IoT platform, which includes firmware that runs on ESP8266 with the help of Wi-Fi as in Fig. 1. It is a platform-based object which connects various objects and transmits data from one to another using Wi-Fi protocol.

ARDUINO UNO

It is a microcontroller board based on ATmega328p as in Fig. 2. It works on an open source platform, and the board is filled with a set of analog and digital inputs or vice versa in case of outputs. It contains a microchip 8 bit AVR chip. The board contains 14 pins of digital inputs along with 6 pins of analog outputs. The board can be charged by using a 9v battery or an USB cable. This board acts as glue in binding all other components of the project in order to maintain balance.

Fig. 1 Pictorial representation of ESP8266

Fig. 2 ATmega328p

RF Technology

This technology uses radio waves to transmit the data from one desired body to another. This technology transmits data over an IoT network by utilizing the transmitter and receiver kit. This technology reduces the inadequacies in the process as it uses radio frequencies.

Data Interpretation Model:

- This module stores the database, and the result analysis is carried out by using a SVM algorithm.
- The SVM algorithm is used in risk factor analysis on the site, and the results obtained are stored in the database.
- The stored data helps in interpreting the risk factor of the site with respect to previous risks taken place in the original site.

4 Architecture Diagram

In the architecture diagram as shown in Fig. 3, the main sensors such as temperature, humidity and emergency switch are connected to a NodeMCU, which is further attached with a RF transmitter through which the RF receiver gets connected to Arduino UNO to obtain the data. The Arduino is further connected to alert system, LCD and a GSM. The GSM module is connected in two ways to the Web interface in order to help with the user experience.

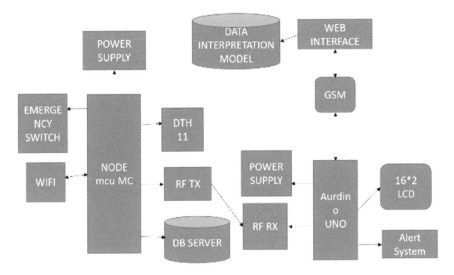

Fig. 3 Architecture diagram

5 Prototype Implementation

This project uses a RF-based tracking system, which helps in transmitting the data over the IoT network and detecting the probability of the safety of workplace environment by calculating the risk factor using SVM algorithm. This project contains two sections, i.e. control room and helmet section. Both are connected through RF-based technology. The control room includes an LCD display and an emergency button. The end project connects both the domains by using a Web interface.

The main sensors such as temperature, humidity and emergency switch are connected to a NodeMCU MC, which is further attached with a RF transmitter through which the RF receiver connected to Arduino UNO gets the data. The Arduino is further connected to an alert system, LCD and a GSM as in Fig. 4. The GSM module is connected in two ways to the Web interface to help with the user experience.

6 Testing and Evaluation

The testing and evaluation of the project has been done by using ThingSpeak platform. This platform ensures to have separate eyes on the required fields which need to be concentrated. The three major fields being temperature, humidity and alert.

In the platform, create a channel specifying the name of the channel. After that, make three fields and label them temperature, humidity, and alert. In the fields, create a graph showing the fluctuations caused to the protocol during different situations at different temperature and humidity ranges.

Fig. 4 Prototype

FIELD 1: Temperature

In this field, we can see the spikes in the graph with fluctuations in the temperature. When the temperature reaches above the limit either of the environment or the protocol, the spikes can be noted and this sends a message to the control room using the database through GSM and Wi-Fi.

FIELD 2: Humidity

In this field, we can see the spikes in the graph with fluctuations in the humidity range. When the humidity reaches above the limit either of the environment or the protocol, the spikes can be noted and this sends a message to the control room using the database through GSM and Wi-Fi.

FIELD 3: Alert

Here in field 3, the graph varies with respect to the protocol usage of the button in the processing unit. This alert system comes with a panic button attached with the helmet. As soon as the protocol uses this button, we see the button turning red in the IDE, further indicating in sending the panic message to the user through GSM and Wi-Fi.

As mentioned in the above fields, all the data is collected from the different fields used such as temperature, humidity and physical alert system as in validated output from Figs. 5, 6, 7 and 8. All this data is stored in the backend of ThingSpeak platform and is connected to the main prototype through GSM module.

The RF technology binds the whole prototype in a single part where one can see the physical alert system, control room and the helmet part.

Fig. 5 Graph of temperature fluctuations

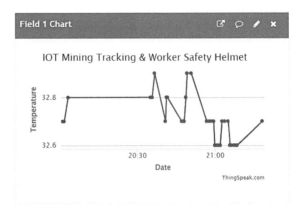

Fig. 6 Graph of humidity fluctuations

Fig. 7 Graph of temperature and humidity

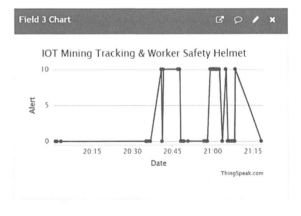

Fig. 8 Alert button

The successful implementation of this prototype suggests user the usage of RF technology is predominant when it comes to accuracy. Apart from the implementation of RF technology, the physical alert button helps in calculating the risks, which humans fail to predict.

7   Conclusion

This project has preferred and successfully used the RF technology when compared to other technologies such as YOLO4, LORAWAN, YOLO5, Zigbee and many others. RF technology shows a greater accuracy of around 95%, whereas the other technologies have an accuracy of around 86%, 84%, 92%, 81%, respectively, as in Fig. 9. Hence, the usage of RF technology enhances the accuracy.

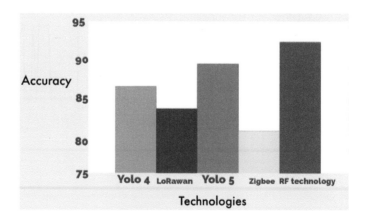

Fig. 9 Graph from previous studies

The usage of RF technology in the project with the help of RF transmitter and RF receiver not only help in increasing accuracy but also it opened a new gateway to record the data in the background with the help of accuracy that can be achieved with the proposed technology.

The constant recording of the data in the background with the help of an interface development not only increase the accuracy of findings but also provide the data to access the forthcoming hazards, thus making workers' life better.

References

1. Kumaravel, A. K., Sravani, K., & Lakshmi Sreenivasa Reddy, L. (2021). Worker safety helmet. In *Proceedings of "International Journal of Advanced Research in Computer and Communication Engineering"*.
2. Pradeepkumar, G., Sanjay Rahul, S., Sudharsanaa, N., Suvetha, S., & Ponnusamy, D. (2021). Smart helmet for the mining industry using LoRaWAN. In *Proceedings of "G Pradeepkumar et al 2021 J. Phys.: Conf. Ser. 1916"*.
3. Tajane, P. S., Shelke, S. B., Sadgzir, S. B., & Shelke, A. N. (2020). Safety helmet from dangers using IOT. In *Proceedings of "International Research Journal of Engineering and Technology (IRJET)"*.
4. Yakub, B., Deepthi, E., Jain, V., Vikram Singh, S., & Ravinder Singh, R. (2019). Iot used Mining tracking system and worker safety helmet. In *Proceedings of "International Journal of Research in Advent Technology*, Special Issue, NCRCEST 2019".
5. Dhanalakshmi, A., Lathapriya, P., & Divya, K. (2017). Smart helmet for improvising safety in mining industries. In *Proceedings of ""* 2017, IISN Journal.
6. Abdulla, R., Nataraj, C., Eldemeresh, T., & Al-Aldie, M. K. A. (2020). IOT based worker safety helmet for mining industry application. In *Proceedings of " 2020 IJSART* (Vol. 29, No. 1).
7. Jayasree, V., & Kumari, N. (2020). IOT based worker safety helmet for construction workers. In *Proceedings of " IEEE 2020 7th International Conference"*.
8. Noorin, M., & Suma, K. V. (2018). IOT based wearable device using WSN technology for miners. In *Proceedings of "IEEE conference in 2018 3rd international conference"*.
9. Mehata, K. M., Shankar, S. K., Karthikeyan, N., Nandhinee, K., & Robin Hedwig, P. (2019). IOT based safety and health monitoring system for construction workers. In *Proceedings of "IEEE 2019 1st International conference on innovation in information and communication technology"*.
10. Mangala Nandhini, V., Padma Priya, G. V., Nandhini, S., & Dinesh, K. (2018). IOT based smart helmet for ensuring safety of industrial workers. In *Proceedings of "IRERT 2018 conference"*.
11. Vishnu Kumar, A., Nirmal Kumar, A., Velumurugan, J., Poornima, S., & Pavithra, K. (2018). For the paper *"Coal Mine Workers Safety Helmet using Li-Fi Data Stored in the Cloud" in the proceedings of "IEEE 2018 conference"* (Vol. 7).
12. Borker, S. P., & Baru, V. B. IOT based smart helmet for underground mines. In *Proceedings of "IJRESM"* (Vol. 1, issue 9).
13. Sujitha, S., Shajilin Loret, J. B., & Merlin Gethsy, D. (2020). IOT based smart mine safety system using Arduino. In *Proceedings of "IJCSMC"* (Vol. 9, issue 5).
14. Ravi Kumar, G., & Keerthi Reddy, B. (2018). IOT based and intelligent helmet for wireless sensor network. In *Proceedings of "IJESRT"*.

Using IT in Descriptive Statistics and Way ANOVA Analysis to Assessment Development in Some Anthropometric Indicators of Thai Ethnic Students Born Between 2003 and 2006 in Thuan Chau District, Son La Province, Vietnam

Tran Thi Minh and Mai Van Hung

Abstract Adolescence is marked by changes in physical, intellectual and social relationships which are getting more complex. According to the World Health Organization (WHO), puberty, which starts from 10 to 19 years old, is a very important transition period in terms of physiological psychology and body and is seen as the preparation for the full development of muscles and the completion of organs and competence. This study looks at the development of some anthropometric indicators of Thai students in the years 2018, 2019 and 2020 to discover the diagrams of growth in height, weight, BMI and influential factors. These factors play a decisive role in the stature of a child as adulthood.

Keywords Ethnic · People · Anthropometry · Indicator · Physical · Height

1 Introduction

The primary goal in the socio-economic development strategy of every country is to improve the quality of life. This is because the idea of *"people development is key to success"* has been recognized on a global scale. In the first Human Development Report in 1990, the United Nations Organization stated: *"People are the real wealth of a nation. People are at the center of the development process"*[1]. This is reaffirmed in the United Nations Human Development Report in 2020, *"The process of development – human development - should at least create an environment for people, individually and collectively, to develop to their full potential and to have a reasonable chance of leading productive and creative lives that they value."* [2].

T. T. Minh
Tay Bac University, Sơn La, Vietnam

M. Van Hung (✉)
Research Center for Anthropology and Mind Development, VNU University of Education, Hanoi, Vietnam
e-mail: hungmv@vnu.edu.vn

© The Author(s), under exclusive license to Springer Nature Singapore Pte Ltd. 2023   149
S. Shakya et al. (eds.), *Sentiment Analysis and Deep Learning*,
Advances in Intelligent Systems and Computing 1432,
https://doi.org/10.1007/978-981-19-5443-6_13

Many policies on care, protection and education aiming at improving the physical condition of Vietnamese people, especially children have been implemented. The identification of biological indicators of different subjects in various localities will contribute to the scientific assessment of the overall development of Vietnamese people in different periods.

Son La is a mountainous province in the northern region of Vietnam, with the majority of students from ethnic minorities such as Thai, Mong, Dao, Xa, Muong. This is also a province with a young population; and thus, its labor resources are abundant. Therefore, there is a need for healthy and intelligent labor.

The study is conducted to analyze the growth of some anthropometric indicators of Thai students born in 2003, 2004, 2005 and 2006 within three continuous years 2018, 2019 and 2020. This is necessary in assessing the growth and also helps to improve the quality of the young population from ethnic minorities today.

2 Methodology

The study was conducted with the participation of 267 Thai students in Son La province, in which 130 are male students and 137 are female students. The participants or research subjects are randomly selected and all the standards for a morphological and anthropological study has been strictly followed [3].

The anthropometric indicators surveyed are the height and weight of Thai students born in 2003, 2004, 2005 or 2006 for the years 2018, 2019 and 2020.

The data analysis process was carried out at the "Anthropology lab" at the Center for Anthropology and Mind Development, Hanoi National University of Education. Data are then processed by software programs Excel 2010 [4] and SPSS 2.0.

- Descriptive statistics used includes: frequency, percentage for qualitative variables; mean values, standard deviations for quantitative variables.
- Student-t test has been applied to compare quantitative variables standard distribution between two independent groups. One way ANOVA analysis is also taken to compare quantitative variables of ≥ 3 groups and post-hoc test is used to compare pairs of two groups. This test compares anthropometric measurements (height, weight, BMI) between age groups and sexes.
- Pearson correlation coefficient (r) is used to measure linear correlation between two variables, with values from $+1$ to -1.

- The closer r goes to 1 or -1, the stronger the linear correlation will be
- The closer r goes to 0, the weaker the linear correlation will be

$r > 0$ indicates a positive correlation between two variables, that is, the increase in the value of one variable will lead to the increase in the value of the other and vice versa. Meanwhile, $r < 0$ indicates a negative correlation between two variables, which means, if the value of one variable increases, the value of the other will decrease and vice versa. When $r = 0$, the two variables have no linear correlation.

– Multi-variable linear regression analysis method:

• The form of a multivariate linear regression equation is: $Y = \beta0 + \beta1 X1 + \beta2X2 + \dots$

  In which: $X1, X2 \dots$ are independent variables.
  Y is the dependent variable; $\beta0, \beta1, \beta2\dots$ are the regression coefficients.

– Multi-variable logistic regression analysis method:

• The form of a multivariate logistic regression equation is:

  $Y = \text{Loge} = = 0 + 1 \times 1 + 2 \times 2 + 3 \times 3 + \dots$
  In which: pi is the probability of the event occurring;
  β_0 is a regression constant; $\beta_1, \beta_2 \dots$ are the regression coefficients.
  $X1, X2, X3\dots$ are independent variables.
  Predictive function model is: $p = \frac{e^y}{1+e^y}$.
  The optimal multivariate logistic model is expected to be:

• a simple model with few prognostic variables and can be applied in reality.
• able to describe the data in a satisfactory way, that is, provide the closest prediction to the actual value of the dependent variable in the model.
• significant in biology and clinical.

3 Results and Discussion

3.1 Standing Height of Thai Ethnic Students

Over the three-year investigation of biological indicators lasting from 2018 to 2020, we have measured the height and weight indices of the Thai student group in Thuan Chau district, Son La province to have a better insight about the health status as well as nutritional condition of the study subjects.

Standing height is one of the most basic biological values that reflects the growth and development of the human body across age groups. Studies on this indicator show that standing height varies according to age, sex and is affected by living environment.

The results from analytic comparison of standing height of Thai male and female students are shown in Tables 1 and 2 and Figs. 1 and 2.

Table 1 shows that there was a gradual increase in the standing height of Thai male students and female students in Son La province during the period from 2018 to 2020. At the same age, the heights of male students and female students are significantly

Table 1 Standing height of Thai students by age and sex in the period 2018 - 2020

Indices			2003 Male (1)	Female (2)	$\overline{X}_1-\overline{X}_2$	2004 Male (1)	Female (2)	$\overline{X}_1-\overline{X}_2$
Standing height	2018	$\overline{X}\pm SD$	1.5401 ± 0.0334	1.5000 ± 0.0172	0.0401	1.4628 ± 0.0328	1.4726 ± 0.0189	−0.0098
		Increase	–	–		–	–	
	2019	$\overline{X}\pm SD$	1.5844 ± 0.0334	1.5174 ± 0.0169	0.067	1.532 ± 0.0329	1.5108 ± 0.0182	0.0212
		Tăng	0.0443	0.0174		0.0692	0.0382	
	2020	$\overline{X}\pm SD$	1.6045 ± 0.0335	1.5405 ± 0.0186	0.064	1.5824 ± 0.0332	1.5242 ± 0.0190	0.0582
		Increase	0.0201	0.0231		0.0504	0.0134	
n			37	44		31	31	

Indices			2005 Male (1)	Female (2)	$\overline{X}_1-\overline{X}_2$	2006 Male (1)	Female (2)	$\overline{X}_1-\overline{X}_2$
Standing height	2018	$\overline{X}\pm SD$	1.4110 ± 0.0447	1.4213 ± 0.0205	−0.0103	1.3283 ± 0.0445	1.3798 ± 0.0211	−0.0515
		Increase	–	–		–	–	
	2019	$\overline{X}\pm SD$	1.4602 ± 0.0423	1.4662 ± 0.0209	−0.006	1.3624 ± 0.0449	1.4347 ± 0.0206	−0.0723
		Tăng	0.0492	0.0449		0.0341	0.0549	
	2020	$\overline{X}\pm SD$	1.5131 ± 0.0431	1.5111 ± 0.0225	0.002	1.4318 ± 0.0420	1.4834 ± 0.0228	−0.0516
		Increase	0.0529	0.0449		0.0694	0.0487	
n			32	30		30	32	

Table 2 Average standing height of Thai ethnic students by age and sex

Age	Sex						p	General $\overline{X} + SD$
	Male			Female				
	$\overline{X}$	SD	n	$\overline{X}$	SD	n		
12	1.3283	0.0445	30	1.3798	0.0211	32	< 0.05	1.354 ± 0.033
13	1.3867	0.0448	62	1.428	0.0206	62	< 0.05	1.4074 ± 0.0327
14	1.4516	0.0390	93	1.4741	0.0209	93	< 0.05	1.4629 ± 0.03
15	1.5284	0.0365	100	1.5073	0.0193	105	< 0.05	1.518 ± 0.0279
16	1.5834	0.0333	68	1.5208	0.018	75	< 0.05	1.552 ± 0.026
17	1.6045	0.0335	37	1.5405	0.0186	44	< 0.05	1.5725 ± 0.026

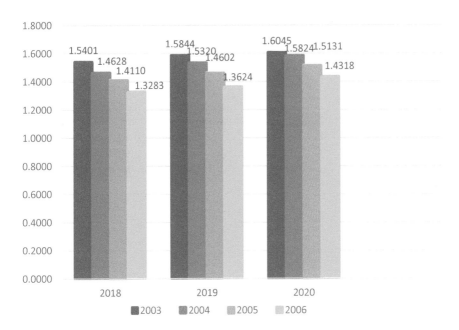

Fig. 1 Height of Thai male students, period 2018–2020

different and increase unevenly. The average height of female students is greater than that of male students across the groups: group of students who were born in 2006 in all 3 years of survey (corresponding to ages 12, 13, 14), group of those born in 2005 in 2 times of measuring in 2018 and 2019 (corresponding to ages age 13, 14) and group of students born in 2004 at the time of measurement in 2018 (age 14 respectively). In the remaining age groups, the height of male students is higher than that of schoolgirls.

Our research results are consistent with the findings in the study of Tran Thi Loan [5], that is, there is relatively large height growth in the period of 12–17 years old.

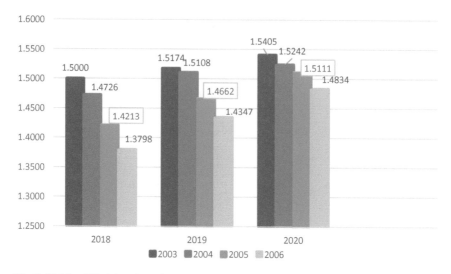

Fig. 2 Height of Thai female students, period 2018–20

Female students grows fastest in height at the age of 12–13 years, which is two years earlier than male students'. The growth rates of height by age for male students and female students are not the same (the difference, $p < 0.05$, has statistical significance); the differences in average standing height between the two sexes in an age group are not significant.

The average standing height of Thai students by age group saw a rapid growth in both sexes, with a fairly big increase in height across all age groups. The average increase in height is 0.0552 m/year for male students and 0.0321 m/year for female students.

The height of Thai male students and female students is shown in Figs. 1 and 2.

Male students grow fastest when they are in the ages of 14 and 15, with an increase of 0.0768 m, while for female students, it is in their 12–13 year old period, and the increase is 0.0482 m. The largest difference in height between male and female students can be seen when they are 17 years old, with the gap of 0.064 m. These differences in height between male and female students have a statistical significance ($p < 0.05$).

In summary, the growth of standing height of Thai students in our study is in accordance with the law of height growth of Vietnamese children. However, the average standing height of Thai students in Thuan Chau district, Son La province is lower than that in the survey conducted by the Ministry of Health in 2003 and then the average height according to WHO statistics in 2007. This is because most ethnic minority families here have difficult economic conditions, so they are not properly aware of the nutritional regime and cannot afford a reasonable care regime to enhance their children's stature. Therefore, measures should be taken to further improve the height of children in the study and in Vietnam in general.

Table 3 Average weight of Thai students by age and sex

Age	Sex						p	General $\overline{X} \pm$ SD
	Male			Female				
	$\overline{X}$	SD	n	$\overline{X}$	SD	n		
12	26.3643	3.7443	30	27.9216	2.0610	32	< 0.05	27.1429 ± 2.9027
13	30.7103	3.6714	62	32.3321	2.5722	62	< 0.05	31.5212 ± 3.1218
14	35.3337	3.9606	93	36.7827	2.6134	93	< 0.05	36.7827 ± 3.2870
15	40.2981	3.7898	100	39.5146	2.8323	105	< 0.05	39.9064 ± 3.3110
16	43.9615	3.5269	68	41.5014	2.6238	75	< 0.05	42.7314 ± 3.0753
17	46.9108	2.8273	37	43.2636	2.6904	44	< 0.05	45.0872 ± 2.7589

3.2 The Weight of Thai Students by Age and Sex

3.2.1 The Weight of Thai Students

Weight is one of the important parameters that reflect the development of the child's body, nutrition regime and lifestyle. Data on weight of Thai students by age and sex are shown in Tables 3 and 4.

The average weight gain for male students is 4.1093 kg/year, and for women is 3.0684 kg/year. Weight growth is relatively fast during the period of 12–17 years old. It rockets when they are 13–14 years old, with the time of rapid weight gain for female students being from 13 to 14 years old, and for male students being between 14 and 15.

The fastest rate of weight gain for male students happens when they reach the period of 14–15 years old, with the increase of 4.9644 kg. For female students, it is when they are 13–14 years old, and the increase is 4.4506 kg. The biggest difference in weight between male and female students is at 17 years old, with a difference of 3.6472 kg. This difference is statistically significant ($p < 0.05$).

3.2.2 Weight of Thai Ethnic Students by Age and Sex

Weight of male and female Thai students over 2018–2020 period is seen in Table 4.

It can be seen from Table 4 that at the same age, the weights of male and female students are significantly different and increases unevenly. The weight of female students is greater than that of male counterparts across the groups: group of students who were born in 2005, 2006 in all three years of survey (corresponding to ages 12, 13, 14), group of those born in 2004 at measuring time in 2018 (corresponding to age 14). In the remaining age groups, the weight of male students is more than that of female students. The rates of weight gain by age for male and female students are not the same (this difference is statistically significant, $p < 0.05$), but the difference in average standing weight between two sexes in an age group is not.

Table 4 Weight of Thai students by age and sex, period 2018–2020

Weight	Indices	2003		$\bar{X}_1-\bar{X}_2$	2004		$\bar{X}_1-\bar{X}_2$	2005		$\bar{X}_1-\bar{X}_2$	2006		$\bar{X}_1-\bar{X}_2$
		Male	Female		Male	Female		Male	Female		Male	Female	
2018	$\bar{X} \pm SD$	41.0503 ± 3.1328	39.9221 ± 2.5543	**1.1282**	35.4116 ± 4.2933	36.0167 ± 2.4678	**−0.6051**	31.5863 ± 3.5984	32.0924 ± 3.1668	−0.5061	26.3643 ± 3.7443	27.9216 ± 2.0610	−1.5573
	Increase	–	–	–	–	–	–	–	–	–	–	–	–
2019	$\bar{X} \pm SD$	44.0343 ± 2.8233	42.0989 ± 0.6206	**1.9354**	39.6416 ± 4.455	38.0342 ± 2.4966	**1.6074**	35.5788 ± 3.7862	36.4764 ± 3.1919	−0.8976	29.8343 ± 3.74433	32.5719 ± 1.9777	−2.7376
	Increase	2.984	2.1768		4.23	2.0175		3.9925	4.384		3.47	4.6503	
2020	$\bar{X} \pm SD$	46.9108 ± 2.8273	43.2636 ± 2.6904	**3.6473**	43.8887 ± 4.2305	40.9039 ± 2.6269	**2.9848**	40.2025 ± 3.7816	40.5877 ± 3.4458	−0.3852	35.0108 ± 3.8024	37.855 ± 2.1806	−2.8442
	Increase	2.8765	1.1646		4.2471	2.8697		4.6237	4.1113		5.1765	5.2831	
N		37	44		31	31		32	30		30	32	

The weights of Thai male and female students are shown in Figs. 3 and 4.

The fastest rate of weight gain for male students happens when they reach the period of 14–15 years old, with the increase of 4.9644 kg. For female students, it is when they are 13–14 years old, and the increase is 4.4506 kg. The biggest difference in weight between male and female students can be seen when they are at the age

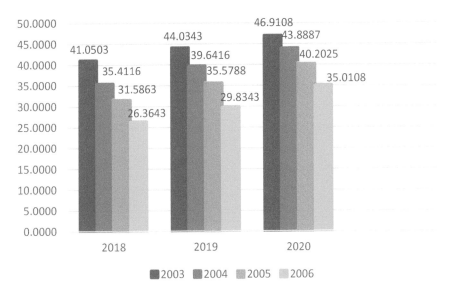

Fig. 3 Weight of Thai male students, period 2018–2020

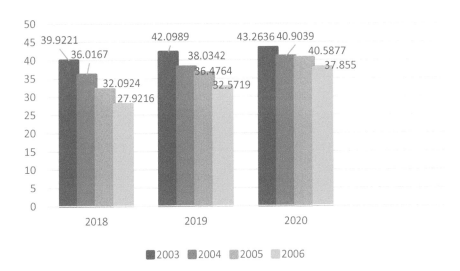

Fig. 4 Weight of Thai female students, period 2018–2020

Table 5 Average BMI of Thai students by age and sex

Age	Sex						p	General $\overline{X} + SD$
	Male			Female				
	$\overline{X}$	SD	n	$\overline{X}$	SD	n		
12	14.9682	2.1491	30	14.6755	1.1551	32	< 0.05	14.8218 ± 1.6521
13	15.9884	1.8873	62	15.8607	1.2816	62	< 0.05	15.9245 ± 1.5845
14	16.7768	1.7895	93	16.9338	1.2515	93	< 0.05	16.8553 ± 1.5205
15	17.2582	1.5766	100	17.4034	1.3430	105	< 0.05	17.3308 ± 1.4598
16	17.5421	1.4303	68	17.9596	1.3001	75	< 0.05	17.7508 ± 1.3652
17	18.2499	1.3981	37	18.2405	1.2589	44	< 0.05	18.2452 ± 1.3285

of 17, with the difference being 3.6472 kg. This difference is statistically significant ($p < 0.05$).

3.3   BMI of Thai Students by Age and Sex

The average BMI indices of Thai students in our study is shown in Tables 5 and 6.

The average BMI of Thai students by age group is different in both sexes (the difference is statistically significant, $p < 0.05$). The average BMI of male students is 0.6564/year, while it is 0.7130/year for females. Besides, the difference in BMI between two sexes in a particular age group is not significant.

Table 6 shows that, at the same age, BMIs of male and female students have different values. BMI of females is greater than that of males in the groups of students who were born in 2006 at the 2020 survey (corresponding to age 14), in 2005 in all three surveys (ages 13, 14, 15 respectively), in 2004 at the 2018 and 2020 surveys (ages 14 and 16, respectively) and in 2003 at the 2018 and 2019 surveys (ages 15, 16, respectively). In the remaining age groups, the BMI of male students is greater than that of female counterparts. The changes of BMI by age for male and female students are not the same (the difference is statistically significant, $p < 0.05$). The BMI of male and Thai students is shown in Figs. 5 and 6.

At the same age, BMIs for male and female students of Thai ethnicity are not the same. To be precise, BMI of male students is lower than that of female students. This is because the female development period begins earlier than that of male. In general, the differences in BMI of male and female students from 12 to 17 years old are statistically significant ($p < 0.05$).

Table 6 BMI of Thai students in the period of 2018–2020

BMI	Indices	2003				2004		
		Male	Female	$\overline{X}_1-\overline{X}_2$		Male	Female	$\overline{X}_1-\overline{X}_2$
2018	$\overline{X}\pm SD$	17.3354 ± 1.5706	17.7545 ± 1.2851	−0.4191		16.5273 ± 1.7212	16.6238 ± 1.3099	−0.0965
	Increase	–	–	–		–	–	–
2019	$\overline{X}\pm SD$	17.5692 ± 1.4118	18.2959 ± 1.2780	−0.7267		16.8724 ± 1.6284	16.6760 ± 1.2449	0.1964
	Increase	0.2338	0.5414			0.3451	0.0522	
2020	$\overline{X}\pm SD$	18.25 ± 1.3981	18.2405 ± 1.2405	0.0095		17.5150 ± 1.44.879	17.6233 ± 1.3222	−0.1083
	Increase	0.6808	−0.0554			0.6426	0.9475	
n		37	44			31	31	

BMI	Indices	2005				2006		
		Male	Female	$\overline{X}_1-\overline{X}_2$		Male	Female	$\overline{X}_1-\overline{X}_2$
2018	$\overline{X}\pm SD$	15.8688 ± 1.6497	15.8941 ± 1.6033	−0.0253		14.9682 ± 2.1491	14.6755 ± 1.1551	0.2927
	Increase	–	–	–		–	–	–
2019	$\overline{X}\pm SD$	16.6841 ± 1.5578	16.9735 ± 1.4971	−0.2894		16.1079 ± 2.1249	15.8273 ± 0.9599	0.2806
	Increase	0.82	1.0794			1.1397	1.1518	
2020	$\overline{X}\pm SD$	17.5669 ± 1.5309	17.7795 ± 1.4991	−0.2126		17.1192 ± 2.0896	17.2040 ± 0.9474	−0.0848
	Increase	0.8828	0.806			1.0113	1.3767	
n		32	30			30	32	

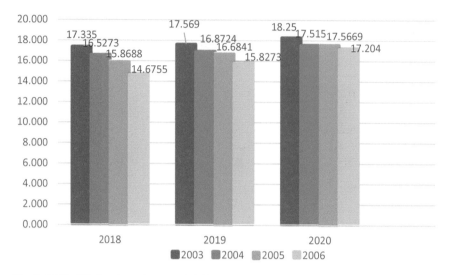

Fig. 5 BMI of Thai male students, period 2018–2020

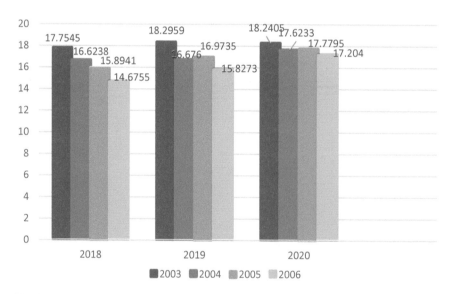

Fig. 6 BMI of Thai female students, period 2018–2020

4 Conclusion

The study results on morphological indicators of students show that the height and weight indices increase with age. However, the growth rate of these 2 indices by age and by sex is uneven or not the same. Rapidly growing period of these indicators appears later in male than in female students, which illustrates the unevenness in the law of growth according to the sex of children.

The average BMI of Thai ethnic students at some secondary schools in Thuan Chau district, Son La province, is mainly in one of 4 levels: constant lack of energy at levels 1, level 2, level 3 and normal level. This reflects that their morphological indicators are far lower than the national and international average. To improve the situation, it is necessary to have a combination of many activities from changing the public's perception of the role of nutrition through communication and education to improving the quality of life of ethnic minorities in Thuan Chau district, Son La province, in particular and the Northwest region in general.

References

1. United Nations. (1990). *Human development report*. United Nations Development Program.
2. United Nations. (2020). *Human development report*. United Nations Development Program.
3. Ministry of Health. (2003). *Normal Vietnamese biological values in the 1990s–20th century*. Medical Publishing House.
4. Dam, D. T. (2003). *Using Microsoft Escel in biological statistics*. Medical Publishing House, Hanoi.
5. Loan, T. T. (2002). *Research on some physical and intellectual indicators of students from 6–17 years old in Cau Giay district, Hanoi*. PhD thesis in Biology, Hanoi Pedagogical University.

LSTM-Based Deep Learning Architecture for Recognition of Human Activities

Pooja Kallepally and M. Rajesh

Abstract This paper uses raw sensor data to develop and train the networks for human activity recognition in a variety of applications, ranging from tracking their fitness and the safety monitoring. The models may simply be expanded to be trained with additional data sources for greater accuracy or classification extension for different prediction classes. This research delves into WISDM's accessible dataset as well as the distinct characteristics of each class for various axes. Furthermore, for the application of human activity identification, the design of a long short-term memory (LSTM) architectural model is defined. In the first 500 epochs of training, an accuracy of over 96 percent and a loss of less than 30% have been achieved.

Keywords Sensors · Long short-term memory (LSTM) · Activity recognition · Acceleration

1 Introduction

The objective of human activity recognition (HAR) is to group practices in light of information gathered from a bunch of sensors in cell phones. Cell phones are currently available to basically everybody in the computerized age, and video surveillance is commonly used [1]. Cell phones have become typical all over the planet. These cell phones have inherent sensors like an accelerometer, sensors, among others. These sensors respond to movement of a cell phone using body. HAR is utilized to comprehend the activities performed by people in such recordings/camera. Given the quantity of cameras, it is outside the real-life possibilities for a solitary individual for gathering to keep track of them 24 * 7 with high exactness while downplaying working expenses in a powerful manner. The speed increase of an accelerometer is

P. Kallepally (✉) · M. Rajesh
Department of Computer Science and Engineering, Amrita School of Engineering, Amrita
Vishwa Vidyapeetham, Bengaluru, India
e-mail: BL.EN.P2CSE20014@bl.students.amrita.edu

M. Rajesh
e-mail: m_rajesh@blr.amrita.edu

© The Author(s), under exclusive license to Springer Nature Singapore Pte Ltd. 2023   163
S. Shakya et al. (eds.), *Sentiment Analysis and Deep Learning*,
Advances in Intelligent Systems and Computing 1432,
https://doi.org/10.1007/978-981-19-5443-6_14

estimated in m/s² along each of the three axes. Gyroscope is one more noticeable sensor for action recognizable proof. It really adds a speed or turn pivot to the information given by the accelerometer. Numerous analysts have utilized information from the two sensors or only one of them. HAR is advancing in applications to make life better with the arrangement of different wearable gadgets that will have a greater number of sensors than these two, for example, pulse using external sensors isn't always a good idea.

2 Literature Review

Bülbül et al. [1] proposed a research article on human activity recognition using smartphones. They introduced SVM, a method for separating examples that makes the greatest use of hyperdimensional planes. Although SVM can be utilized with or without supervision, supervised SVM is frequently faster and more successful. An accuracy of 87 percent was attained when SVM with a cubic polynomial kernel used to identify the dataset. The SVM strategy isn't proper for enormous informational collections. At the point, when given information is more, SVM doesn't perform well.

Paper [2], "MEx: Multimodal Exercises Dataset for Human Activity Recognition," by Anjana Wijekoon and Wen-chao Xu investigated several methodologies, with CNN, being the most typically utilized to process visual pictures. They are frequently referred to as move invariant or space invariant artificial neural networks, and they primarily focus on translation and common design invariance. The vector was created by flattening the frame; then, the frames were processed to create timing window. For one path, a single-dimensional display with a frame width vector of length frame width height/frame height/frame per second. 2D method primarily employed two-dimensional convolution and total pooling levels.

Paper [3] "Subject Cross-Validation in Human Activity Recognition," by Akbar Dehghaniet al., proposed a model that uses k-fold cross-validation as the primary methodology for testing model output and adjusting hyperparameters. This means that measurements are both random and identically distributed, which means that data points are taken from the same source separately. As a result, k-fold cross-validation may worsen activity recognizer performance, especially when overlapping sliding windows are utilized. The influence of subject cross-validation on the performance of HAR was investigated.

Paper [4], "Dynamic Vision Sensors for Human Activity Recognition," by Stefanie Anna Baby, proposed a model that included a dynamic vision sensor that processes only the foreground objects captured by the camera while completely ignoring the unnecessary processing of backgrounds, reducing processing requirements, and greatly improving detection efficiencies. Because every pixel's intensity difference is exclusively determined by the texture edge, the DVS output is extremely sparse. Various portions of the DVS data are also represented as motion maps, which are then utilized to define aspects of interest while employing the sparse data and segments of movement acquired by the DVS to increase confidence in activity detection.

Ghazal [5] employed an SVM classifier to extract human skeleton information from the camera input, such as the placement of joints in the human body structure. Based on the input collected from the camera feed, this data may be utilized to estimate the human stance and to detect the activity connected with the movements of these data points.

Ding [6] built a model to identify different human activities using a decision tree-based technique. The notion of bagging, which helps randomize the selection of the partitioning data nodes in the tree construction, improves the classification accuracy of a single-tree classifier. The vector was allocated to the supplied class based on a majority vote based on the decision tree provided. This procedure was repeated for each tree in the forest. To attain the needed high accuracy, this procedure needed a vast quantity of tagged data.

Studies from [7–14] have also shed light on various methods for detection of the activities.

3   Proposed System

The model is built with the LSTM architecture and trained on accelerometer data to classify six activities. This model has two fully connected and two LSTM layers, each with 64 units. The initiation work for the hidden layers is ReLU. With regards to neural network, the rectifier direct unit is an initiation as illustrated mathematically in,

$$F(x) = \max(0, x) \tag{1}$$

ReLU is a very popular activation function since it does not activate all neurons at the same time. At the output layer, SoftMax is used as the activation function because the SoftMax layer provides the desired probability distribution for each classification. Using the basic LSTM cell method from Google's TensorFlow library, the deep learning algorithm is optimized by creating a loss function to evaluate the difference between the model predictions and the desired label and a set of weights that the neural network can use to generate an accurate prediction, resulting in a loss function with a lower value.

ADAM optimizer is used in the LSTM model as Fig. 1. Adam optimization is a deep learning model optimization training technique that replaces stochastic gradient descent. Adam optimizer combines the finest features of the AdaGrad and RMSProp methods to provide an optimization technique for noisy issues with sparse gradients.

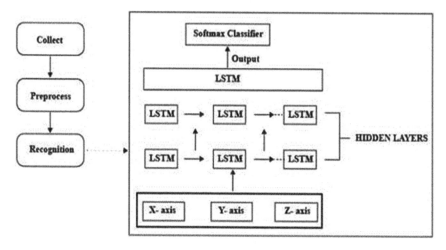

Fig. 1 HAR using LSTM

4 LSTM Architecture

LSTM deep learning approaches are employed to train a model using a publically available dataset containing just accelerometer data. The learned model will then attempt to anticipate activities in real time one by one. LSTM can be utilized for assignments like unsegmented, connected, discourse acknowledgment, and inconsistency location in network traffic or interruption recognition frameworks, for instance interruption identification frameworks.

Figure 2 consists of a typical LSTM unit, with an input gate, an output gate, and forget gate. The LSTM resolves a huge issue with intermittent neural networks' short memory. The LSTM utilizes a progression of GATES, each with its own RNN, to keep, neglect, or overlook information that focuses relying upon a probabilistic model. Recurrent networks in which each neuron is replaced by a memory unit are known as LSTMs. A real neuron with a recurrent self-connection is housed in the memory unit. The cell state of the LSTM network refers to the activations of those neurons within the memory units.

Furthermore, LSTMs might be used to settle issues with unstable and vanishing angles. In layman's terms, these worries are the consequence of intermittent weight

Fig. 2 LSTM network

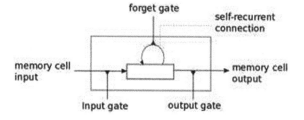

changes in a neural network training. Inclinations increment or reduction as ages pass, and with each change, it becomes more straightforward for the slopes of the framework to duplicate in whichever way. Because of this compounding, the slopes are either too large or excessively minuscule. While conventional RNNs have serious restrictions, for example, detonating and vanishing slopes, the LSTM configuration fundamentally mitigates the problems.

5 Methodology

This model as shown in Fig. 3 considers six activities that are running, sitting, standing, walking upstairs, and walking downstairs. These exercises are accessible in raw data information given by WISDM and are ideal to use as a base for research. A large portion of the exercises includes rehashed movement making it simpler to prepare and perceive. The x-axis addresses the flat movement which is toward the left and right sides of the legs and hands. This is utilized to catch the level movement of leg. The y-axis distinguishes vertical movement which is the course to the top and lower part of the human, while z-axis addresses the movement into and out of the screen, which can be utilized to catch legs in reverse and forward movement.

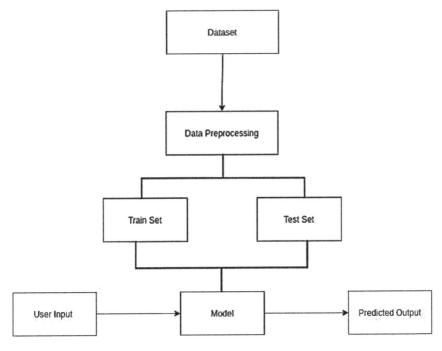

Fig. 3 System architecture

The data collected by the accelerometer are used to perform various activities, such as walking, sitting, and standing. The data are recorded at 50 Hz. Due to the nature of the data collected, it is not possible to access the raw data. Instead, a preprocessing version has been created. The preprocessing version is able to split the data into 128 data points, with 50% overlap. It then splits the data into two components: the body motion components and the gravitational force.

6 Results

The dataset for jogging activity in Fig. 4 shows the pattern created by the user, which is different from the other activities that are recorded. The frequency lines are not aligned like those used in other activities. The position of the user is also determined by the y-axis, which fluctuates around 10–20°. This means that the person is in the right position while performing the activity. The z-axis here is similar to walking. It ranges in the negative values because of the user activity (jogging) which requires a lot more force.

In the walking pattern shown in Fig. 5, the frequency is similar in all the axis but has own force and radius. The peaks represent the human walking pattern in a straight line.

The output in Figs. 6 and 7 shows a similar pattern as the motion of the user is in constant motion. The x-axis represents the motion of the leg, and since there is no movement, the graph is constant. The y-axis shows the gravitation pull toward the surface, and the z-axis is in perpendicular motion, and so, the graph is constant in both the axes.

The LSTM model appears to be well trained with an accuracy of over 96%, with cross-entropy losses well below 0.2 in both training and validation data. The cross-entropy loss is 0.13 which is shown in Fig. 8.

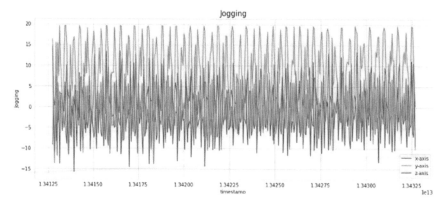

Fig. 4 Jogging pattern

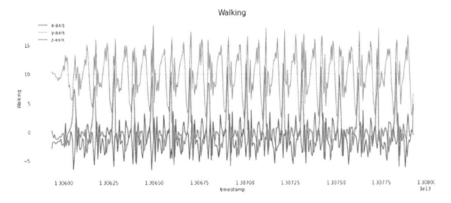

Fig. 5 Walking pattern

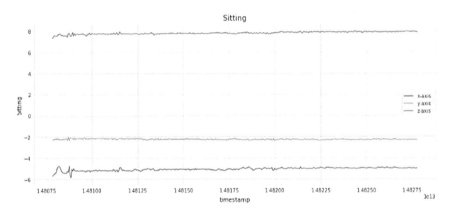

Fig. 6 Sitting pattern

As can be seen from Fig. 9 confusion matrix, the two most common activities in the dataset, i.e., walking and jogging, are correctly classified with very high accuracy. Although sitting and standing are minority classes, the model is still able to distinguish them correctly. Accuracy is not as high as other classes for upstairs and downstairs operations. This is to be expected because the two operations are very similar; hence, the underlying data may not be enough to accurately distinguish them.

7 Conclusion and Future Work

The movement of the sensor all along x, y, and z (3-dimensional) axes, as well as the timestamp at which the readings are obtained, has been studied. Based on the raw time series data, a simple LSTM network is built. The LSTM-based model accomplishes

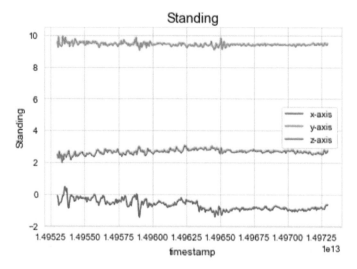

Fig. 7 Standing pattern

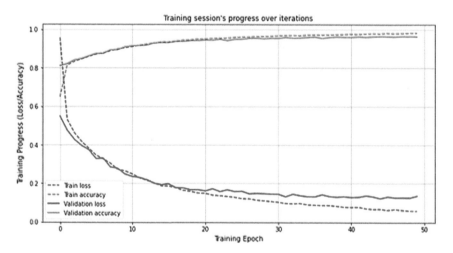

Fig. 8 Cross-validation loss

classification accuracy of 96%. This is done by combining the tri-pivotal data of the accelerometer with the tri-axial data. It is possible to detect comparison actions like standing and sitting, as well as perceive various workouts having substantially very high precision.

The future work may focus on working on newly developed datasets suited to real-world scenarios, including more operations for prediction and more participants. If not, data collection efforts can also be attempted with a larger number of participants with different lifestyles, behaviors, ages, genders, etc. Moreover, predicting unusual

Fig. 9 Confusion matrix

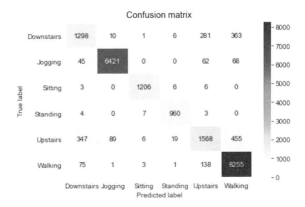

human behaviors can be performed. Using a functional activity recording system, one can monitor dependents such as the elderly in smart homes and assess how active they are for health services. Additionally, for all occupants living in smart buildings, human activity recognition can be used to check their comfort levels on factors such as temperature and humidity.

References

1. Bülbül, E., Çetin, A., & Doğru, I. A. (2018). *Human activity recognition using smartphones "Gazi".* University Ankara, Turkey, IEEE.
2. Wijekoon, A., Wiratunga, N., & Cooper, K. (2019). *MEx: Multi-modal exercises dataset for human activity recognition.* arXiv:1908.08992v1[cs.CV]. Aug 13, 2019.
3. Dehghani, A., Glatard, T., & Shihab, E. (2018). *Subject cross-validation in human activity recognition,* Conference, July 17, 2018, Washington, DC, USA.
4. Baby, A. A., Vinod, B., Chinni, C., & Mitra, K. (2018). *Dynamic vision sensors for human activity recognition.* arXiv:1803.04667v1[cs.CV]. 13 Mar 2018.
5. Ghazal, S., & Khan, U. S. (2018). Human posture classification using skeleton information. In *2018 international conference on Computing, Mathematics and Engineering Technologies—iCoMET 2018.*
6. Ding, G., Tian, J., Wu, J., Zhao, Q., & Xie, L. (2018). Energy-efficient human activity recognition using wearable sensors.. In *2018 IEEE Wireless.*
7. Meghana, C., Himaja, M., & Rajesh, M. (2020). Communications and networking conference workshops (WCNCW): IoT-Health 2018: IRACON workshop on IoT enabling technologies in healthcare. In *Smart attendance monitoring system using local binary pattern histogram.* Information and Communication Technology for Competitive Strategies (ICTCS 2020).
8. Sreevidya, B., Rajesh, M., & Sasikala, T. (2018). Performance analysis of various anonymization techniques for privacy preservation of sensitive data. In *International conference on intelligent data communication technologies and internet of things.*
9. Subbiah, U., Kumar, D. K., Thangavel, S. K., & Parameswaran, L. (2020). An extensive study and comparison of the various approaches to object detection using deep learning. In *2020 International Conference on Smart Electronics and Communication (ICOSEC).*
10. Senthilkumar, T., & Narmatha, G. (2016). Suspicious human activity detection in classroom examination. In *Computational intelligence, cyber security and computational models. Advances in intelligent systems and computing* (Vol. 412). Springer.

11. Venkata Krishna Rao, B., Gopikrishna, K. S., Sukrita, M., & Parameswaran, L. (2021). Activity recognition using LSTM and inception network. In *Soft computing and signal processing. Advances in intelligent systems and computing* (Vol. 1325), Springer.
12. Kavya, J., & Geetha, M. (2016). An FSM based methodology for interleaved and concurrent activity recognition. In *2016 International Conference on Advances in Computing, Communications and Informatics (ICACCI).*
13. Bhambri, P., Bagga, S., Priya, D., Singh, H., & Dhiman, H. K. (2020). Suspicious human activity detection system. *Journal of IoT in Social, Mobile, Analytics, and Cloud, 2*(4), 216–221.
14. Tripathi, M. (2021). Sentiment analysis of Nepali COVID19 tweets using NB, SVM AND LSTM. *Journal of Artificial Intelligence, 3*(03), 151–168.

A Deep Learning Framework for Classification of Hyperspectral Images

Likitha Gongalla and S. V. Sudha

Abstract The emergence of deep learning models has paved way for solving many real world problems. Hyperspectral image classification has its significance in very critical applications associated with defence and agricultural research. The existing methods showed that there is need for improving quality in hyperspectral image classification for computer vision applications. In this paper, we proposed a framework known as Hyperspectral Image Classification Framework (HICF) with underlying mechanisms. We proposed an algorithm known as deep learning-based hyperspectral image classification (DL-HIF) for improving classification of hyperspectral images. A prototype is built using Python data science platform for evaluating the performance of the proposed algorithm. Our experimental results are compared with two existing Machine Learning (ML) models such as Support Vector Machine (SVM) and Neural Network (NN). The proposed DL-HIF outperforms the state-of-the-art models.

Keywords Machine learning · Deep learning · Hyperspectral image classification · Deep learning framework

1 Introduction

Hyperspectral imaging taken from satellites or through remote sensing has its significance in many computer vision applications. The emergence of deep learning models has paved way for solving many real world problems. Hyperspectral image classification has its significance in very critical applications associated with defence and agricultural research across the electromagnetic spectrum hyperspectral imaging technology can read. Therefore, it plays crucial role in many real world applications. With the advancements in ML in the form of deep learning, it is made possible to have improved quality in the classification of hyperspectral imagery.

L. Gongalla (✉) · S. V. Sudha
School of Computer Science and Engineering, VIT-AP University, Amaravati, Andhra Pradesh, India
e-mail: liks.201819@gmail.com

© The Author(s), under exclusive license to Springer Nature Singapore Pte Ltd. 2023   173
S. Shakya et al. (eds.), *Sentiment Analysis and Deep Learning*,
Advances in Intelligent Systems and Computing 1432,
https://doi.org/10.1007/978-981-19-5443-6_15

Many approaches came into existence as per the literature. Many of the approaches used CNN-based models. Alipourfard et al. [3] combined CNN and subspace-based feature extraction for leveraging accuracy of the model. Zhao et al. [4] used multiscale approach in deep learning with feature extraction. Yue et al. [5] used deep CNN model for hyperspectral image classification. Zhang et al. [6] introduced multi-metric active learning concept for classification. Chen et al. [7] used feature extraction and CNN for classification of hyperspectral imagery. Wang et al. [13] used the concept of fast dense spectral approach with CNN. Wang et al. used, however, the problem with existing models is that there is need for more appropriate configuration of layers in CNN. Our contributions in this paper are as follows.

1. We proposed a framework known as Hyperspectral Image Classification Framework (HICF) with underlying mechanisms.
2. We proposed an algorithm known as Deep Learning-based hyperspectral image classification (DL-HIF) for improving classification of hyperspectral images.
3. A prototype is built using Python data science platform for evaluating the performance of the proposed algorithm.

The remainder of the paper is structured as follows. Section 2 focuses on literature review on deep learning methods for hyperspectral image classification. Section 3 presents the proposed framework based on deep learning. Section 4 presents experimental results of the proposed framework and Sect. 5 concludes the paper.

2 Related Work

This section presents review of literature on deep learning models for hyperspectral image classification. Liu et al. [1] proposed a deep learning concept known as active deep learning for classification. Yue et al. [2] used spatial pyramid pooling in the proposed deep learning framework to improve performance. Alipourfard et al. [3] combined CNN and subspace-based feature extraction for leveraging accuracy of the model. Zhao et al. [4] used multiscale approach in deep learning with feature extraction. Yue et al. [5] used deep CNN model for hyperspectral image classification. Zhang et al. [6] introduced multi-metric active learning concept for classification. Chen et al. [7] used feature extraction and CNN for classification of hyperspectral imagery. Deep learning-based frameworks are explored in [8–10].

Gao et al. [11] used deep induction network-based approach for classification to have better performance. Fabelo et al. [12] used deep learning for hyperspectral images associated with human brain. Wang et al. [13] used the concept of fast dense spectral approach with CNN. Zhong et al. [14] used spectral spatial residual network concept for classification of images. Sun et al. [15] defined concept of fusion of deep features for efficient classification. Gap et al. [16] used spatial classification approach. Singh et al. [17] used deep learning for classification of imagery. Wang et al. used CNN along with alternately updated spectral spatial approach for classification.

Multiple deep belief networks and 3D deep learning approaches are other important methods found in the literature.

3 Proposed Methodology

We proposed a deep learning-based framework known as Hyperspectral Image Classification Framework (HICF). It is meant for improving efficiency in hyperspectral image classification. It is based on Convolutional Neural Network (CNN) which has different layers such as convolutional layers, pooling layers and fully connected layers. Finally, the fully connected layers transform the maps into feature vectors that will help in prediction of class labels.

Each convolutional layer has many learnable filters. It makes use of ReLU as activation function which is used to converge in terms of gradient vanishing. The results of pooling layers are given to fully connected layers where they transform the maps into feature vectors that will help in prediction of class labels.

Algorithm 1 Deep Learning-based Hyperspectral Image Classification (DL-HIF)

Algorithm: Deep Learning-based Hyperspectral Image Classification (DL-HIF)

Inputs: Hyperspectral imagery

Outputs: Classification results

1. Start
2. Apply convolutional layer 1
3. Apply pooling layer 1
4. Apply convolutional layer 2
5. Apply pooling layer 3
6. Apply fully connected layer 1
7. Apply fully connected layer 2
8. Apply softmax layer
9. $M = \text{TrainModel}(feature\ maps)$
10. $M' = \text{FitModel}(M)$
11. $P = \text{DetectResults}(M')$
12. Return $P/$

As presented in Algorithm 1, it takes spectral images as input and perform classification based on the deep learning framework presented in Fig. 1.

The confusion matrix shows in Fig. 2 is used to derive performance metrics. In this research, accuracy of the models is evaluated using Eq. 1.

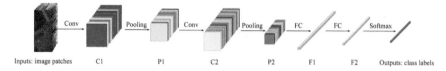

Fig. 1 Proposed Hyperspectral Image Classification Framework (HICF)

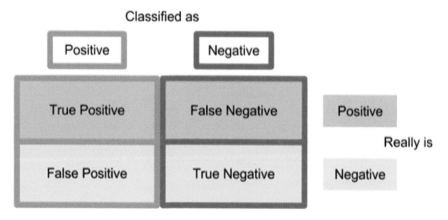

Fig. 2 Confusion matrix

$$\text{Accuracy} = \frac{TP + TN}{TP + TN + FP + FN} \qquad (1)$$

The accuracy value can be in the range from 0 to 1. The value 1 indicates highest performance.

4 Experimental Results

Experiments are made with the prototype built using data set collected from. It shows the data representation of input spectral samples, performance of the classification models in terms of accuracy for each input sample and average accuracy for all input samples.

Figure 3 shows the representation of nine input samples where the number of iterations is shown in horizontal axis while the vertical axis shows corresponding value for spectral image in terms of mean_spectrum, mean_spectrum − std_spectrum and mean_spectrum + std_spectrum.

As presented in Table 1, the accuracy % of different classification models such as the proposed DL-HIF, SVM and NN are provided.

As presented in Fig. 4, different hyperspectral samples are provided in horizontal axis while vertical axis show accuracy of each classification model.

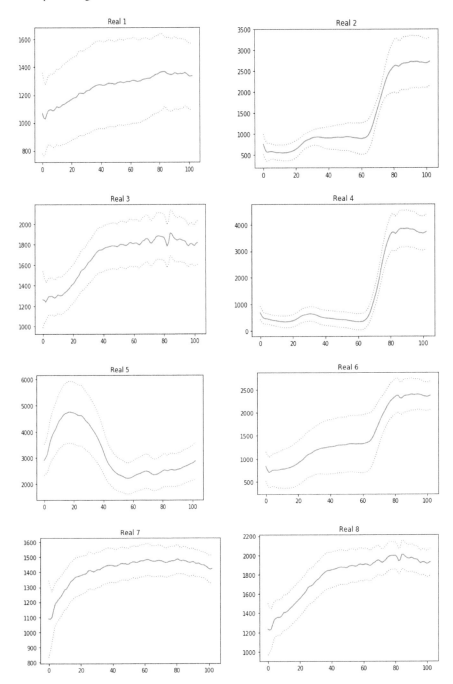

Fig. 3 Results representing input hyperspectral samples

Fig. 3 (continued)

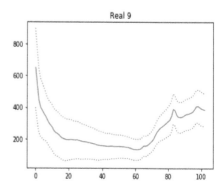

Table 1 Performance of different classification models

Hyperspectral sample #	Accuracy (%)		
	SVM	NN	DL-HIF (Proposed)
1	0.89178	0.909115	0.94188
2	0.94188	0.902702	0.98196
3	0.7515	0.814325	0.92184
4	0.91182	0.929455	0.98196
5	0.9018	0.922742	0.98196
6	0.82164	0.913223	0.94188
7	0.61122	0.845989	0.96192
8	0.84168	0.870638	0.9519
9	0.89178	0.998593	0.97194

Fig. 4 Performance of different classification models

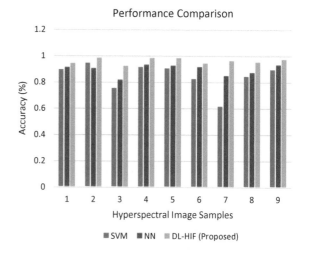

Table 2 Average accuracy of classification models for 9 hyperspectral samples

Classification models	Average accuracy (%)
SVM	0.840567
NN	0.89296
DL-HIF (Proposed)	0.959693

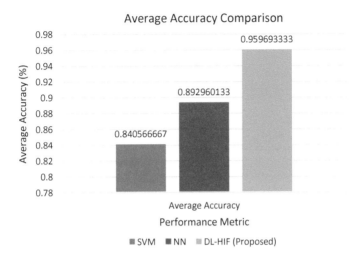

Fig. 5 Average accuracy of classification models for 9 hyperspectral samples

As presented in Table 2, the average accuracy (%) of each model is presented. The average accuracy of 9 hyperspectral samples is considered here for performance comparison.

As presented in Fig. 5, the average accuracy (%) of each model is presented. The average accuracy of 9 hyperspectral samples is considered here for performance comparison. The average accuracy of the SVM is 84.0566667%, the NN is 89.2960133% and DL-HIF is 95.9693333%. The results revealed that the proposed model outperforms the existing models. As deep learning-based model has more sophistication in learning process, the proposed method showed highest performance.

5 Conclusion

In this paper, we proposed a framework known as Hyperspectral Image Classification Framework (HICF) with underlying mechanisms. We proposed an algorithm known as Deep Learning-based Hyperspectral Image Classification (DL-HIF) for improving classification of hyperspectral images. Our experimental results are compared with two existing Machine Learning (ML) models such as Support Vector Machine (SVM) and Neural Network (NN). The average accuracy of the SVM is

84.0566667%, the NN is 89.2960133% and DL-HIF is 95.9693333%. The results revealed that the proposed model outperforms the existing models. As deep learning-based model has more sophistication in learning process, the proposed method showed highest performance. In future, we intend to use Generative Adversarial Network (GAN) architecture-based deep learning to improve hyperspectral image classification performance further.

References

1. Liu, P., Zhang, H., Eom, K. B. (2016). Active deep learning for classification of hyperspectral images. *IEEE Journal of Selected Topics in Applied Earth Observations and Remote Sensing*, 1–13.
2. Yue, J., Mao, S., & Li, M. (2016). A deep learning framework for hyperspectral image classification using spatial pyramid pooling. *Remote Sensing Letters, 7*(9), 875–884.
3. Alipourfard, T., Arefi, H., & Mahmoudi, S. (2018). A novel deep learning framework by combination of subspace-based feature extraction and convolutional neural networks for hyperspectral images classification. In *IGARSS 2018—2018 IEEE International Geoscience and Remote Sensing Symposium [IEEE IGARSS 2018—2018 IEEE International Geoscience and Remote Sensing Symposium, Valencia, Spain* (2018.7.22–2018.7.27)] (pp. 4780–4783).
4. Zhao, W., Guo, Z., Yue, J., Zhang, X., & Luo, L. (2015). On combining multiscale deep learning features for the classification of hyperspectral remote sensing imagery. *International Journal of Remote Sensing, 36*(13), 3368–3379.
5. Yue, J., Zhao, W., Mao, S., & Liu, H. (2015). Spectral–spatial classification of hyperspectral images using deep convolutional neural networks. *Remote Sensing Letters, 6*(6), 468–477.
6. Zhang, Z., & Crawford, M. M. (2017). A batch-mode regularized multimetric active learning framework for classification of hyperspectral images. *IEEE Transactions on Geoscience and Remote Sensing, 55*(11), 6594–6609.
7. Chen, Y., Jiang, H., Li, C., Jia, X., & Ghamisi, P. (2016). Deep feature extraction and classification of hyperspectral images based on convolutional neural networks. *IEEE Transactions on Geoscience and Remote Sensing, 54*(10), 6232–6251.
8. Chen, Y., Lin, Z., Zhao, X., Wang, G., & Gu, Y. (2014). Deep learning-based classification of hyperspectral data. *IEEE Journal of Selected Topics in Applied Earth Observations and Remote Sensing, 7*(6), 2094–2107.
9. Audebert, N., Le Saux, B., & Lefevre, S. (2019). Deep learning for classification of hyperspectral data: A comparative review. *IEEE Geoscience and Remote Sensing Magazine, 7*(2), 159–173.
10. Li, J., Zhao, X., Li, Y., Du, Q., Xi, B., & Hu, J. (2018). Classification of hyperspectral imagery using a new fully convolutional neural network. *IEEE Geoscience and Remote Sensing Letters*, 1–5.
11. Gao, K., Guo, W., Yu, X., Liu, B., Yu, A., & Wei, X. (2020). Deep induction network for small samples classification of hyperspectral images. *IEEE Journal of Selected Topics in Applied Earth Observations and Remote Sensing, 13*, 3462–3477.
12. Fabelo, H., Halicek, M., Ortega, S., Shahedi, M., Szolna, A., Piñeiro, J., Sosa, C., O'Shanahan, A., Bisshopp, S., Espino, C., Márquez, M., Hernández, M., Carrera, D., Morera, J., Callico, G., Sarmiento, R., & Fei, B. (2019). Deep learning-based framework for in vivo identification of glioblastoma tumor using hyperspectral images of human brain. *Sensors, 19*(4), 1–25.
13. Wang, W., Dou, S., Jiang, Z., & Sun, L. (2018). A fast dense spectral-spatial convolution network framework for hyperspectral images classification. *Remote Sensing, 10*(7), 1–19.
14. Zhong, Z., Li, J., Luo, Z., & Chapman, M. (2017). Spectral-spatial residual network for hyperspectral image classification: A 3-D deep learning framework. *IEEE Transactions on Geoscience and Remote Sensing*, 1–12.

15. Sun, G., Zhang, X., Jia, X., Ren, J., Zhang, A., Yao, Y., & Zhao, H. (2020). Deep fusion of localized spectral features and multi-scale spatial features for effective classification of hyperspectral images. *International Journal of Applied Earth Observation and Geoinformation, 91*, 1–10.
16. Gao, F., Wang, Q., Dong, J., & Xu, Q. (2018). Spectral and spatial classification of hyperspectral images based on random multi-graphs. *Remote Sensing, 10*(8), 1–20.
17. Singh, S., & Kasana, S. S. (2018). Efficient classification of the hyperspectral images using deep learning. *Multimedia Tools and Applications*, 1–14.

Improved Security on Mobile Payments Using IMEI Verification

M. Kathiravan, M. Sambath, B. Bhuvaneshwari, S. Nithya Krishna, W. Jeshwin, and Nikil Babu

Abstract The conventional financial frameworks such as the exchanges over the bank counter with a passbook are upgraded by adding the components of electronic banking, all the bank exchanges/transaction are made possible over the Internet with the progression of data and the corresponding innovation. Nonetheless, there are significant risks associated with this innovation, such as banking fraud. Hence, to fabricate trust among banking clients, it is vital to build security systems. In electronic banking, security systems basically centre on demonstrating a solid framework for the web-based exchange, particularly client validation. Numerous experts have proposed various models for web-based client access validation in the financial industry. A large portion of the research depends on the conventional type of username and secret key, yet with the differing systems of secret key structures. For example, secret phrase in erratic worth, secret key in erratic worth via obtained channel, and biometric preferred finger impression acknowledgement, voice acknowledgement, or retina acknowledgement. This paper has examined the existing client access for Internet banking based on mobile phones. The elements of the SIM card and the attributes portable of the SIM card are portrayed. Cloning, a danger to the SIM card is additionally portrayed. Discoveries show that all the security models for online client access contain secret words such as worth, biometric, or PIN. In this manner, none of the current client accesses proposed the possibility of the client access in the light of the International Mobile Equipment Identity (IMEI) number to strengthen the security of the client access. The proposed research work is more focused on security the mobile payment by utilizing the IMEI.

Keywords Mobile payment · IMEI · Cryptography · Security · Hash algorithm

M. Kathiravan (✉) · M. Sambath · S. Nithya Krishna · W. Jeshwin · N. Babu
Computer Science and Engineering, Hindustan Institute of Technology and Science, Chennai, India
e-mail: mkathiravan@hindustanuniv.ac.in

B. Bhuvaneshwari
Department of Information Technology, Panimalar Engineering College, Chennai, India

N. Babu
Chitkara University Institute of Engineering and Technology, Chitkara University, Punjab, India

1 Introduction

Mobile payment is the payment (transfer of funds in return for a good or a service), where the mobile phone is involved in the initiation and confirmation of the payment. Although there are many developed encryption algorithms, they cannot be used directly on mobile phones. There are three main reasons for this problem. First, most mobile phones do not have powerful CPUs and have a limited amount of internal memory. Secondly, wireless networks have lower bandwidth and less reliability as compared to wired networks. Finally, cryptography is necessary for establishing security in mobile payments. However, the complexities of cryptographic algorithms are not usually specified for users. When these complexities become revealed, the acceptance of these algorithms will be difficult.

The portable number (that is enlisted at significant financial organizations ahead of time) is utilized in relation to the encryption techniques utilized in exchanges. Regardless of whether PCs or frameworks are linked to wired networks, portable frameworks use battery power with limited energy. Such issues generally lead us to another security calculation in view of cell phone assets. This calculation utilizes the following terms:

- The serial number of the IMEI International Mobile Equipment Identity (IMEI) is an alphanumeric code used to identify GSM and CDMA mobile phones. If you know where to look, you can usually find it in the phone's battery compartment. The IMEI number is used by the GSM organization to distinguish genuine gadgets from bogus ones and, along these lines, might be utilized for avoiding taking cell phones to get to corporations in a certain nation. For suggested computation, secret phrase generator software is stashed away in the handset; consequently, IMEI assumes a major role in handset confirmation.
- Get back to the basics. After the client authenticates the username and password, the server disconnects from the client and tries to communicate with it directly. While using Windows Server, this approach is used to confirm the identity of dial-up users in the system. Each client's phone number is included in his or her information. In the event that a client account is stolen or someone else tries to connect to the server, he is merely prepared to share the line via that number that had registered on the server, as a precautionary measure.
- In order to generate a unique password for each exchange, the IMEI number of the device will be used. On the server, the exact confidential phrase is stored and generated using the same code-generating application, making it easy to cross-check.

2 Literature Survey

E-ticketing, mobile e-commerce, and e-payment are just a few of the numerous services that newer types of mobile phones are capable of providing. In the modern day, NFC-enabled mobiles are of special relevance, since NFC technology enables convergence of service spanning several apps into a single mobile device. NFC payment systems now employ the EMV protocol; however, this protocol has two severe weaknesses that might put customers at danger between their payment devices as well as merchant's POS. To ensure the safety of NFC-based mobile payment transactions, these two security flaws must be addressed. In this research, a new security layer is presented to enhance security of EMV. It has been shown that EMV's security mechanism is impenetrable to well-known security exploits [1].

In electronic commerce, digital payment methods have critical role in ensuring transaction security. Electronic payment systems based on rigorous math calculations would be susceptible to quantum computers. To combat the threat posed by quantum computers, electronic payment methods are being enhanced using quantum cryptography. However, even if quantum cryptography is believed as completely safe, logical flaws are expected as usually to create major issues. 6 For the first time, we used formal analysis to validate security of quantum cryptography-based electronic payment systems. It is possible to find weaknesses in protocols using formal analysis techniques, which may then be used to make the protocols themselves more secure. An electronic payment system that relies on quantum cryptography is not really secure, according to our findings. There is a lack of accountability and justice in initial procedure, according to our findings. In order to ensure accountability and fairness, we suggested better approach [2, 3], which we tested using formal analysis.

In this research [4], we provide statistical modelling of digital money billing methods, depending upon graph analytics approaches. In this way, authors' crash experiments may be used as the basis for system stability analysis. In these crash-tests, participants in the digital money billing system (participants of graph) are simulated to be destroyed by both equipment and hardware failures and getting hacked. Integral estimates of system steadiness are constructed by combining a number of graph centrality metrics in a certain sequence. An economic security matrix is used to calculate crucial degree of destruction of network centrality metric as well as aggregate metrics during destruction in the article. Mobile phones, which are becoming more popular due to the proliferation of wireless networks and portable gadgets like them, provide a tremendous potential to empower mobile devices as payment devices. There are, however, numerous concerns which must be addressed for making mobile phone payments work, such as the fact that mobile phones have a limited amount of electricity and a less capable CPU. Several public-key encryption techniques for mobile payment have recently been introduced.

In mobile networks, however, the restricted abilities of portable devices as well as wireless communications render these algorithms inappropriate. Using the call back approach and OTP based on the IMEI number, novel mobile phone security mechanism is presented in this work. In the light of resource constraints, this method offers necessary security for mobile payment. The suggested algorithm's efficiency has been shown via simulations using prototype data. The expansion of information and technology to different service sectors, particularly product marketing companies and product's trade, is a result of this evolution. The rise of e-commerce has resulted in a slew of new developments in the worlds of business and marketing. As a result, modifications to payment mechanism must follow the e-commerce model, which is to say, Internet-based transactions. As an essential component of online commerce and electronic payment facilitates non-cash transactions while also making the process of acquiring goods more convenient for customers. In any nation or socioeconomic setting, money may be defined as any commodity or verifiable record universally recognized in payment for services and goods as well as debt repayment [5].

The era of information and communication technology (ICT) and advanced development lead to dynamic changes in the business climate, where deals keep on moving from cash-based exchanges to electronic-based exchanges [6].

In early processing days with a couple of PC frameworks and a little select gathering of clients, this model demonstrated viable. With the appearance of the Internet, online business, and the multiplication of PCs in workplaces and schools, the client base has become both in number and in segment base. Individual clients never again have single passwords for single frameworks, yet are given the test of recollecting various passwords for various frameworks, from email, to web accounts, to banking and monetary administrations. This paper presents a calculated model portraying how clients and frameworks cooperate in this capacity and looks at the outcomes of the growing client base and the utilization of secret word memory helps. A framework model of the dangers related with secret word-based confirmation is introduced according to a client driven perspective including the develop of client secret phrase memory helps. When gone up against with a lot of information to recollect, clients' foster memory helps to help them in the errand of recalling significant snippets of data. This client secret word memory helps structure a scaffold between in any case detached frameworks and meaningfully affects framework level security across various frameworks interconnected by the client. A fundamental examination of the ramifications of this client driven interconnection of safety models is introduced [7, 8].

It will probably be impervious to these days too successive phishing and pharming assaults, and furthermore to additional traditional ones like social designing or man-in-the-centre assaults. The central issue of this model is the requirement for multifaceted shared confirmation, rather than essentially putting together the security with respect to the advanced authentication of the monetary substance, since generally speaking clients cannot perceive the legitimacy of a testament, and may not focus on it. By adhering to the guidelines characterized in this proposition, the security level of the web banking climate will expand and clients' trust will be upgraded, hence permitting a more gainful utilization of this help [9].

In the new day's, web applications use is expanding by banking, monetary establishments and wellbeing or emergency clinic the executives frameworks, for example, on the web or net banking and portable banking, online business applications, news channels, ongoing and short term data, and so on. This multitude of online applications [10, 11] need help for security properties like validation, approval, information secrecy, and touchy data spillage. The most generally acknowledged normal strategy for confirmation for a web-based application is to utilize a mix of alphanumeric with exceptional characters usernames and passwords. Net or online applications ought to help a solid secret key, (for example, a mix of alphanumeric with unique characters).

In the past, late investigations uncover that the end clients today have on a normal roughly 10–15 passwords to safeguard their Internet-based records to do their genuine exchanges. By and large, a typical web (web) client having one secret key might be not difficult to recollect, however, controlling numerous passwords for various web or web applications is tedious errand and a security danger. Generally, passwords are not gotten by any means as they can be guessable or someone (a malignant client) can be taken. To conquer this, passwords should be more grounded verification arrangements. To tackle the issues that dangerously login online business, web banking framework and conventional powerful secret phrase framework needs outer gear, a better personality verification plot in view of PKI-SIM card [12] is proposed in this paper.

These days web-based banking [13] is getting prominence due for its few potential benefits, like simplicity of activity and virtual money related exercises, i.e. client's actual presence at various bank areas is not any more required. Nonetheless, as a result of shortage of openness at various areas, cost, and slow speed, number of individuals utilizing web correspondence is a little level of the cell phone clients, particularly in the non-industrial nations. Additionally, security in web-based banking is as yet a main issue. A large portion of the monetary establishments has their own space and deals electronic exchange offices. In this paper, we propose a plan to use the web-based existing financial office by means of short message administration (SMS) accessible in cell phone innovation without having the web association. Considering guaranteeing an elevated degree of safety, we present two-level security conspires: voice confirmation and advanced watermarking.

The advancement of the Internet and the appearance of web-based business encouraged digitalization [14, 15] in the instalment processes by giving an assortment of electronic instalment choices including instalment cards (credit and charge), advanced and portable wallets, electronic money, contactless instalment strategies, and so forth. Portable instalment administrations with their expanding fame are as of now under the period of change, heading towards a promising future of speculative conceivable outcomes alongside the development in innovation. In this paper, we will assess the present status and development of portable instalments and other electronic instalment frameworks in business sectors all over the planet and investigate the eventual fate of this industry.

The worldwide spread and utilization of the web and cell phone has added to the improvement of advanced instalments. In spite of its development potential, up to this point there is an absence of examination giving an exhaustive amalgamation and investigation of elements influencing the utilization, reception, and acknowledgement of computerized instalment strategies. This study means to address this hole by giving an exhaustive audit of the connected writing recovered from Scopus and Web of Science information bases. Following a methodical strategy, a last example of 193 examination articles was recognized and broke down. The outcomes features that a solitary hypothesis has neglected to make sense of the mind boggling nature of electronic instalment completely reception [16, 17].

Presently, a day we have n number of exchange techniques to do yet we definitely disapprove of extortion SMS and phone calls. That they are calling from bank representative. This call is in regards to for bank check without realizing we are offering every one of the subtleties to them, then we are losing our sum. The existing system uses OTP and 3D security-based check UPI instalments. The disadvantage is that the open and vote-based nature of the web opens purchasers to make easy money contributions from corrupt sellers. Anybody with even humble information on site creation can construct a web-based store with compelling reason need to guarantee the accessibility of stock or a method for offering types of assistance [18, 19].

In the proposed system, the view of the telephone IMEI number this exchange works. By utilizing the communication API, we will get the IMEI number. To make safer, we are utilizing SHA 256 algorithm. On the off chance that any approved individual gains admittance to the secret word maker programme, he/she is simply ready to have a correspondence with the server and cannot play out the transactions. As a consequence of affirmation, the past exchanges secret phrase will be sent and the exchange will be finished.

3 Methodologies

Using a theoretical model, the framework design explains how a framework is built, how it is used, and several other aspects of how it is used. A design representation is a traditional portrayal and depiction of a framework, organized in such a manner as to facilitate understanding of the framework's structures and methods of operating as show in Fig. 1.

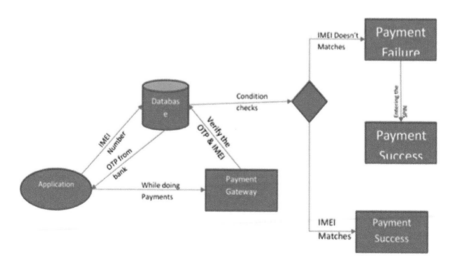

Fig. 1 Architecture diagram

Interoperability among QR-based instalment results of various organizations: RuPay, Visa, MasterCard, and AmEx. Substitute for PoS terminals, low cost acknowledgement instalment solution push-based exchange, for example, starts with the card holder. Do not bother offering the actual card to the trader.

- Open Android Studio with the project to which you want to add the module.
- From the Android Studio menu, click File > New > Import Module. Select the
- Eclipse ADT project folder by the AndroidManifest.xml file and click Ok.
- If necessary, change the module name and click Next.
- Importing to Android Studio encourages users to migrate a certain library and project dependency to Android Studio as well as add a dependency declaration to the build gradle file during the import process. See create an Android Library for additional information on transferring project and library dependencies. As a result, you will no longer have to manually manage any well-known binary libraries, source libraries, or JAR files that have recognized Maven coordinates with Maven dependencies. Your workspace directory as well as any real path maps may also be entered to resolve any unsolved comparative paths, variable paths, and linked resources.
- Click Finish.

4 Implementation and Result Analysis

We achieved authentic results using the IMEI method, which is very much more secure, compared to the traditional phone number method.

A login page is a passage page to a site that requires clients to distinguish between proof and validation, represented in Fig. 2.

A registration page in Fig. 3 enables users and organizations to independently register and gain access to your system.

The IMEI transaction page as in Fig. 4, which provides the various options for money transactions.

Figure 5 shows how, after selecting the mode of transaction, the payment gateway navigates to the page where the user enters their bank information.

Figures 2, 3, 4 and 5 are described in detail. The page that asks for your login credentials to access a certain programme is known as a "login application." You could have seen it if you logged into Facebook, Twitter, or another social media site. Generate a login screen and learn how to deal with security issues when a bogus attempt is made in this chapter.

To begin with, you must specify that the input type for both text views be password. Security experts use the terms "login" and "sign-in" interchangeably to describe the procedure whereby a person authenticates and verifies their identity in order to obtain accessibility to a mobile computer system. Users often have a username and a password, and these credentials are often referred to as a "login" in the context of online services (or logon, or sign-in, or sign-in). When it comes to current secure systems, mail or SMS validation is typically required for additional protection. The user may

Fig. 2 Login page

Fig. 3 Registration page

log out if they no longer need access (log off, sign out, or sign off). A verification code (OTP) is sent to the user's mobile phone during registration. There is no longer a chance of a person enrolling with a false mail address or phoney mobile number. This module verifies if an email address or mobile number exists and may be accessed by a user. The free module comes with ten free email addresses as well as ten free SMS transactions. Email, SMS, and OTP verifications are all available in this

Fig. 4 IMEI transaction page

Fig. 5 Adding user bank details

module. A telephony manager is available to get information about the phone's telephony services. It is possible for applications to utilize procedures in this category for determining the status of telephonic services and subscriber information. According to the National Institute for Standards and Technology, secure hash algorithms are a group of cryptographic hash functions that are part of the federal government's data processing standards, including: SHA-0: A retronym for the initial 160-bit hash function released in 1993 under the title "SHA."

5   Conclusions

We have proposed a straightforward and effective IMEI-based validation component alongside a callback method for electronic exchange frameworks. The secret word created by the client's cell phone is checked by the instalment server, utilizing directly in correspondence with the client. Along these lines, it gives far more noteworthy security than the charge card-based validation components. With the far reaching utilization of GSM innovation around the world, the proposed calculations can be executed to give sufficient security for the greater part of the cell phones that have just restricted equipment assets.

Acknowledgements Thanks to the Department of Computer Science and Engineering, Hindustan Institute of Technology and Science, Chennai, India.

References

1. Hassan, M. A., & Shukur, Z. (2022). Device identity-based user authentication on electronic payment system for secure E-wallet apps. *Electronics, 11*, 4.
2. Hammood, W. A., Abdullah, R., Hammood, O. A., Asmara, S. M., Al-Sharafi, M. A., & Muttaleb Hasan, A. (2020). A review of user authentication model for online banking system based on mobile IMEI number. *IOP Conference Series Materials Science and Engineering , 769.*
3. Ghiduk, A. S. (2012). Design and implementation of the user interface and the application for mobile devices. *International Journal of Computer Applications,* ISSN 0975-8887.
4. Marcus, P., & Peitek, N. (2016). Retrofit love working with APIs on android future studio.
5. Dewi, P. (2010). Pemodelan process Bisnis Menggunakan Activity Diagram UML dan BPMN. *Jurnal Fakultas Teknology Industri Universitas Kristen Petra Surabaya.*
6. Ramos, M., & Tulio, M. (2018). Angular JS performance. *IEEE Computer Society,* ISSN 0740-7459.
7. Ruby, S., & Richarson, L. (2007). *Restful web service.*
8. Samisa. (2008). *Restful, PHP, web service.* Packet Publishing.
9. Sharfina, Z., & Santoso, B. (2016). An Indonesian adaptation of the system usability scale.
10. Fatonah, S., Yulandari, A., & Wibowo, P. W. (2018). A review of E-payment in E-commerce. *Journal of Physics Conference Series.*
11. Wirdasan, D. (2019). Teknology e-commerce Dalam process Bisnis. 7(2).

12. Alam, S. B., et al. (2010). A secured electronic transaction scheme for mobile banking in Bangladesh incorporating digital watermarking. In *2010 IEEE International Conference on Information Theory and Information Security.*
13. Chowdhury, M. M., & Noll, J. (2007). Distributed identity for secure service interaction. In *2007 Third International Conference on Wireless and Mobile Communications (ICWMC'07).*
14. Yin, X., et al. (2009). An improved dynamic identity authentication scheme based on PKI-SIM card.
15. Priya, R., Tamil Selvi, V., & Ramesh Kumar, G. (2014). A novel algorithm for secure internet banking with finger print recognition. In *International Conference on Embedded Systems (ICES).*
16. Basavala, S. R., Kumar, N., & Agarrwal, A. (2012). Authentication: An overview, its types and integration with web and mobile applications. In *2nd IEEE International Conference on Parallel, Distributed and Grid Computing.*
17. Hammood, O. A., Kahar, M. N. M., Mohammed, M. N., & Hammood, W. A. (2017). Issues and challenges of video dissemination in VANET and routing protocol: Review. *Journal of Engineering and Applied Sciences, 12*(11), 9266–9277.
18. Conklin, A., Dietrich, G., & Walz, D. (2004). Password-based authentication: A system perspective. In *37th Annual Hawaii International Conference on System Sciences.*
19. Martino, A. S., & Perramon, X. (2008). A model for securing E-banking authentication process: Anti phishing approach. In *IEEE Congress on Services-Part I.*

Evaluating the Effectiveness of Classification Algorithms for EEG Sentiment Analysis

Sumya Akter⬥, **Rumman Ahmed Prodhan**⬥, **Muhammad Bin Mujib**⬥, **Md. Akhtaruzzaman Adnan**⬥, **and Tanmoy Sarkar Pias**⬥

Abstract Electroencephalogram (EEG) signals from the brain provide additional information about emotional states that we may be unable to convey verbally. Machine learning algorithms can effectively predict the emotion from brain waves. So, we design a research to evaluate the effectiveness of multiple machine learning techniques - Naive Bayes, Logistic Regression, XGBoost, SVM, Decision Tree, Random Forest, KNN, and deep learning models—CNN, LSTM, and Bi-LSTM for classifying sentiment from brain signals. In our experiment, the DEAP dataset is used as a collection of brain signals representing different human sentiments. The Fast Fourier transformation (FFT) which shifts the data to the frequency domain is used to extract features from the time series EEG data. Among all the algorithms CNN, KNN and Random Forest achieved the highest accuracy of 96.64%, 95.8%, and 95.28% respectively on the binary classification of valence. The results demonstrate that it is possible to attain accuracy comparable to or even outperform some of the deep learning models by combining appropriate feature extraction techniques (in this case FFT) with machine learning algorithms.

Keywords Support vector machine · Random forest · K-nearest neighbors · Long short-term memory · Convolutional neural network · Naive Bayes · Electroencephalogram · Fast fourier transform

S. Akter · R. A. Prodhan · M. B. Mujib · Md. A. Adnan
University of Asia Pacific, Dhaka 1205, Bangladesh
e-mail: 18101029@uap-bd.edu

R. A. Prodhan
e-mail: 18101018@uap-bd.edu

M. B. Mujib
e-mail: 18101017@uap-bd.edu

Md. A. Adnan
e-mail: adnan.cse@uap-bd.edu

T. S. Pias (✉)
Virginia Tech, Blacksburg, VA 24061, USA
e-mail: tanmoysarkar@vt.edu

1 Introduction

Emotion is a behavioral and physiological condition characterized by a range of feelings, thoughts, and actions. The ability to recognize and analyze the emotions of others is known as emotion recognition. This may be accomplished by using brain signals [1], facial expressions, body language, and voice tone [2]. It is a vital social skill since it helps us to react correctly to other people's emotions. The capacity to recognize emotions from brain activity is known as emotion recognition from brain signals [3]. Many authors [4], have worked on emotion recognition and have used a variety of ways to do so, but the most effective of them is to detect emotions from brain signals.

1.1 Background

Numerous researches on emotion recognition have been conducted in recent years utilizing the DEAP dataset, which is publicly available. In the DEAP dataset emotion are categorized into valence, arousal, dominance, liking, and familiarity. It has been demonstrated that a variety of different modeling approaches are used in previous studies. Table 1 depicts several literature reviews on the previous study on basis of different modeling techniques. Zheng et al. [17] analyze the performance of a variety of common feature extraction, feature selection, and classification algorithms in a systematic manner. They compared six different features such as differential entropy (DE), asymmetry(ASM), differential asymmetry (DASM), rational asymmetry (RASM), power spectral density(PSD) and differential causality (DCAU). In recent times FFT [8, 10, 11] and DWT [11] methods are mostly used to achieve higher accuracy on EEG signal. In this study, FFT is used for feature extraction. There are also many different machine learning and deep learning algorithms that are used to detect emotion from EEG signals. Parui et al. [14] used XGBoost algorithm and achieved 75% accuracy in different emotion labels. Their proposed algorithm was also compared with Bayesian, KNN, Decision Tree, and Random Forest algorithms. Doma et al. [12] performed this by utilizing basic machine learning techniques such as SVM, Naive Bayes, Decision Trees, Logistic Regression, KNN, and LDA, all of which obtained an accuracy of at least 80%. Many deep learning algorithms like Convolution neural network(CNN) or Long short-term memory (LSTM) can be used to identify emotions. Many studies combined CNN and LSTM to acquire high performance [5]. Cui et al. [6] merged the DE feature extraction method with Bi-LSTM and CNN and achieve an average of 94% accuracy on the DEAP dataset. There are many other different technique like GCNN [13], SVM [12, 15, 16], logistic regression [12, 17], decision trees [14] have also been explored previously. To detect emotion more accurately, this study will compare 10 different machine learning and deep learning algorithms and which are Naive Bayes, Logistic regression, XGBoost, SVM, Decision tree, Random forest, KNN, CNN, LSTM, Bi-LSTM with Fast Fourier feature extraction technique. The chart in Fig. 1 represents the average accuracy vs algorithm.

Table 1 Overview of the literature review based on the different algorithms

Studies	Year of publication	Feature	Algorithm	Performance
Arjun et al. [5]	2022	Autoencoder	CNN, LSTM	Arousal: 69.5%, Valence: 65.9%
Cui et al. [6]	2022	DE	DE-CNN-BiLSTM	Arousal : 94.86%, Valence: 94.02%
Ma et al. [7]	2021	Wavelet	SVM	Avg Arousal and Valence: 89.72%
Hasan et al. [8]	2021	FFT	CNN	Valence:96.63%, Arousal: 96.17%
Yin et al. [9]	2021	DE	GCNN, LSTM	Valence: 90.45%, Arousal: 90.60%
Maeng et al. [10]	2020	Hjorth, DWT, FFT	LSTM	Arousal: 94.8%, Valence: 91.3%
Luo et al. [11]	2020	DWT, FFT	SNN	Arousal: 74%, Valence: 78%, Dominance:80%, Liking: 86.27%
Doma et al. [12]	2020	PCA	Naive Bayes, Logistic regression, KNN, SVM, Decision Tree, LDA	Avg of accuracy: 75%
Wang et al. [13]	2019	Time-frequency domain	P-GCNN, GCNN, SVM, DBN	Arousal: 77.03%, Valence: 73.31%, Dominance: 79.20%
Parui et al. [14]	2019	Raw	Bayesian, KNN, C4.5, Decision Tree, Random Forest, XGBoost	Avg 75%(XGBoost)
Liu et al. [15]	2018	EMD, SBS	KNN, SVM	Valence: 86.46%, Arousal: 84.90%
Raghav et al. [16]	2018	Wavelet Decomposition, PCA	SVM, KNN, Naive Bayes, Random Forest	Max accuracy: 75% (Random Forest)
Zheng et al. [17]	2017	PSD, DE, DASM, RASM, ASM, DCAU	KNN, SVM, GELM, Logistic regression	Avg Arousal and Valence: 69.67%

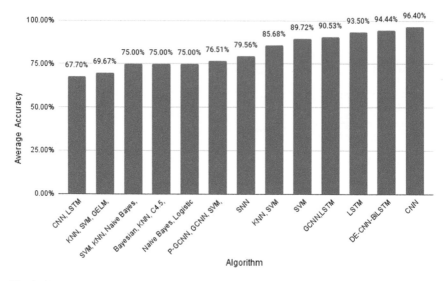

Fig. 1 Average accuracy versus algorithm

1.2 Motivation

The study of emotional brain signals is a complex and complicated endeavor. Electroencephalography (EEG) is a method for determining the brain's electrical activity. The electroencephalogram (EEG) signal may be used to identify electrical activity linked with various emotions. It is very important to know how human emotions vary from one state to other. The study shows the machine learning algorithms with extracted features can effectively recognize emotions compared to deep learning algorithms.

1.3 Objective

The goal of this paper is to detect emotion from brain signal using machine learning and deep learning algorithms. A comprehensive analysis between machine learning and deep learning algorithm is also added in this paper. This is advantageous in a variety of applications, including identifying emotions during human-computer interaction, monitoring emotions during different activities, and assisting those who suffers from emotional illnesses like depression and sadness. Deep learning is more used for this. Deep learning is a very effective technique for discovering complicated patterns in data [18]. But Machine learning and feature extraction is also effective if done together correctly. In this paper, the author shows how machine learning algorithms work with feature extraction, much like deep learning. This study makes the following significant contributions:

- On the DEAP dataset, we evaluate the performance of several machine learning techniques for detecting sentiment.
- After applying fast Fourier transform (FFT) as feature extraction on EEG data, we compare Machine Learning algorithms with Deep Learning algorithms.
- We prove that by using fast Fourier transform (FFT) as feature extraction, machine learning can perform almost similar to or even exceeds deep learning models.

2 Methods

Figure 2 illustrates the whole workflow process to evaluate different machine learning algorithms. Firstly, the DEAP dataset is selected as the raw EEG data. Then various data preprocessing techniques are applied. Following that, the preprocessed data is segmented using FFT in the feature extraction stage. Then using the same feature extracted data all the experiment is conducted.

2.1 Dataset

There are many publicly available dataset on EEG signals for emotion recognition including DEAP [19], SEED [20], LUMED [21], DREAMER [22] and AMIGOS [23]. Among these, the DEAP and the SEED are the most popular datasets. The DEAP dataset is widely used for evaluating different algorithms, thus it is selected for this study. The DEAP dataset is a multimodal dataset that is primarily used to analyze human affective states. The dataset team chose 120 music videos with a runtime of one minute. Table 2 provides an overview of the DEAP dataset. The raw EEG of a single participant is shown in the first step of Fig. 2. The EEG data of subjects 1–22 and 23–32 are recorded in Twente and Geneva respectively. For each participant, 40 trials are recorded showing 40 videos out of the selected 120 videos. The duration of each recording is 63 s.

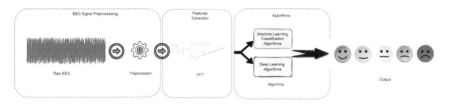

Fig. 2 Workflow diagram

Table 2 A synopsis of the DEAP dataset

Dataset type	Multimodal
Subjects	32
Videos	40
EEG Channels	32
Labels	Valance, Arousal, Dominance, Liking and Familiarity
Original sample rate	512 Hz
Preprocessed downsampled rate	128 HZ
Rating Values	Between 1 to 9 in continuous number (Valance, Arousal, Dominance and Liking). Between 1 and 5 in discrete number (Familiarity)

2.2 Preprocessing

The 512 Hz EEG data are downsampled to 128 Hz. Blind source separation is used to reduce eye artifacts. A bandpass filter of 4.0–45.0 Hz frequency range is used. The data is averaged according to a commonly used reference. Since the EEG data was collected in two distinct places, the order of EEG channels is changed to match the Geneva sequence. Each trial contains 60 s of data and a three-second pre-trial baseline. The pre-trial section is then eliminated. Additionally, the trials are reordered to compare the video presentation order to the video order.

2.3 Feature Extraction

Feature extraction is critical to minimizing signal loss, reducing overfitting, and simplifying implementation [24]. Generally, rather than utilizing raw data, any feature extraction approach may provide superior classification results. Frequency-domain characteristics are used in this research, which is primarily used to break down signal data into the sub-bands. Wavelet transform (WT), fast Fourier transform (FFT), and Equivocator methods (EM) are often used to extract EEG characteristics. From these feature extraction approaches to the EEG data, the fast Fourier transform is selected. FFT is used for computing the discrete Fourier transform or inverse discrete Fourier transform of a sequence. The discrete Fourier transform (DTF) is denoted by the following notation:

$$x[k] = \sum_{n=0}^{N-1} x[n] e^{\frac{-j2\pi kn}{N}} \qquad (1)$$

Here, n denotes the domain's size. To compute the DFT of a discrete signal $x[n]$, each of its values should be multiplied by e power to some function of n. Then, the results acquired for a certain n should be summed. Calculating the DFT of a signal has $O(N^2)$

complexity. As the name implies, the fast Fourier transform (FFT) is substantially quicker than the discrete Fourier transform (DFT). The Fourier transform decreases the complexity from $O(N^2)$ to $O(NlogN)$.

For this study, 14 EEG channels are selected. Those EEG channels are Fp1, AF3, F3, F7, FC1, P3, PO3, Fp2, Fz, F4, F8, C4, P4, PO4 and their numbers are 1, 2, 3, 4, 6, 11, 13, 17, 19, 20, 21, 25, 29 and 31 respectively. FFT breaks the raw EEG signals into 5 different sub-bands shown in Fig. 3. Using this algorithm the raw EEG data is segmented using a sliding window of size 256 and each sliding step size is 16. After sliding through the entire raw EEG signal, it is converted from the time domain to the frequency domain. Then the frequency domain signal is divided into 5 different sub-bands which are delta (4–8), theta (8–12), alpha (12–16), beta (16–25), gamma (25–45) (Fig. 3).

2.4   Machine Learning Algorithms

For evaluating the effectiveness, several machine learning algorithms are employed to learn the correlation between the data and the label. In this study, 7 different types of classifiers are used to evaluate the effectiveness of the machine learning algorithm. Selected machine learning algorithms are Support vector machine(SVM), Logistic regression, Decision Tree, Random Forest, K-Nearest Neighbor (KNN), Gaussian Naive Bayes, and XGBoost.

2.4.1   Support Vector Machine

The Support Vector Machine (SVM) [20] technique is a kind of supervised machine learning that may be used for classification and regression applications. The approach is based on determining the optimum hyperplane for dividing a dataset into two groups. Considering training examples $x_i \in R, i = 1, \ldots, n$ and label $y \in \{1, -1\}^n$, the objective of this classifier is to estimate parameters $w \in R^p$ and $b \in R$ such that the $\text{sign}(w^T \phi(x) + b)$ predicts the class of a sample. SVMs perform better when there is a distinct boundary between the two classes.

$$\min_{w,b,\varsigma} \frac{1}{2} w^T w + C \sum_{i=1}^{n} \varsigma_i \qquad (2)$$

subject to $y_i (w^T \phi(x_i) + b) \geq 1 - \varsigma_i, \varsigma_i \geq 0, i = 1, \ldots, n$

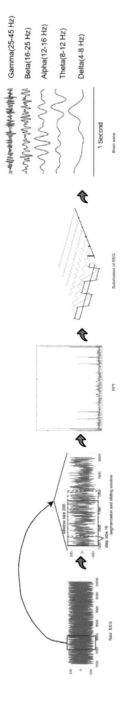

Fig. 3 Using FFT for feature extraction

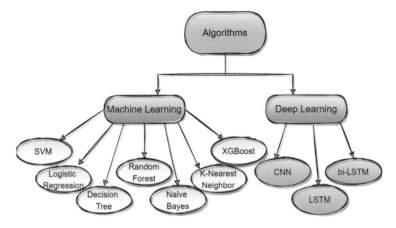

Fig. 4 Classification algorithm tree

2.4.2 Logistic Regression

Logistic regression is a statistical technique for predicting the probability of a binary outcome [25]. A binary result is a variable that may take on just two values: true or false, yes or no, or 0 or 1. The form of logistic function is given below:

$$p(x) = \frac{1}{1 + e^{-(x-\mu)/s}} \tag{3}$$

Here, location parameter is μ (the curve midpoint, where $p(\mu) = 1/2$) and scale parameter is s.

2.4.3 Decision Tree

A decision tree is a decision-support tool that creates a tree-like graph or model of actions and their associated consequences, which may include random event outcomes, resource costs, and utility [26]. At node m let the data be described by Q_m with consisting of N_m samples. For every candidate split $\theta = (j, t_m)$ consisting of a feature and threshold t_m, partition the data into left and right subsets like the following:

$$Q_m^{\text{left}}(\theta) = \{(x, y) \mid x_j <= t_m\} \tag{4}$$

$$Q_m^{\text{right}}(\theta) = Q_m \setminus Q_m^{\text{left}}(\theta) \tag{5}$$

2.4.4 Random Forest

A random forest is a supervised learning technique for classification and regression that is used in machine learning[27]. It is a decision tree-based ensemble learning system. It then takes the predictions from each tree and averages them to get the final classification. The random forest training method employs the general approach of bootstrap aggregating, or bagging, to train tree learners. Given a training set of $X = x1, \ldots, x_n$ along with responses $Y = y1, \ldots, y_n$, bagging repeatedly (B times) takes a random sample from the training set with replacement and fits trees to these samples:
For $b = 1, \ldots, B$:

  1. n training samples from X, Y; denote them as X^b, Y^b
  2 On X^b, Y^b, train a classification or regression tree f^b.

After training, it is possible to make predictions for unknown samples x' by summing the predictions from all the separate regression trees on x':

$$\hat{f} = \frac{1}{B} \sum_{b=1}^{B} f_b(x') \tag{6}$$

or, in the case of classification trees, by majority voting.

2.4.5 K-Nearest Neighbor

K-Nearest Neighbor (KNN) is a machine learning technique that classifies data points in feature space based on their neighbors [28]. KNN is a non-parametric and slow method for learning. Non-parametric means that no assumptions regarding the distribution of the underlying data are made.

At test time, locate the nearest example in the training set and return the corresponding label.

$$\hat{y}(x) = y_{n^*}, \text{ where } n^* = \arg\min_{n \in D} \text{dist}(x, x_n) \tag{7}$$

dist returns the distance between two points x and x_n. There are many configuration possibilities for this function. Manhattan distance, Euclidean distance, Cosine distance, Jaccard distance.

2.4.6 Gaussian Naive Bayes

Gaussian Naive Bayes classifiers are probabilistic machine learning models that assume each feature's value follows a Gaussian distribution. This means that the model may be used to estimate the probability of a data point belonging to a certain class given the values of the features. The features likelihood is assumed like the following:

$$P(x_i \mid y) = \frac{1}{\sqrt{2\pi\sigma_y^2}} \exp\left(-\frac{(x_i - \mu_y)^2}{2\sigma_y^2}\right) \tag{8}$$

Occasionally, make an assumption of variance is independent of Y (i.e., σi) or independent of Xi (i.e., σk) or both (i.e., σ).

2.4.7 XGBoost

XGBoost is a machine learning technique for classification and regression problems [29]. It is an implementation of a gradient boosting algorithm. In terms of accuracy, XGBoost has been found to surpass other machine learning algorithms. Additionally, XGBoost is renowned for its speed and efficiency. It is capable of handling large datasets and rapidly training models.

At iteration t, the objective function (loss function and regularization) that we need to minimize is as follows:

$$\mathcal{L}^{(t)} = \sum_{i=1}^{n} l(y_i, \hat{y}_i^{(t-1)} + f_t(x_i)) + \Omega(f_t) \tag{9}$$

Where, y_i = Actual value(label) extracted from the training set. Can be written as $f(x + \delta)$ where $x = \hat{y}_i^{(t-1)}$

2.5 Deep Learning Algorithms

Three distinct kinds of models are used in this paper to evaluate deep learning algorithms. To analyze deep learning methods, convolutional neural networks (CNN), long short-term memory (LSTM), and bidirectional long-short term memory (bi-LSTM) models are used. The architectures of these 3 models are shown in Fig. 5. Log loss is a loss function that is commonly used in classification tasks and is one of the most frequently utilized metrics in deep learning [30]. It is just a logarithmic variant of the likelihood function.

$$-(y \log(p) + (1 - y) \log(1 - p)) \tag{10}$$

2.5.1 Convolutional Neural Network

Convolutional neural networks (CNN) are a kind of deep learning neural network that is often used to evaluate visual imagery and signals [31]. CNN is composed of neurons with learnable weights and biases.

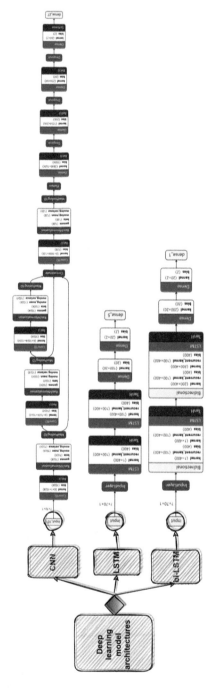

Fig. 5 Deep learning model architectures

2.5.2 Long Short-Term Memory

The long short-term memory algorithm is a deep learning technique for predicting the next value in a series [32]. It is a sort of recurrent neural network that is taught by the use of backpropagation.

2.5.3 Bidirectional Long Short-Term Memory

Bidirectional Long short-term memory is a kind of deep learning algorithm that is capable of learning from both historical and future data [33]. It is a recurrent neural network (RNN) architecture that has been taught to recall long-term dependencies.

2.6 Data Preparation

After applying FFT for feature extraction, each of the 32 EEG files contains 19,520 rows and 4 columns, From each of the files 14,640 rows are selected for training, and 4880 rows are selected for testing. Thus the data was split into 75% into training and 25% into testing. Valence is divided into binary class, as it is the most popular classification to compare model performance on EEG signals. 1–4.9 in one category and 5–9 in another category.

2.7 Evaluation Metrics

To maintain consistency in the evaluation, the experiment results on the valence label are shown as all of the labels perform almost similarly. Also, the same feature extraction technique and data preparation are used for every model. Most of the default parameters are used where possible. To measure the performance of an algorithm testing accuracy is taken from the testing data. Precision, recall, f1 score, and accuracy is used to determine the performance of each model.

The precision of a machine learning algorithm is a metric that indicates how near the algorithm's predictions are to the actual values. Additionally, this Table 3 represents the precision of each algorithm.

$$Precision = \frac{TP}{TP + FP} \tag{11}$$

The recall is a measure that indicates how well a model is at properly identifying positive samples from the test set. This Table 3 also represents the recall values.

Table 3 Classification report of different algorithms for binary classification

Algorithms name	Classes	Precision	Recall	F1
Naive Bayes	0	0.49	0.44	0.47
	1	0.6	0.65	0.62
Logistic Regression	0	0.55	0.29	0.38
	1	0.6	0.82	0.69
XGBoost	0	0.55	0.29	0.38
	1	0.6	0.82	0.69
SVM	0	0.75	0.65	0.70
	1	0.76	0.83	0.79
Decision Tree	0	0.81	0.81	0.81
	1	0.85	0.85	0.85
Random Forest	0	0.95	0.94	0.95
	1	0.95	0.97	0.96
KNN	0	0.95	0.95	0.95
	1	0.96	0.96	0.96
CNN	0	0.96	0.97	0.96
	1	0.97	0.97	0.97
LSTM	0	0.75	0.72	0.73
	1	0.78	0.83	0.8
Bi-LSTM	0	0.74	0.74	0.74
	1	0.8	0.8	0.8

$$\text{Recall} = \frac{\text{TP}}{\text{TP} + \text{FN}} \tag{12}$$

The $F1$ score is a metric that indicates the accuracy of a machine learning model. It is the harmonic mean of the precision and recall of the model which is represented by Table 3. The higher the $F1$ score, the more accurately the model can distinguish between positive and negative samples.

$$F1 = \frac{2 * (\text{Precision} * \text{Recall})}{(\text{Precision} + \text{Recall})} \tag{13}$$

3   Results and Discussions

The optimum hyperplane is estimated by the specific machine learning algorithm e.g. for CNN the complex non-linear hyperplane is determined by a multilayer perceptron using deep features from convolutional layers. To illustrate the findings in Fig. 6, the testing precision of each method is plotted. Random forest, KNN, and CNN

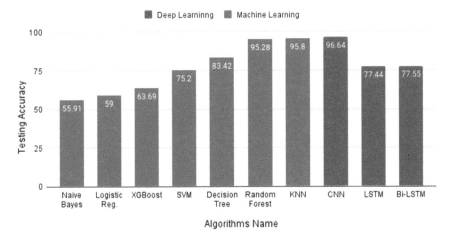

Fig. 6 Testing accuracy versus algorithms name

exhibit exceptional performance. Comparing the performance of machine learning algorithms with deep learning algorithms is fascinating for identifying sentiment from brain signals, a typically difficult problem. Although the accuracy of the CNN model is the highest. The majority of scholars use CNN for this activity. As FFT is used for feature extraction, Random forest and KNN have achieved outstanding precision. They almost equal the accuracy of the CNN model while greatly outperforming LSTM and Bi-LSTM. Consequently, despite the great accuracy of the CNN model, machine learning algorithms perform well with feature extraction.

Table 3 represents the classification report of different machine learning and deep learning algorithms for binary classification. The classes column shows that DEAP dataset labels are divided into 2 classes which are 0 and 1. Class 0 contains data from 1–4.99 and 1 contains data from 5–8.99 for more accurate calculation handling.

4   Conclusion and Future Research Direction

Using FFT and machine learning models, it is feasible to attain accuracy that matches or even exceeds that of deep learning models, according to the aforementioned trials. It is difficult to extract characteristics from raw EEG data, but this can be accomplished by FFT. In the future, the author wants to explore the correlation between the different labels of emotions and how to recognize emotions more precisely from brain signals.

5   Data and Code Availability

The DEAP dataset upon which this study is based is available on their official website. All the experiment's code is available on Github. The code is written in Python using the Google Colab platform.

References

1. Hassan, R., Hasan, S., Hasan, M. J., Jamader, M. R., Eisenberg, D., & Pias, T. (2020). Human attention recognition with machine learning from brain-eeg signals. In *2020 IEEE 2nd Eurasia Conference on Biomedical Engineering, Healthcare and Sustainability (ECBIOS)*, pp. 16–19. https://doi.org/10.1109/ECBIOS50299.2020.9203672
2. Shakya, S., & Smys, S. (2021). Big data analytics for improved risk management and customer segregation in banking applications. *Journal of ISMAC, 3*(03), 235–249.
3. Sungheetha, A., & Sharma, R. (2020). Transcapsule model for sentiment classification. *Journal of Artificial Intelligence, 2*(03), 163–169.
4. Apu, M. R. H., Akter, F., Lubna, M. F. A., Helaly, T., & Pias, T. S. (2021). Ecg arrhythmia classification using 1d cnn leveraging the resampling technique and gaussian mixture model. In *2021 Joint 10th International Conference on Informatics, Electronics Vision (ICIEV) and 2021 5th International Conference on Imaging, Vision Pattern Recognition (icIVPR)*, pp. 1–8. https://doi.org/10.1109/ICIEVicIVPR52578.2021.9564201
5. Arjun, Rajpoot, A. S., & Panicker, M. R. (2022). Subject independent emotion recognition using eeg signals employing attention driven neural networks. *Biomedical Signal Processing and Control, 75*, 103547 (2022). https://doi.org/10.1016/j.bspc.2022.103547, https://www.sciencedirect.com/science/article/pii/S1746809422000696
6. Cui, F., Wang, R., Ding, W., Chen, Y., & Huang, L. (2022). A novel de-cnn-bilstm multifusion model for eeg emotion recognition. *Mathematics, 10*(4) (2022). https://doi.org/10.3390/math10040582, https://www.mdpi.com/2227-7390/10/4/582
7. Ma, X., Liu, P., Wang, X., & Bai, X. (2021). EEG emotion recognition based on optimal feature selection. *Journal of Physics: Conference Series, 1966*(1), 012043 (2021). https://doi.org/10.1088/1742-6596/1966/1/012043, https://doi.org/10.1088/1742-6596/1966/1/012043
8. Hasan, M., Rokhshana-Nishat-Anzum, Yasmin, S., & Pias, T. S. (2021). Fine-grained emotion recognition from eeg signal using fast fourier transformation and cnn. In *2021 Joint 10th International Conference on Informatics, Electronics Vision (ICIEV) and 2021 5th International Conference on Imaging, Vision Pattern Recognition (icIVPR)*, pp. 1–9. https://doi.org/10.1109/ICIEVicIVPR52578.2021.9564204
9. Yin, Y., Zheng, X., Hu, B., Zhang, Y., & Cui, X. (2021). EEG emotion recognition using fusion model of graph convolutional neural networks and LSTM. *Applied Soft Computing, 100*, 106954.
10. Maeng, J. H., Kang, D. H., & Kim, D. H. (2020). Deep learning method for selecting effective models and feature groups in emotion recognition using an asian multimodal database. *Electronics, 9*(12). https://doi.org/10.3390/electronics9121988, https://www.mdpi.com/2079-9292/9/12/1988
11. Luo, Y., Fu, Q., Xie, J., Qin, Y., Wu, G., Liu, J., et al. (2020). EEG-based emotion classification using spiking neural networks. *IEEE Access, 8*, 46007–46016. https://doi.org/10.1109/ACCESS.2020.2978163.

12. Doma, V., & Pirouz, M. (2020). A comparative analysis of machine learning methods for emotion recognition using EEG and peripheral physiological signals. *Journal of Big Data,7*. https://doi.org/10.1186/s40537-020-00289-7

13. Wang, Z., Tong, Y., & Heng, X. (2019). Phase-locking value based graph convolutional neural networks for emotion recognition. *IEEE Access, 7,* 93711–93722. https://doi.org/10.1109/ACCESS.2019.2927768.

14. Parui, S., Bajiya, A., Samanta, D., & Chakravorty, N. (2019). Emotion recognition from eeg signal using xgboost algorithm, pp. 1–4. https://doi.org/10.1109/INDICON47234.2019.9028978

15. Liu, Z. T., Xie, Q., Wu, M., Cao, W. H., Li, D. Y., & Li, S. H. (2019). Electroencephalogram emotion recognition based on empirical mode decomposition and optimal feature selection. *IEEE Transactions on Cognitive and Developmental Systems, 11*(4), 517–526. https://doi.org/10.1109/TCDS.2018.2868121.

16. Raghav, G., Nongmeikapam, K., Dixit, A., Bose, S., & Singh, D. (2018). Evaluating classifiers for emotion signal on deap dataset

17. Zheng, W. L., Zhu, J. Y., & Lu, B. L. (2019). Identifying stable patterns over time for emotion recognition from EEG. *IEEE Transactions on Affective Computing, 10*(3), 417–429. https://doi.org/10.1109/TAFFC.2017.2712143.

18. Pias, T. S., Kabir, R., Eisenberg, D., Ahmed, N., & Islam, M. R. (2019). Gender recognition by monitoring walking patterns via smartwatch sensors. In *2019 IEEE Eurasia Conference on IOT, Communication and Engineering (ECICE)*, pp. 220–223. https://doi.org/10.1109/ECICE47484.2019.8942670

19. Koelstra, S., Muhl, C., Soleymani, M., Lee, J. S., Yazdani, A., Ebrahimi, T., et al. (2012). Deap: A database for emotion analysis; using physiological signals. *IEEE Transactions on Affective Computing, 3*(1), 18–31. https://doi.org/10.1109/T-AFFC.2011.15.

20. SEED dataset. https://bcmi.sjtu.edu.cn/home/seed/. Accessed: 12 April 2022.

21. Loughborough University EEG based Emotion Recognition Dataset(LUMED). https://www.dropbox.com/s/xlh2orv6mgweehq/LUMED_EEG.zip?dl=0. Accessed: 11 April 2022.

22. Katsigiannis, S., & Ramzan, N. (2018). Dreamer: A database for emotion recognition through EEG and ECG signals from wireless low-cost off-the-shelf devices. *IEEE Journal of Biomedical and Health Informatics, 22*(1), 98–107. https://doi.org/10.1109/JBHI.2017.2688239.

23. Miranda-Correa, J. A., Abadi, M. K., Sebe, N., & Patras, I. (2021). Amigos: A dataset for affect, personality and mood research on individuals and groups. *IEEE Transactions on Affective Computing, 12*(2), 479–493. https://doi.org/10.1109/TAFFC.2018.2884461.

24. Jenke, R., Peer, A., & Buss, M. (2014). Feature extraction and selection for emotion recognition from EEG. *IEEE Transactions on Affective Computing, 5*(3), 327–339. https://doi.org/10.1109/TAFFC.2014.2339834.

25. LaValley, M. P. (2008). Logistic regression. *Circulation, 117*(18), 2395–2399.

26. Song, Y. Y., & Ying, L. (2015). Decision tree methods: Applications for classification and prediction. *Shanghai Archives of Psychiatry, 27*(2), 130.

27. Pal, M. (2005). Random forest classifier for remote sensing classification. *International Journal of Remote Sensing, 26*(1), 217–222.

28. Zhang, Z. (2016). Introduction to machine learning: K-nearest neighbors. *Annals of Translational Medicine, 4*(11).

29. Mitchell, R., & Frank, E. (2017). Accelerating the XGBoost algorithm using GPU computing. *PeerJ Computer Science, 3,* e127.

30. Sarif, M. M., Pias, T. S., Helaly, T., Tutul, M. S. R., & Rahman, M. N. (2020). Deep learning-based bangladeshi license plate recognition system. In *2020 4th International Symposium on Multidisciplinary Studies and Innovative Technologies (ISMSIT)*, pp. 1–6. https://doi.org/10.1109/ISMSIT50672.2020.9254748

31. Pias, T. S., Eisenberg, D., & Islam, M. A. (2019). Vehicle recognition via sensor data from smart devices. In *2019 IEEE Eurasia Conference on IOT, Communication and Engineering (ECICE)*, pp. 96–99. https://doi.org/10.1109/ECICE47484.2019.8942799

32. Graves, A. (2012). Long short-term memory. *Supervised Sequence Labelling with Recurrent Neural Networks*, 37–45.

33. Zhang, S., Zheng, D., Hu, X., & Yang, M. (2015). Bidirectional long short-term memory networks for relation classification. In *Proceedings of the 29th Pacific Asia Conference on Language, Information and Computation*, pp. 73–78.

Analytics and Data Computing for the Development of the Concept Digitalization in Business and Economic Structures

Mykhailo Vedernikov⬤, **Lesia Volianska-Savchuk**⬤, **Maria Zelena**⬤, **Nataliya Bazaliyska**⬤, and **Juliy Boiko**⬤

Abstract This work examines the evolution of business and economic structures under the influence of digitalization, updating the introduction of digital technologies in business and economic structures and assessing the current level of automation of enterprise processes, as well as identifying key opportunities for digitalization and digitization for enterprises. The use of quantitative dependence of indicators to accelerate the development of business and economic structures in the context of global transformation is proposed. For many companies, this is a revolutionary opportunity. With this in mind, innovative enterprise development strategies must take into account the requirements of digitalization, which involves the use of integrated mobile applications, social networks, analytics, and cloud technologies. It is necessary to improve and modernize the traditional system of management of business and economic structures and best of all, to replace it with an integrated cloud platform, which will create a digital infrastructure of the enterprise. It is also worth mentioning the long-term strategy, which must be carefully thought out and adapted to the evolutionary and technological realities of today. It is necessary to make innovation as a key strategy in the management of business and economic structures. The methods of using the relationship of statistical indices to analyze the impact of staffing factors and their average productivity on net income from sales in the forecast period are considered. Substantiation on the basis of the index method of forecasts (plans, tasks, or standards) of macro-and microeconomic phenomena and processes in the preparation of management decisions allows to improve the level of use of available statistical information. It should also be noted that the stated methodological provisions of the index analysis of the dynamics of net income from sales can be used at the regional level to assess changes in the volume of gross regional product or products sold in various economic activities.

Keywords Data analytics · Digitalization · Digital technologies · Statistical indexes · Net income · Labor productivity · Staff

M. Vedernikov · L. Volianska-Savchuk · M. Zelena · N. Bazaliyska · J. Boiko (✉)
Khmelnytskyi National University, 11, Instytuts'ka Str, Khmelnytskyi 29016, Ukraine
e-mail: boiko_julius@ukr.net

1 Introduction, Background, Motivation and Objective

The proposed work covers the topical issues of introducing modern intelligent decision-making concepts in the personnel management system and assessing the areas of application the comprehensive digitalization of business structures in the context the globalization of the economy.

The world is constantly evolving, the performance of work tasks becomes automated, and jobs occupy people's jobs [1, 2]. All these changes, that pose new organizational challenges to HR services, push for unexpected decisions, and force them to work in conditions of destabilization and social upheaval [3–6]. At the same time, the impact of digital technologies on personnel management is increasing and accelerating, which requires the definition and evaluation of personnel administration processes to create an appropriate HR strategy [7, 8].

Progress never stands still, and digital technologies are increasingly penetrating various areas of business [9–11]. We are witnessing fundamental changes in the labor market. The digitalization of the economy significantly transforms all the traditional management functions of the organization, and above all, the field of personnel management [12]. To date, everyone must, first of all, define role as a team one that helps management and employees quickly adapt to the digital way of thinking. It is vital to create a team, that not only knows all the processes and functions of business and economic structures, but also be able to monitor, analyze and implement new technologies, that appear almost every day.

Every year, more and more companies automate HR functions and thus simplify and make more efficient work of HR managers and HR department in general [13, 14]. Research shows that there is a significant gap between those companies, that actively implement change, confidently use new resources and technologies to improve HR processes, and those, who are in a waiting position. The passivity of many companies leads to significant losses of competitiveness, as they not only lose the opportunity to invest in their own capital but also risk losing access to a workforce with unique opportunities.

Digital HR brings together social networks, mobile applications, cloud technologies, augmented reality and is a new platform to improve the work of both employees and candidates, improve and process their experience. Digital solution developers provide the technical component of digital HR [15], while company management and HR departments must build their own integrated digital personnel management strategies and programs. The transition to digital transformation should be based on an in-depth analysis of internal and external factors, the study of advantages and disadvantages, as well as industry constraints and potential partners.

The relevance of implementing digital HR capabilities is indisputable, especially important for companies in the growth and expansion, large and international companies, with high importance of investment attractiveness, a large number of vacancies, or those that pay special attention to quality selection and adaptation of employees. In other cases, the various capabilities of digital HR can significantly improve the performance of any HR functions, increase the efficiency of human capital, and develop a corporate culture of HR [16].

2 Methods the Implementing Digital HR Capabilities

2.1 Implementation of Statistical Assessment of Functional Interactions of Users of Social and Economic Phenomena and Processes

Unfortunately, in the system of statistical methods of analysis, the index method is used to perform statistical evaluation of the functional relationships of social and economic phenomena and processes. Statistical indices allow us to identify the available reserves for improving the use of factors that shape the change in the overall indicator of the index model. Therefore, the methodology of index analysis of the influence of factors on the formation of net income from sales of machine-building plants needs further development. Taking into account the needs of economic practice, it is necessary to choose from the accumulated wealth of index methodology, theoretically sound and practically the most acceptable methods of index calculations, and present them in a form accessible to practical needs.

Statistical index is a relative indicator that quantifies the change of the studied phenomenon in dynamics or in comparison with the standard, based on the characteristics of functional connections of factors that form its level. It can be some value of the plan, norm, forecast or territorial comparison). The components of such general indicator are a homogeneous or an inhomogeneous population. The application of the methodology of statistical indices to the analysis of inhomogeneous aggregates is hindered by the problem of finding a common measure for the components of such a phenomenon. It can be, for example, the volume of trade in different types of goods. This, in particular, concerns the change in net income from sales due to extensive (quantitative volume of goods and services, number of staff) and intensive (prices and productivity of staff) form of net income from sales [17].

The general theory of statistics identifies the synthetic and analytical concepts at the basis of the index method. The use of the synthetic form of statistical indices allows you to evaluate the relationship between the dynamics of one characteristic from the change in time of several others based on the calculation of individual indices (synthetic concept). The analytical concept is based on the construction and evaluation of statistical index systems [18]. If the index analysis of changes in net income from sales in the dynamics involves identifying the degree of influence of factors, such as the number of employees (number of employees), productivity, and prices, we'll characterize the impact of these factors on changes in net income from sales in reporting period, compared to the base, and apply the following system of statistical indices:

Table 1 Initial data for calculating the change in net income from sales of machine-building plant products for 1–5 years

Year	Net income from sales of products p_1q_1, thousand UAH	Price index for region i_p, %	Net income from sales of products (at comparable prices) $\frac{p_1q_1}{i_p} = p_0q_1$, thousand UAH	Average number of staff T, people	Productivity (at comparable prices) $\frac{p_0q_1}{T}$, thousand UAH
1	12,205	110.2	11,075.32	2237	4.95
2	10,363	100.2	10,342.32	2310	4.48
3	323,789	99.4	325,743.46	2359	138.09
4	458,202	111.,8	409,840.79	2314	177.11
5	273,792	148.5	184,371.72	2271	81.19

$$\frac{\sum p_1q_1}{\sum p_0q_0} = \frac{\sum T_1}{\sum T_0} \cdot \left(\frac{\sum p_0q_1}{\sum T_1} \div \frac{\sum p_0q_0}{\sum T_0} \right) \cdot \frac{\sum p_1q_1}{\sum p_0q_1}$$

where q_0 and q_1—the number of products, sold in the base and reporting periods, respectively; p_0 and p_1—unit price of sold products in the base and reporting periods, respectively; T_0 and T_1—number of employees (average number of staff and persons).

The initial data analytics for the index analysis of changes in net income from sales of products on the example of the data of the machine-building plant are given in Table 1.

Using the data analytics in Table 1, determine the indices of net income from sales, number of employees and their productivity for 2–5 years, the values of which are given in Table 2.

The influence of individual factors on the change in net income from sales in the dynamics of the example of the machine-building plant in 2 year will be considered using Table 3.

Data analysis Table 3 shows that in 2 year compared to 1 year, with a total decrease in the year of net income from sales by 1842 thousand UAH, the increase due to changes in prices amounted to 20.7 thousand UAH or 1.1% of total growth. Due to the decrease in the level of labor productivity in 2 year compared to 1 year, there was a decrease in net income from sales by 2261.0 thousand UAH, which is 122.7% of the total increase. And, the increase in the number of staff led to an increase in net income from sales by 398.3 thousand UAH (21.6% of the total increase in net income from sales in 2 year).

Consider the dynamics (2–5 years) of the influence of factors on the amount of net income from sales, constructing Table. 4. Its analysis shows, that although, given the data in Table 2, the amount of net income from sales of machine-building plant products for the five years studied increased 22.4 times (the ratio of the corresponding values for 5 and 1 years).

Table 2 Dynamics of indices of net income from sales, number of employees, their productivity and prices for 2–5 years

Year	Formula for calculation	Index, %	Magnification (+), reduction (−), %
Total net income from sales of products			
2	$\dfrac{\sum p_1 q_1}{\sum p_0 q_0}$	84.9	− 15.1
3		31.2	31.2
4		141.5	41,5
5		59.8	− 40.2
Number of staff			
2	$\dfrac{\sum T_1}{\sum T_0}$	103.3	3.3
3		102.1	2.1
4		98.1	− 1.9
5		98.1	− 1.9
Staff productivity			
2	$\left(\dfrac{\sum p_0 q_1}{\sum T_1} \Big/ \dfrac{\sum p_0 q_0}{\sum T_0}\right)$	82.1	− 17.9
3		30.8	30.8
4		129.0	29.0
5		41.0	− 59.0
Price			
2	$\dfrac{\sum p_1 q_1}{\sum p_0 q_1}$	100.2	0.2
3		99.4	− 0.6
4		111.8	11.8
5		148.5	48.5

However, it is necessary to describe the annual dynamics for 4–5 years in more detail (Table 4).

The total decrease in net income from sales in 5 year compared to 4 year is 184,410 thousand UAH and is formed mainly by reducing the level of staff productivity (265,315.7 thousand UAH) and their number (8514.6 UAH). However, the increase in prices for the company's products allowed to increase the overall indicator (net income from sales) by 89,420.3 thousand UAH.

It should also be noted a positive trend to increase net income from sales due to prices. Thus, in 5 year the corresponding value of the share of this indicator is 48.5% of the total increase.

In addition to the use of statistical indexes, the use of index analysis in setting economic projections (objectives, plans, and standards) is also important in order to analyze the influence of factors on the formation of net income from the sale of products in dynamics.

Improving the use of available statistical information in the planning and management of macro-and microeconomic processes occurring at both the national and

Table 3 Influence of certain factors on the increase (decrease) of net income from sales of machine-building plant products for 2 year

Unit	Increase (+), decrease (-) in net income from sales	Including as a result of change		
		Number of staff	Labor productivity	Price
	$\sum p_1q_1 - \sum p_0q_0$	$\left(\sum T_1 - \sum T_0\right) \cdot \frac{\sum p_0q_0}{\sum T_0}$	$\left(\frac{\sum p_0q_1}{\sum T_1} - \frac{\sum p_0q_0}{\sum T_0}\right) \cdot \sum T_1$	$\sum p_1q_1 - \sum p_0q_1$
Thousand UAH	$10,363 - 12,205 = -1842$	$(2310 - 2237) \times (12,205/2237) = 398.3$	$((10,342.32/2310) - (12,205/2237)) \cdot 2310 = -2261.0$	$10,363 - 10,342.32 = 20.7$
%	100	21.6	-122.7	1.1

Table 4 The impact of certain factors on the annual increase (decrease) in net income from sales of machine-building plant products for 2–5 years

Year	Unit	Total annual increase (+), decrease (-) of net sales revenue products $\sum p_1 q_1 - \sum p_0 q_0$	Including due to changes		Price
			Number of staff $(\sum T_1 - \sum T_0) \cdot \frac{\sum p_0 q_0}{\sum T_0}$	Labor productivity $(\frac{\sum p_0 q_1}{\sum T_1} - \frac{\sum p_0 q_0}{\sum T_0}) \cdot \sum T_1$	$\sum p_1 q_1 - \sum p_0 q_1$
2	Thousand UAH	– 1842.0	398.3	– 2261.0	20.7
	%	– 100.0	21.6	– 122.7	1.1
3	Thousand UAH	313,426.0	219.8	315,160.6	–1954.5
	%	100.0	0.1	100.6	–0.6
4	Thousand UAH	134,413.0	–6176.6	92,228.3	48,361.2
	%	100.0	–4.6	68.6	36.0
5	Thousand UAH	– 184,410.0	–8514.6	– 265,315.7	89,420.3
	%	– 100.0	–4.6	– 143.9	48.5

regional levels, largely depends on the level of application of statistical methods for its analysis in the preparation of management decisions.

This problem can be solved using the index method, which allows you to provide a statistical assessment of factors in substantiating forecasts (plans, objectives, or standards) of economic phenomena.

In order to estimate forecasts for indicators such as net sales income, wage fund, etc., it is necessary to identify the relationship and quantify the relationship of factors determining their development in the forecast future based on normative information (predicted or planned values of indicators).

Consider ways to use the relationship of statistical indices to analyze the impact of staffing factors and their average productivity on net income from sales in the next period. As a basis of comparison it is necessary to accept initial actual levels, and instead of the reporting data—t values of the forecasted (planned) number of workers and their forecasted (planned) average level of labor productivity and the prices of the future period:

$$\frac{\sum p_{ij} q_{ij}}{\sum p_o q_o} = \frac{\sum T_{ij}}{\sum T_o} \cdot \left(\frac{\sum p_o q_{ij}}{\sum T_{ij}} \div \frac{\sum p_o q_o}{\sum T_o} \right) \cdot \frac{\sum p_{ij} q_{ij}}{\sum p_o q_{ij}}$$

Let's analyze the impact of the main factors that determine the development of projected net income from sales (Table 5).

Using the data in Table 5, we will define by means of an index method the factors influencing increase in the forecasted net income from sale of production, in comparison with actually reached volume of the basic microeconomic indicator of development of machine-building plant (Table 6).

Table 5 Initial data for the calculation of the projected net income from sales of machine-building plant products for 6–7 years

Year	Net income from sales of products, thousand UAH	Price index for region,%	Net income from sales (in comparable prices), thousand UAH	Average number of staff, persons	Labor productivity (in comparable prices), thousand UAH
1	2	3	4	5	6
1	12,205	110.2	11,075.32	2237	4.95
2	10,363	100.2	10,342.32	2310	4.48
3	323,789	99.4	325,743.46	2359	138.09
4	458,202	111.8	409,840.79	2314	177.11
5	273,792	148.5	184,371.72	2271	81.19

Estimated data for 6–7 years in columns 2, 3, 5, obtained on the basis of regression equations of the trend for the last five years (1–5)

6	506,974.1	140.5	360,887.03	2320	155.57
7	604,075.4	149.3	404,605.09	2327	173.87

Table 6 The impact of certain factors on the increase (decrease) in the projected net income from sales of machine-building plant in 6 year compared to the actual data for 5 year

The factor of change in the amount of net income from sales	Increase (+), decrease (−)	
	Thousand UAH	Percent (%)
Total increase/decrease: $\sum p_{ij}q_{ij} - \sum p_o q_o$	$506,974.1 - 273,792 = 233,182.1$	100
Including because of: • changes in average productivity: $\frac{\sum p_o q_{ij}}{\sum T_{ij}} - \frac{\sum p_o q_o}{\sum T_o} \cdot \sum T_{ij}$	$\left(\frac{360,887.03}{2320} - \frac{273,792}{2271}\right) \cdot 2320 =$ $81,211.7$	34.83
• changes in the number of staff: $\left(\sum T_{ij} - \sum T_o\right) \cdot \frac{\sum p_o q_o}{\sum T_o}$	$(2320 - 2271) \cdot \frac{273,792}{2271} = 5883.3$	2.52
• price level changes: $\sum p_{ij}q_{ij} - \sum p_o q_{ij}$	$506,974.1 - 273,792 = 146,087.1$	62.65

Data analysis Table 6 shows that the main factor in the formation of the projected growth of net income from sales in the amount of 233,182.1 thousand UAH is an increase in the projected price level of 146,087.1 thousand UAH from the total increase (the share of this factor is 63% and is predominant compared to other factors).

We will also consider the scientific and methodological provisions of the application of statistical indices in the analysis and forecasting of the influence of factors on the formation of the wage bill on the example of the initial data of the machine-building plant in the dynamics for 1–5 years with the following system of statistical indices:

$$\frac{\sum f_1 T_1}{\sum f_0 T_0} = \frac{\sum T_1}{\sum T_0} \cdot \left(\frac{\sum f_0 T_1}{\sum T_1} \div \frac{\sum f_0 T_0}{\sum T_0}\right) \cdot i_p$$

where T_0 and T_1—average number of staff (persons) in the base and reporting periods, respectively; f_0 and f_1—average annual salary of one employee in the base and reporting periods, respectively.

The initial data for the index analysis of the change in the amount of the wage fund on the example of the data of the machine-building plant are given in Table 7.

According to Table 7, we define indices of the volume of the wage fund, the number of employees and real annual wages for 2–5 years. The results of the calculations are given in Table 8.

Data analysis Table 8 shows, that in 2 years compared to 1 year, with a total increase over the year in the amount of the wage fund by 5717.1 thousand UAH, the increase due to changes in prices amounted to 132.6 thousand UAH or 2.3% of the total increase.

Due to the increase in the level of real wages in 2 year, compared to 1 year there was an increase in the salary fund by 3603.2 thousand UAH, which is 63.0% of the

Table 7 Initial data for calculating the change in the volume of the payroll of the machine-building plant for years 1–5 years

Year	Remuneration fund, thousand UAH	Price index for Khmelnytsky region,%	Remuneration fund (in comparable prices), thousand UAH	Average number of staff, people	Real annual salary, thousand UAH
	$f_1 T_1$	i_p	$\frac{f_1 T_1}{i_p} = f_0 T_1$	T	$\frac{f_0 T_1}{T}$
1	60,713.60	110.2	55,094.01	2237	24.63
2	66,430.70	100.2	66,298.10	2310	28.70
3	70,106.90	99.4	70,530.08	2359	29.90
4	81,017.00	111.8	72,466.01	2314	31.32
5	104,458.30	148.5	70,342.29	2271	30.97

Table 8 Dynamics of wage fund indices, number of employees, real annual wages and inflation for 2–5 years

Year	Formula for calculation	Index, %	Increase (+), reduction (−),%
The total amount of the payroll			
2	$\frac{\sum f_1 T_1}{\sum f_0 T_0}$	109.4	9.4
3		105.5	5.5
4		115.6	15.6
5		128.9	28.9
Number of staff			
2	$\frac{\sum T_1}{\sum T_0}$	103.3	3.3
3		102.1	2.1
4		98.1	−1.9
5		98.1	−1.9
Real staff salary			
2	$\left(\frac{\sum f_0 T_1}{\sum T_1} / \frac{\sum f_0 T_0}{\sum T_0} \right)$	105.7	5.7
3		104.0	4.0
4		105.4	5.4
5		88.5	−11.5
Inflation (prices)			
2	$\frac{\sum f_1 T_1}{\sum f_0 T_1}$	100.2	0.2
3		99.4	−0.6
4		111.8	11.8
5		148.5	48.5

total increase, and an increase in the number of staff led to an increase in the salary fund in 1981, UAH 3000 (34.7% of the total increase in the wage bill for 2 year).

Data analysis Table 8 shows that although, given the data of Table 7, the volume of the wage fund of the machine-building plant for the five studied years has increased 1.72 times. The ratio of the corresponding values for 5 and 1 years. However, it is necessary to describe the annual dynamics for 4–5 years in more detail.

The total increase in the volume of the wage fund in 5 year compared to 4 year—23,441.3 thousand UAH, mainly formed due to a decrease in the level of real wages of staff (9169.2 thousand UAH) and their number (1505.5 thousand UAH). However, the adjustment of wages in accordance with the level of inflation (prices) made it possible to obtain an increase in the total indicator (volume of the wage fund) by 34,116.0 thousand UAH.

Consider in the dynamics 2–5 years, the influence of factors on the volume of the wage fund, building Table 9.

Substantiation on the basis of the index method of forecasts (plans, tasks, or standards) of macro-and microeconomic phenomena and processes in the preparation of management decisions allows to improve the level of use of available statistical information. It should also be noted, that the stated methodological provisions of the index analysis of the dynamics of net income from sales can be used at the regional level to assess changes in the volume of gross regional product or products sold in various economic activities.

Ensuring high rates of development and competitive advantages of the enterprise depends on the effective implementation of its organizational measures. Making effective management decisions is not possible without the formation and assessment of the level of corporate culture. Some companies have already gained extensive experience in practical work and information about their corporate culture. However, how to use it properly in the preparation of strategic management decisions, management does not know. In this regard, the question of the choice of tools for assessing the level of corporate culture, which would quickly identify internal opportunities and weaknesses in the activities of enterprise personnel, is particularly acute.

2.2 Influence of Enterprise Corporate Culture Components on Individual Components of Enterprise Personnel Motivation

Our general and element-by-element assessment is the basis for the improvement of corporate culture and the formation of appropriate programs and events at enterprises. In order to obtain the completeness of information regarding staff motivation for effective activity, it is advisable to provide the experts of the focus group, in addition to the proposed (existing), to determine elements and sub-elements of corporate culture, which are relevant, in their opinion, for them personally and enterprise personnel. The new concept of personnel management implies complication of mechanisms of

Table 9 The impact of certain factors on the annual increase (+), decrease (−) of the remuneration fund of the machine-building plant for 2–5 years

| Year | Unit | Total annual increase (+), decrease (−) of the salary fund $\sum p_1q_1 - \sum p_0q_0$ | Including due to changes | | Price |
			Number of staff $\left(\sum T_1 - \sum T_0\right) \cdot \frac{\sum p_0q_0}{\sum T_0}$	Real wages $\left(\frac{\sum p_0q_1}{\sum T_1} - \frac{\sum p_0q_0}{\sum T_0}\right) \cdot \sum T_1$	$\sum p_1q_1 - \sum p_0q_1$
2	UAH	5717.1	1981.3	3603.2	132.6
	%	100.0	34.7	63.0	2.3
3	UAH	3676.2	1409.1	2690.2	−423.2
	%	100.0	38.3	73.2	−11.5
4	UAH	10,910.1	−1337.4	3696.5	8551.0
	%	100.0	−12.3	33.9	78.4
5	UAH	23,441.3	−1505.5	−9169.2	34,116.0
	%	100.0	−6.4	−39.1	145.5

labor motivation, continuous improvement of employee's skills, creation of a system of continuing professional education, shift of emphasis towards individual forms of work activity [19].

To illustrate the economic and mathematical model of the dependence of personnel performance indicators on certain elements (attributes and factors) of corporate culture, a study was conducted of the influence of factor characteristics on the results of activities for engineering enterprises "A", "B", "C", and "D". The result characteristic is personnel productivity and profitability of products. The factor characteristic is the proportion of employees, whose motive for effective activity is professional development and vocational training, which determine career growth. The share of people covered by vocational training and advanced training.

The results of the study show that to determine the effectiveness of the implementation of proposals for improving the corporate culture, mainly use the assessment of changes in labor productivity through corporate events. To do this, we take a photograph of the working day of employees with the determination of the labor intensity of work for each function, operation, etc., involved in this process, before and after taking corporate measures. The change in labor productivity as a result of the introduction of corporate measures is calculated according to the Formula:

$$\Delta P = \frac{T_1 - T_2}{T_2} \cdot 100\%$$

where T_1—labor intensity of operation before taking measures, minutes; T_2—labor intensity of the operation after (during) taking measures, minutes.

Staff motivation is part of the corporate culture and is a tool for creating the desired attitude to the company's activities and job responsibilities. If it takes into account the standards and institutions of the corporate culture, staff members will have a clear understanding of the norms of conduct and how these norms will be promoted.

We note that increasing staff motivation has the most positive impact on the development and quality of corporate culture of the enterprise, which, in turn, improves the image and business reputation of the company and its competitiveness.

As practice shows, if the staff motivation system does not take into account the peculiarities of the corporate culture, its effectiveness is significantly reduced. Building such a motivation system will be difficult, if human resources professionals lack the necessary competence and a clear understanding of the functions and tasks performed by the corporate culture. They simply do not see the relationship between the concepts of "the peculiarity of the enterprise's corporate culture" and "the peculiarity of staff motivation".

Many employers who face the challenges of the crisis, decide to optimize staff costs by reducing and abandoning effective motivation programs. This, of course, saves significant funds, but increases production risks. Firstly, even if the country is in crisis, sooner or later it will end. And the enterprise will have to work according to old methods and technologies, which will reduce its competitiveness. Secondly, personnel, whose development and training are not engaged in loses competence, reduces the quality of work and productivity. At the same time, a well-formed and developed corporate culture allows even in a period of financial instability to train employees using in-house staff training.

Therefore, staff motivation is an integral part of the corporate culture, developing which, the employer must take into account and purposefully use the corporate culture prevailing in the enterprise.

In modern market conditions, there was a need to develop the motivation of the company's personnel. All this is due to the fact that the implementation of the economic growth of the enterprise also occurs through the motivational influence on the staff for effective activities, which requires new approaches. It is argued, that the basis for the formation and development of motivation of enterprise personnel should be a systematic approach. According to it, motivation is presented as a set of main subsystems, aimed at certain results of activity.

Staff motivation is a set of different forms and methods that, when actively interacting with internal and external environments and information support, changes structure in order to purposefully influence the interests, behavior, and activities of employees, while maintaining integrity.

Effective motivation of the personnel will become only if it fully covers the personnel of the enterprise and its individualization. To improve the quality of life of staff, the use of methods of motivation of staff and forms of corporate culture, except for the proper level of remuneration, is justified. In this regard, it is advisable to consider a staff motivation model, the main idea of which is a combination of economic and social methods and forms of corporate culture. This model corresponds to the modern requirements of science and business practice and is characterized by the integrity of the functioning of a certain set of motivational components, related and coordinated with the corporate mission, external and internal factors and information support.

The systematization of conceptual approaches to the problem of motivation has become the basis for the formation and development of motivation of personnel at industrial enterprises. On this basis, there is a structural model of motivation process, which reveals content of stages and interconnection of the main categories of motivation, as well as contributes to correct understanding of occurrence and satisfaction of need and allows to coordinate efforts on achievement of high-end results of work of the enterprise through activation of personnel activity [17]. Examining the motivation of staff, we note that the most significant motivation factors are job security, high pay, a bonus system, and social benefits. Consequently, material interest continues to be one of the main motivating factors of workers' work.

Let's define external and internal factors of corporate culture influence on staff motivation (according to Table 10).

The formation of corporate culture depends on the conditions, in which the enterprise operates. And, it's aimed at solving two main problems: adaptation to the external environment and internal integration of all components.

It is worth noting, that the role of motivation in the formation of corporate culture occupies a fundamental place. The basic motivation tools are well known to almost all HR managers. They include the most popular examples of motivation that have an impact on corporate culture, namely: the provision of an official car, payment for

Table 10 Impact of corporate culture on staff motivation

Main direction of influence content	Effect result
Internal factors	
Provide a positive moral and psychological climate, cohesion, and strengthen motivation	Increasing productivity
Create a clear understanding of formal rules among workers	Improving the quality of communications
Promote creative decision-making and teamwork	Development of teamwork skills and creative initiative of staff
Factors of external impact	
Promote consumer, supplier and business partner cohesion	Perceiving business partners as business colleagues and reducing transaction costs
Form the image and authority of the enterprise in the business community and among consumers	Goodwill growth and company reputation growth

mobile communications, personal development training, numerous bonus programs, life insurance, non-state pension programs, etc.

Each institution provides the employee with only the amount of incentives (material and moral) that it is able to resolve financially. But you should not forget that effective work and high loyalty of staff requires effective motivation that can meet all four levels of employee needs.

The HR manager should build the strategy and tactics of individual motivational tools together with the CEO, because the main principle of motivation is effectiveness (as each result should have an impact on the implementation of the strategic goal of the enterprise, organization, or institution) and fairness (as any activity should be encouraged depending on the employee's contribution).

A powerful motivating element is usually the corporate culture of the organization or the reform changes in its functioning [20]. In order to ensure the competitiveness of the enterprise in local, regional and state markets, it is necessary to make certain adjustments to the personnel management templates, used from time to time.

If the company does not want to accept the latest developments, technologies and ideas, it will be waiting for a period of stagnation and, as a result, the final exit from the market.

Guided by the experience of foreign enterprises, we model a certain list of recommendations regarding effective corporate culture as a component of the motivational system of the enterprise.

Therefore, first of all, leaders should maintain an appropriate socio-psychological climate in the team in such a way, that it itself acts as a powerful motivator for further action. Such an approach requires a change in priorities the development of a new vision of the enterprise, taking as a basis the need for transformation.

In order to successfully modernize the existing corporate culture of the enterprise, the core corporate values must be reviewed. In order for an enterprise, institution or organization, that aims to improve such a motivational element, as a corporate culture to take a leading position in the market for goods and services. It should first change the behavior of personnel. Therefore, each employee must feel fit and properly encouraged for further career accomplishments. And the motivational mechanism must be fundamentally different from the existing one. The renewal of corporate values must be carried out because the previous ones no longer have the maximum influence on workers and have lost motivating role.

Figure 1 shows the sequence of actions that allows you to form a corporate culture, oriented toward a victorious end result, due to the clear interaction of all structural divisions of an enterprise, organization or institution.

We describe the main stages of building an effective corporate culture.

So, the first stage is the construction of an enterprise vision, that is, the formation of a strategic picture of promising changes in the behavior of the team. The second stage is behavior that allows you to form the corporate values of an enterprise, organization or institution. More details of the formation of corporate principles were discussed above. The third phase includes an assessment of the quality of work and the results of the individual contribution of each employee to the achievement of the strategic goal.

The basis of Fig. 1 is the established effective corporate culture of the enterprise, capable of resisting external influences and guaranteeing the loyalty of employees.

As a result of the evaluation, existing management competencies are analyzed and staff are given the opportunity to evaluate management according to a number of criteria. The manager, who has scored the minimum number of points according to the selected parameters must upgrade qualifications. The last stage, the fourth one, is characterized by a motivational component. In general, staff motivation includes a number of components [20]: motivation of work activity, that is, motivation of staff to work effectively, providing the necessary remuneration and meeting existing needs; motivation for stable and productive employment; motivation to develop the

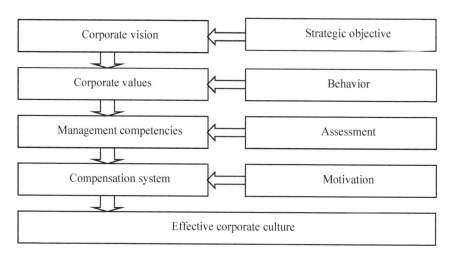

Fig. 1 Algorithm for formation of effective corporate culture

employee's competitiveness; motivation to own the means of production; motivation to choose a new place of work and the like. In other words, the result will be a compensation system that can become a good basis for the formation of a stable socio-psychological climate in the team.

Profound changes in the structure and quality of the labor force, in the content of labor, the exhaustion of reserves for increasing labor efficiency due to the physical capabilities of a person, require unconventional approaches to strengthening labor motivation and improving the corporate culture of staff. At the same time, the task is to give effect to those opportunities of a person that are related to his skill, education and training, instructions, development of labor potential, desire for creativity and self-realization.

Managers of all levels have long known that an increase in wages motivates an employee only for the first three months. Therefore, for increase in efficiency, in view of a long-term outlook, it is necessary to use such non-material means of motivation as:

- constant filling of the social package;
- providing new opportunities for professional development and career growth;
- conducting team building trainings outside the organization;
- celebration of professional holidays, following corporate traditions, etc.

Practical measures to put in place new reserves to increase the labor activity of employees are directly related to the humanization of labor. The latter, as experience shows, is now the main link of intangible motivation for work [20].

It is clear, that the main goal of such programs and methods of motivation is to maximize profit. However, one cannot fail to see the positive moment, that new approaches to motivation bring to the activities of employees, to meet their needs. They promote the development of creative potential of workers, improve the quality of working life, lead to democratization of production management, meet the needs of higher order needs of belonging and involvement, in recognition, self-affirmation and self-expression. Next, the content of individual components of labor humanization programs will be considered in more detail, the implementation of which is aimed at strengthening the intangible motivation of work.

During the development of the communication model for the analysis of motivational factors the most important ones were chosen: salary, attitude of managers, moral and psychological climate, social benefits. All this is connected with each other, for example, the moral and psychological climate—with the attitude of the leadership. But only one factor is related to several: this is the relationship of employees, and social benefits, and salary, when it is motivational. In general, all the indicated factors that affect the employee's culture sometimes act in the opposite direction. Their accounting is necessary when analyzing the cultural state in the organization.

In enterprises, in which a value corporate culture is formed, the motivational mechanism will reflect a harmonious combination of stimulating and motivating events. Such enterprises will not make the typical error of dominating only incentive measures, because this leads to the fact that the employee's needs for self-expression, etc., are not met, or only motivating, since in this case the turnover of employees can increase due to uncompetitive wages. In a combination of motivational and stimulating measures, motivational measures will still remain the main ones. A corporate culture that includes a staff motivation management mechanism is a stronger foundation than a material incentive. Such an enterprise will be able to survive in difficult crisis times, which is unlikely to succeed in an enterprise, where only high salaries and bonuses are the basis of interest in the labor of workers. In addition, the practical experience of the most successful companies in the field of personnel strategy confirms that corporate culture and value orientation are much more important than material rewards and other incentive funds [21].

Enterprise corporate culture is a system of collectively divided values, beliefs, traditions and standards of behavior of employees. It is expressed in the symbolic means of the spiritual and material environment of people working in a certain organization [22].

Corporate culture as the basis of the motivational mechanism contributes to the development of the employee's dedication. If employees set their own goals and do everything possible to implement them, they will work more fruitfully and energetically. If the manager pays attention to this and formulates goals with employees, the achievement of which is measured by the level and timing of implementation, this significantly increases the productivity of the department or enterprise as a whole. As a result, there is an increase in the efficiency of strategic management of the enterprise [23].

The culture of the organization is one of the factors of competitiveness of a commercial organization; success, efficiency, and sometimes survival. The culture of the organization is becoming increasingly a matter of concern to managers around the world. The culture of the organization is a complex phenomenon. It includes norms, principles, rules, values, ideals, language, jargon, the history of the organization, legends, images, symbols, metaphors, ceremonies, rituals, forms of rewards and rewards, location, house, and environment. Unlike the national culture, which is quite inertial in nature, the culture of the organization can change significantly in a short time (for example, months or years).

3   The Results of the Analysis the Components of Corporate Culture Motivation for Personnel

Methods of creating a positive culture of the organization, as a rule, are informal in nature. But, despite it, there are numerous examples of a powerful and purposeful change in the culture of many organizations. The modern period of development of practice and management theory is increasingly called the "cultural revolution" in management.

It should also be remembered, that since the main goal of corporate culture as the basis for the formation of a motivational mechanism is to ensure the profitability of the enterprise by optimizing personnel management, ensuring the loyalty and tolerance of employees, raising the perception of the enterprise as its own home, resolving issues without conflicts. It is important that all changes that occur at the enterprise and directly concern employees are taken with their direct participation in them. This will contribute to the constant optimization of the motivational mechanism of the enterprise. Indeed, under such conditions, it is possible to create a state of the enterprise that will have independent development. A significant motivating tool for the formation of active innovative behavior of staff should be an updated corporate culture, taking into account the peculiarities of the national labor mentality, historical memory, and patriotism.

In order to strengthen the influence of corporate culture on the harmonization of the development of social and labor relations, a system of tools has been developed, in particular: (1) organizational and methodological impact; (2) economic influence; (3) socio-psychological impact; (4) information influence (oriented to the internal environment of the enterprise); (5) communication impact oriented to the external environment (in particular, interaction with stakeholders to take into account their interests in the formation of positive values of corporate culture).

To overcome certain obstacles in the development of a positive corporate culture, in particular, strengthening its motivational functions, a methodological approach has been developed to assess the impact of the components of corporate culture on the components of motivation for work (Table 11).

The main dynamic parameter of motivation is its effectiveness, which is determined by the extent to which a person's internal motives for activity will coincide with the proposed incentives. When constructing the motivation system, it is necessary to proceed from the principle of achieving optimal correspondence between the means of internal and external motivation. This means that the motivation system must be specific, adapted to the conditions of a particular organization.

This principle has the responsibility not only to identify the mistakes made, but also to analyze them in-depth and draw conclusions about the possible consequences. The purpose of motivation is to improve the implementation of management activities based on the interests of the person and business.

Table 11 Influence of enterprise corporate culture components on individual components of enterprise personnel motivation

Component of corporate culture	Component of the staff motivation system
Human resources management	Pay
Organization of production	Working conditions
Labor discipline	Career opportunity
Innovative development of the enterprise	Material encouragement
Socio-psychological climate	Moral encouragement
Image and authority of the enterprise	Participation in enterprise management
Ethics of professional activity	Equity participation
Symbols, ideas, traditions	Social protection

The principle of the effectiveness of motivation is the basis of the whole system of motivation by taking appropriate measures.

The management staff of enterprises should see a perspective for the organization as a whole and for the organization's relations with the external environment. Managers need to know not just their own work, but also how their own work and the work of other members of the organization affects the goals, that the organization seeks to achieve. They must take into account the influence of the external environment on the organization, and vice versa. The use of systems theory for management made it easier for managers to see the organization in the unity of its components, closely intertwined with the outside world. From these positions, effectiveness can be determined as the degree of achievement of the goals set for the system, the degree of completion of the "necessary" work.

To assess the degree of effectiveness, at least three criteria are needed: (1) do we do "necessary" things according to predefined requirements (as quality of products or services), (2) do we do all "necessary" things? (as quantity), and (3) do you do the "right" things on time? (as timeliness).

Substantiation of directions to increase the effectiveness of staff motivation components will be carried out on the basis of determining the level of influence of motivation components. The Motivation Component Influence Indicator (MCII) describes the measure in which the Motivation Component will provide the desired competitive advantage over its actual level and is defined as:

$$MCII = MC_{actual} \div MC_{desired}$$

where MC_{actual} and $MC_{desired}$—actual and desired levels of influence, provided by a certain motivational component.

And the desired competitive advantage was determined on the basis of expert assessments, as well as on the basis of the desired positions of location in relation to the leader. Calculation of actual level of influence of motivational components for machine-building enterprise D is given in Table 12.

The graphical model of the formed motivation profile of the components for the enterprise "D" is shown in Fig. 2.

The level of competitive advantage is quantified as an integral value of the coordinates of the location of the enterprise on the quality scale and price scale. Both scales are based on the principle: from the minimum level 0 (corresponds to the position of the outsider) to the maximum level (corresponds to the position of the leader).

The above calculations indicate the possibility of significant improvement in the effectiveness of the motivational components "Career Growth Opportunity" and "Social Contacts" at the machine-building enterprise "D", when performing the components of the corporate culture "Personnel Management", "Production Organization", as well as "Labor Conditions" and "Remuneration" when performing the component "Innovative development of the enterprise".

It should be borne in mind, that the efficiency of motivational components can be enhanced by attracting certain productive resources, the use of which should be rational. That is, such as would ensure the achievement of motivational goals, but not

Table 12 Calculation of the actual level of influence of motivational components for machine-building enterprise "D"

Component of corporate culture	Motivational component			
	Working conditions	Pay	Career opportunity	Social protection
Human resources management	2.2/2.8 = 0.79	4.0/4.4 = 0.91	2.5/3.8 = 0.66	2.2/4.0 = 0.55
Organization of production	3.2/4.2 = 0.76	4.2/5.2 = 0.81	4.4/5.4 = 0.81	2.3/4.5 = 0.64
Labor discipline	2.9/3.9 = 0.74	4.5/5.6 = 0.80	3.0/4.0 = 0.75	2.6/3.6 = 0.72
Innovative development of the enterprise	2.8/3.4 = 0.82	3.6/5.9 = 0.61	2.9/3.2 = 0.91	3.1/5.2 = 0.60
Ethics of professional activity	2.2/3.7 = 0.59	2.7/5.8 = 0.47	2.7/5.0 = 0.54	2.0/4.4 = 0.45

Fig. 2 Profile of motivational components at the enterprise "D"

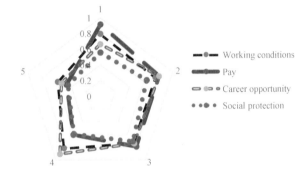

lead to the deterioration of other results of the enterprise. To do this, the enterprise needs to introduce a scale of levels of utility indicators of motivational components, which management can take into account when developing corporate events with the appropriate attraction of resources to increase the motivation of management personnel. Such a scale may be periodically adjusted. In addition, as noted earlier, the weight of motivational components may change over time. And therefore, it will be necessary to calculate for other more significant components of motivation.

For machine-building enterprise "D", the proposed scale of influence levels of motivational components is given in Table 13.

The above scale has the following practical application. In an enterprise, there is no urgent need to attract additional resources to increase the utility of motivational components if the level of the utility indicator is in a high range. But there is an urgent need to attract additional resources if it is revealed that the level of utility of the motivational component is in a low range. Finally, if the level of motivational

234 M. Vedernikov et al.

Table 13 Scale of levels of influence of motivational components at machine-building enterprise "D"

Motivational component	Impact level		
	Low	Average	High
Working conditions	Less than 0.50	0.50–0.75	Above 0.75
Pay	Less than 0.50	0.50–0.75	Above 0.75
Career opportunity	Less than 0.50	050–0.80	Above 0.80
The social protection	Less than 0.50	0.50–0.70	Above 0.70

component utility is in the middle range, management can make decisions to improve the corporate culture at its discretion, based on internal intuition and professional experience.

For the machine-building enterprise "D", it is necessary to attract resources to increase the effectiveness of the motivational components "Remuneration" and "Social Protection", when performing the component of the corporate culture "Personnel Management".

The calculation of the utility indicators of motivational components for the machine-building enterprise "C" was carried out according to the components that were previously determined as the most significant for this enterprise: conditions and remuneration for labor, the possibility of career growth, social protection. Results of calculation of utility indices of motivational components are given in Table 14.

The graphical model of the generated profile of motivational components at the enterprise "C" is shown in Fig. 3.

Table 14 Calculation of the actual level of influence of motivational components for machine-building enterprise "C"

Component of corporate culture	Motivational component			
	Working conditions	Pay	Career opportunity	Social protection
Human resources management	3.2/3.3 = 0.97	5.4/6.0 = 0.90	2.8/3.8 = 0.74	4.2/4.2 = 1.00
Organization of production	1.2/2.2 = 0.55	4.2/5.0 = 0.84	2.4/3.4 = 0.71	2.2/4.2 = 0.52
Labor discipline	2.3/2.9 = 0.79	3.5/3.6 = 0.97	3.1/4.4 = 0.70	3.6/4.6 = 0.78
Innovative development of the enterprise	2.3/3.5 = 0.66	3.6/3.9 = 0.92	1.9/3.2 = 0.59	4.3/5.2 = 0.83
Ethics of professional activity	1.1/2.5 = 0.44	5.7/5.8 = 0.98	1.7/3.2 = 0.53	2.1/3.4 = 0.62

Fig. 3 Profile of motivational components at the enterprise "C"

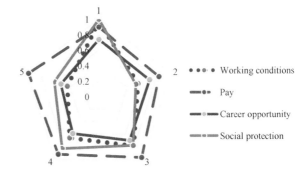

Table 15 Scale of levels of influence of motivational components at machine-building enterprise "C"

Motivational component	Impact level		
	Low	Average	High
Working conditions	Less than 0.50	0.5–0.70	Above 0.75
Pay	Less than 0.70	0.70–0.80	Above 0.80
Career opportunity	Less than 0.50	0.5–0.80	Above 0.80
Ethics of professional activity	Less than 0.55	0.55–0.75	Above 0.75

At machine-building enterprise "C", the effectiveness of such motivational components as "Working Conditions" and "Social Protection", the usefulness of which is the lowest, can be increased in the performance of the function "Organization of Production", as well as the components "Working Conditions", "Career Growth Opportunity" and "Social Protection", which show low levels of utility. For machine-building enterprise "C", a scale of levels of utility indicators of motivational components is proposed, which is given in Table 15.

Based on the calculations and the proposed scale of the motivational components impact levels for the machine-building enterprise "C", the following recommendations can be provided:

(1) it is necessary to attract additional resources to increase the effectiveness of the motivational factor "Working Conditions" in the performance of the managerial function "Professional Ethics", since its value of the utility indicator is in the low range (0.44);

(2) It is necessary to attract additional resources to increase the effectiveness of the motivational factor "Career Opportunity" when performing the function "Professional Ethics", since the value of the utility indicator is in the low range (0.53).

The calculation of the indicators of influence of motivational components for machine-building enterprise "A" was carried out according to the components that were previously determined as the most significant for this enterprise: power and influence, self-improvement, desire for achievements and for obtaining results (Table 16).

The graphical model of the formed profile of motivational components at the enterprise "A" is shown in Fig. 4.

At machine-building enterprise "B", the effectiveness of such a motivational component as "Social Protection", the level of which is the lowest, can be increased, when performing the function "Ethics of professional activity". For machine-building enterprise "A" a scale of influence levels of motivational components is proposed, given in Table 17.

Based on the calculations and the proposed scale of motivation factors for machine-building enterprise A, the following recommendations can be provided:

Table 16 Calculation of influence levels of motivational components for machine-building enterprise "A"

Component of corporate culture	Motivational component			
	Working conditions	Material encourage-ment	Moral encourage-ment	Social protection
Human Resources Management	3.4/5.2 = 0.65	7.1/7.5 = 0.95	5.1/5.3 = 0.96	7.2/7.8 = 0.92
Organization of production	4.2/5.6 = 0.75	3.2/4.5 = 0.71	6.8/7.2 = 0.94	7.9/8.0 = 0.99
Labor discipline	2.9/4.1 = 0.79	4.0/6.3 = 0.63	3.9/4.2 = 0.92	7.1/7.5 = 0.95
Innovative development of the enterprise	5.8/6.1 = 0.80	5.1/6.4 = 0.80	6.1/6.9 = 0.88	5.4/6.2 = 0.87
Ethics of professional activity	3.4/4.4 = 0.65	3.4/5.2 = 0.65	5.1/5.9 = 0.86	5.2/9.1 = 0.57

Fig. 4 Profile of motivational components at the enterprise "A"

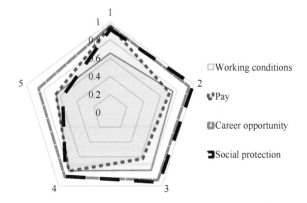

□ Working conditions

⌐• Pay

◖ Career opportunity

⌐ Social protection

Table 17 Scale of influence levels of motivational components at machine-building enterprise "A"

Motivational component	Impact level		
	Low	Average	High
Working conditions	Less than 0.65	0.65–0.75	Above 0.75
Level of education	Less than 0.65	0.65–0.75	Above 0.75
Service experience	Less than 0.85	0.85–0.90	Above 0.90
Social contacts	Less than 0.60	0.60–0.80	Above 0.80

(1) it is necessary to attract additional resources to increase the level of influence of the motivational component "Social Protection" in the performance of the managerial function "Professional Ethics", since the value of the utility indicator is in the low range (0.57);

(2) it is necessary to attract additional resources to increase the level of influence of the motivational component "Material Incentive" when performing the function "Labor Discipline", since the value of the influence level indicator is in the low range (0.63).

4 Discussions

Thus, motivation problems that actively influence the formation of corporate culture are especially acute in personnel management today. A corporate culture can survive under fierce competition only if its personnel correctly and on time assess the environment and trends of social development. The rapid response to changing consumer demand and the actions of competitors, as well as to all social changes, depends on the competence of management personnel, methods and style of the leader. Therefore, the development and implementation of staff incentive systems as the most active catalyst in the formation of a corporate culture, assistance in the achievement of personal goals and goals of the enterprise are relevant today.

Most scientists who study the problems of corporate culture as an effective tool for influencing the motivation of enterprise personnel perceive the manager only from one point of view—the desire for achievement. The author holds the view that the motivational model of the human personality is symmetrical. As the need is met, the positive motivation to achieve naturally goes into negative motivation—to prevent. Negative does not mean bad; it is another form, another stage of personality development.

5 Conclusions

The digital transformation of personnel administration begins with a change in thinking within HR. For many organizations, both in HR and throughout the organization, this is a revolutionary opportunity. In view of this, innovative development strategies should take into account the requirements of digital HR management, which includes the use of integrated mobile applications, social networks, analytics, cloud technologies and VR.

Today, everyone must, first of all, define role as a team one that helps management and employees quickly adapt to the digital way of thinking. It is imperative to improve and modernize the traditional HR solution system, and at best replace it with an integrated cloud platform that will create a digital enterprise infrastructure. It is also worth mentioning the long-term strategy that must be carefully considered and adapted to the evolutionary and technological realities of today. Innovation needs to be made a key strategy within HR. It is vital to create a team that not only knows in-depth all the processes and functions of HR, but also can monitor, analyze and implement new stack technologies that appear almost every day.

An equally important topic is the attitude toward workers in a digitised environment. It is necessary to build relationships with employees as if they are your customers. This will allow, first of all, to increase the involvement of employees, contribute to their career growth and comprehensive development.

It is necessary to monitor other companies in the market, what technologies they use, how they will help them improve their performance and create a favorable working environment for workers, because workers are the greatest value of the company, on which its future largely depends.

Summarizing the results of the corporate culture study as an effective tool for influencing the motivation of enterprise, personnel allow us to highlight the most significant ones:

1. The motivational spectrum of each person is not only unique, but also constantly changing, sometimes taking the most unexpected form. The motivation system at the enterprise should be as complete as possible—formed on the principles of the following functional motivation systems, which in the complex "cover" the entire motivational issues of personnel the system: remuneration and material incentives; social protection and material support; development of skills; career planning; moral stimulus; supporting innovation; competitions and assessments of professionalism; participation in strategic decision-making; participation in entrepreneurial risks.
2. The staff motivation system creates a kind of motivational corset that affects the formation of corporate culture.
3. Considering the motivational sphere of the identity of the worker, it is necessary to mark out its specific properties: aspiration to the power and achievement of leadership, realization of inclinations of the leader, persistence, persistence, activity, ability to intellectual work, brainwork, tendency to reasonable risk, ability to make decisions, readiness to be responsible (not only for the actions, but also for

actions of others), psychological stability, high IQ, and erudition and all-round education.

4. Improving the use of employees' abilities in accordance with the goals of the enterprise and society, since effective human management is directly related to motivation, to organizing external influences on employees, based on knowledge of their needs, motives, interests, values, ideals, and goals.

5. Strategic personnel management is implemented within the framework of the organization, which is not so much a technical and economic system as a social system. In order to obtain the maximum and necessary at a certain moment of return from managerial personnel and in order for their motivation to work to be maximum, a set of measures is needed, among which the creation of an effective corporate culture occupies a very important (as if not the main) place.

6. Corporate culture is becoming an important strategic tool that allows you to focus all divisions of the enterprise on common goals, mobilize employee initiative for creative attitude to labor and strengthen competitive advantages. So, on the one hand, corporate culture, being an integral attribute of any organization, embodies a system of motivation of personnel (at least its individual elements that form the basis of this system), and on the other hand, it meets the strategic requirements of the enterprise.

References

1. Genennig, S. M. (2020). *Realizing digitization-enabled innovation*. Springer.
2. Hulla, M., et al. (2019). A case study-based digitalization training for learning factories. *Procedia Manufacturing, 31*, 169–174.
3. Vedernikov, M., et al. (2020). Management of the social package structure at industrial enterprises on the basis of cluster analysis. *TEM Journal, 1*(9), 249–260.
4. Salun, M., & Palyanychka, Y. (2018). Features and principles of monitoring of industrial enterprise competitiveness. *Economics of Development, 17*(3), 74–82.
5. van Assen, M. F. (2018). The moderating effect of management behavior for Lean and process improvement. *Operations Management Research, 11*, 1–13.
6. Haustein, E., Luther, R., & Schuster, P. (2014). Management control systems in innovation companies: A literature-based framework. *Journal of Management Control, 24*, 343–382.
7. Claus, L. (2019). HR disruption—Time already to reinvent talent management. *BRQ Business Research Quarterly, 22*(3), 207–215.
8. Banerjee, A. (2015). Integrating human motivation in service productivity. *Procedia Manufacturing, 3*, 3591–3598.
9. Groesser, S. N., & Jovy, N. (2016). Business model analysis using computational modeling: A strategy tool for exploration and decision-making. *Journal of Management Control, 27*, 61–88.
10. Lisbeth, C. (2019). HR disruption—Time already to reinvent talent management. *BRQ Business Research Quarterly, 22*(3), 207–215.
11. Vedernikov, M., et al. (2019). Specificity of corporate culture modeling at industrial enterprises in conditions of digital business transformation. In *2020 10th International Conference on Advanced Computer Information Technologies (ACIT)*, Germany, September (pp. 595–600). IEEE.
12. Fischer, M., et al. (2021). Strategy archetypes for digital transformation: Defining meta objectives using business process management. *Information and Management, 57*(5), 103262.

13. Dang, Q. T., et al. (2020). International business-government relations: The risk management strategies of MNEs in emerging economies. *Journal of World Business, 55*(1), 101042.
14. Wang, S. (2018). *Microeconomic theory.* Springer.
15. Klikauer, T. (2014). Morality 1: Disciplinary action, obedience, and punishment. In *Seven moralities of human resource management.* Palgrave Macmillan.
16. Bruns, H. J. (2014). HR development in local government: How and why does HR strategy matter in organizational change and development? *Business Research, 7,* 1–49.
17. Vedernikov, M., et al. (2020). Modeling of controlling activity as an instrument of influence on motivation in the personnel management system of industrial enterprises. In *2020 10th International Conference on Advanced Computer Information Technologies (ACIT)*, Germany, September (pp. 601–606). IEEE.
18. Boiko, J., Volianska-Savchuk, L., Bazaliyska, N., Zelena, M. (2021). Smart recruiting as a modern tool for HR hiring in the context of business informatization. In *2021 11th International Conference on Advanced Computer Information Technologies (ACIT)*, Germany, September (pp. 284–289). IEEE.
19. Mikhaylov, F., Julia, K., & Eldar, S. (2014). Current tendencies of the development of service of human resources management. *Procedia—Social and Behavioral Sciences, 150,* 330–335.
20. Idris, S. A. M., Wahab, R. A., & Jaapar, A. (2015). Corporate cultures integration and organizational performance: A conceptual model on the performance of acquiring companies. *Procedia—Social and Behavioral Sciences, 172,* 591–595.
21. Ibragimova, R. S., Golovkin, D. S. (2021). Approaches to the formation of an industrial enterprise management system. In S. Ashmarina, V. Mantulenko, & M. Vochozka (Eds.), *Engineering economics: decisions and solutions from eurasian perspective. ENGINEERING ECONOMICS WEEK 2020. Lecture Notes in Networks and Systems* (Vol. 139). Springer.
22. Andi, H. K. (2021). Construction of business intelligence model for information technology sector with decision support system. *Journal of Information Technology, 3*(4), 259–268.
23. Thando, D., Eck, R. V., & Zuva, T. (2020). Review of technology adoption models and theories to measure readiness and acceptable use of technology in a business organization. *Journal of Information Technology, 2*(04), 207–212.

The Social Hashtag Recommendation for Image and Video Using Deep Learning Approach

Priyanka Panchal and Dinesh J. Prajapati

Abstract There has been a lot of interest in the recent year in recommending hashtags for images/videos or posts on social media. Several researchers have researched the impact from numerous perspectives. In this paper, we enhance tag recommendation by recommending suitable hashtags considering both contents of the image/video and users' history of the hashtag. On the social media image/video-sharing websites (such as Facebook, Instagram, Flickr, and Twitter), users can upload images or videos and tag them with tags. The proposed method generates candidate keywords, i.e., hashtag by combining techniques for textual tags, image and video activity/object recognition content, and acoustic data. To this end, this paper examines different methodologies that associate information that is multi-modal and suggests hashtags for image or video uploader users to generate tags for their images or videos. Although a substantial amount of study has been carried out on item/product recommendations for E-commerce websites, video recommendations for YouTube and Netflix, and friend suggestions on social media websites, research has not been carried out as much on hashtag recommendations for images/video on social media platform/app/websites, which have now turned out to be a vital role of these social media platforms. Here, in this paper, glance at hashtag recommendations for image/video has been carried.

Keywords Tag recommendation · Social network · Video content · Tag (hashtag) · Deep learning

P. Panchal (✉)
Gujarat Technological University, Ahmedabad, Gujarat, India
e-mail: prinankoo.16@gmail.com

D. J. Prajapati
A. D. Patel Institute of Technology, CVM University, Anand, Gujarat, India
e-mail: it.djprajapati@adit.ac.in

1 Introduction

In the social media platform, whenever we post an image/video, a hashtag is a keyword that is to be added to a part of social media image/video content. The user can add tags to their image/video post and using that hashtag another user can find the content related to a topic of image/video. del.icio.us is a social bookmarking website and Flickr is an image-sharing website that popularized social media hashtags at the beginning of the twenty-first century. There are other additional social networking platforms, including Facebook, Instagram, and Tumblr, that have adopted this method of indicating a tag by utilizing the #symbol. Today, most large social media platform websites/apps contain a certain kind of tagging functionality. A hashtag recommendation can be used to improve a social image/video-sharing site's tagging functionality by offering a hashtag recommendation methodology, in preparation for a new upload and by assisting users in agreeing on the subject of a lexicon for specific topics.

Hashtags, which are commonly used in the domain of microblogs [1, 2], images [3–10], videos [11–17], and news [4, 18–20], are a succinct description of the actual contents and the primary source of information for search engines to identify the necessary content [4]. Social networking services (SNS) have grown in importance component of individuals' everyday lives. The majority of image-sharing platforms like Facebook, Instagram, Twitter, Flickr, etc, … They enable the individuals; tags can be added to the image/videos by users. Because including popular tags will aid people to annotate, categorize, and searchable, tags are an essential component of photos/videos' popularity. However, tagging process is taking a lot of time, and most individuals are unsure of what type of tag to use [9]. Tags have a close relationship to the popularity of social content; together the number and tag quality can be a good indicator of future popularity.

The following are the primary contributions of this paper: (i) define a literature survey of hashtag recommendations for images or videos, (ii) present a filtering method for tag recommendation, (iii) we have defined a comparative analysis of different techniques for tag recommendation, (iv) evaluation parameter for hashtag recommendation, and (v) challenges and trends happens for tag recommendation.

2 Literature Survey

However, users often incorrectly use hashtags. The issue is that an incorrectly tagged post will not appear when others search for correctly tagged posts. Furthermore, it is difficult for the system to suggest tags to use within the context of a short message [21], for example, as a 280-character message on Twitter™ or an image/video on Facebook/Instagram/YouTube. Every day, thousands of images, text messages, micro-videos, and microblogs are spread across the internet. Hashtags help to make a good judgment of all that information. They do not just facilitate us in categorizing information from user posts that are related contexts but as well in identifying and classifying the data and making searching easier. Those who are not connected to our social media network then also with the help of hashtags allowed to view the caption. They help to capture new trending topics that would otherwise go unnoticed [22]. Figure 1 represents the hashtags of numerous forms. Figure 2 is a representative of various hashtag applications.

Tag recommendation for image/videos methodology focuses on different features of images/videos which include the contents of image/videos and the relation between users and the content of images/videos. We distinguish two types of tag recommendation tasks: object-centered and personalized. The goal of the former is to create and

Fig. 1 Various types of hashtags

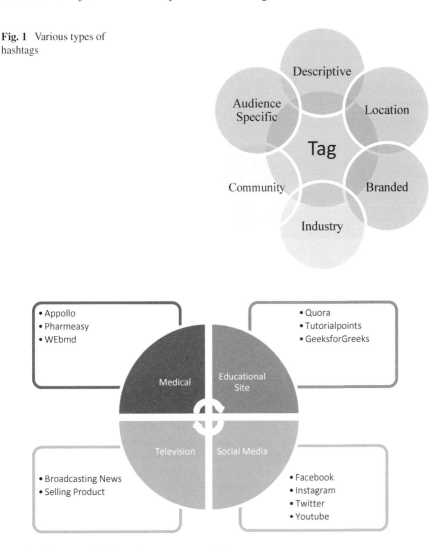

Fig. 2 Different applications of tag recommendations

prioritize candidate tags based on their applicability to the target item, alternatively on how closely the tag is associated with or characterizes the target item [23].

Tag recommendations based on objects, that do not change depending on the intended user, aim to improve the quality of the tag and, as a result, automatically improve the performance services for retrieving information like searching, categorization, and item retrieval recommendation, which use tags as sources of information, in contrast, personalized tag recommendation algorithm considers not just the intended product but also the intended user, with the goal of adding tags that are relevant to the intended object as well as the people. As a result, personalized tag recommenders may return varied outcomes for various users, and improve capturing the user's interests, background, and profile. "In applications where multiple users can assign tags to the same object, such as Last.FM, a personalized tag recommender is not only useful for the individual user (e.g., for content organization), but also in a collective sense," according to [8]. This is due to, when various users' tags were recommended are combined, they may give a more detailed description of the thing, which benefits for services for searching and recommending."

Here, the comparative study of hashtag recommendations is defined below.

2.1   Tag Recommendation for Micro-video

Liu et al. proposed technique for hashtag recommendation which is the User-guided Hierarchical Multi-head Attention Network (UHMAN), wherein they consider user-side data for combining image-level and video-level descriptions of micro-videos as show in Tables 1 and 2. The proposed method was performed on the micro-video-sharing platform Musical.ly from where the dataset has been collected. The experimental findings show that the proposed strategy is successful [11]. The proposed methodology invented based on a neural network, LOGO (short for "muLti-mOdal-based hashtaG recOmmendation"), which uses multiple modalities for micro-video to recommend hashtags. The suggested LOGO approach used the sequential and attentive features simultaneously to integrate all modalities via a multi-view representation learning framework. Extensive studies on the range of both overall performance comparison and micro-level assessments have well justified the efficacy and logic of our suggested strategy [12].

Li et al. [21] proposed method recommend hashtags for micro-videos by presenting a novel multiview representation graph-based information propagation with interactive embedding model. By jointly considering the combination of sequential feature learning, the video-user-hashtag interaction, and the hashtag correlations, it automatically can improve the performance of micro-videos hashtag recommendation. The proposed method performs the experiments on constructed datasets that perform state-of-the-art baselines [13]. Yang et al. [22] proposed the senTiment enhanced multi-mOdal Attentive haShtag recommendaTion (TOAST) model for micro-video hashtag recommendation. The content-based common space learning features are multi-modal sentiment features and multi-modal content features which used methodology as sentiment attention and self-attention respectively. Also, an attention neural network used embedding by an attention neural network for differing the importance of dynamically captured content features and multi-modal sentiment features. The proposed methodology experiment performed on a real-world dataset gives an effective recommendation [14].

Wei et al. [11] proposed a graph convolution network-based personalized hashtag recommendation (GCN-PHR) model, which jointly combines the interaction among < users, hashtags, micro-videos > using advanced GCN techniques and understand their representations. In the proposed model graph, there are three types of nodes are there the users, hashtags, and micro-videos. Proposed modal experiments have been performed on two real-world micro-video datasets which carried out the state of the art by a large margin. Singh et al. [13] proposed the methodology in mobile computing environment for video-sharing websites just like YouTube to improve the popularity of videos. To discover videos of users' interests, combining the tags and titles is beneficial to increase their tendency to watch more videos. The proposed methods applied to gain higher popularity which recommend the tags for the video to combine the tags and titles of video uploaders based on a novel hybrid method that includes a multi-modal content analysis method as well. By combing the video content and textual semantic analysis of the original tag, the proposed method generates the candidate keywords, i.e., tags. The proposed method experiments with the demonstration of videos to improve social popularity effectively as well as extend the length of video viewing time per playback.

Wei et al. [12] propose to exploit user-item interactions to guide the representation learning in each modality and additional personalized micro-video recommendations. User preferences can be defined based on the model-specific representations of the users which build upon the message-passing concept of GNN using multimodal GNN. TikTok, Kwai, and MovieLens datasets that are publicly available, on which significant experimentation have been conducted. The proposed methodology for multi-modal recommendation methods gives a significantly better state of the art.

2.2 Social Tag Recommendation for Images

Zheng et al. [16] proposed technique which presents a tag recommendation system, called PROMPT, which combines the social contents and uploaded photo leveraging to recommend the personalized tag. For recommending the personalized keywords, one important thing needs to understand and determine the similar tagging approaches of the user who belongs to similar groups. Afterward, tags of photos, textual content, and visual metadata are joined together to find the most suitable candidate tags and from that most N tags need to recommend. Using the asymmetric tag co-occurrence, probabilities initialize the score of the candidate's tags. Then after neighbor voting is used to normalize the scores. At last, diminish inaccessible and many nearest neighbors tag have been promoted using random walk methodology. The proposed methodology recommends the top N tags/keywords to the uploaded photo. The proposed methodology demonstrates on the Flickr dataset which photos (that contain test set as 46,700 and train set as 28 million photos). The proposed methodology confirms the outperform state of the art with 1540 unique user keywords/tag. Nguyen et al. proposed method which presents a personalized content-aware image tag recommendation approach that combines both the features of images and users' historical tagging data/information in a factorization model. Deep learning state-of-the-art technique used to extract powerful and important features of images using the object detection and image classification the images. To recommend the tags, an adaptive factorization model is used for combing the features of users tagging behaviors history and image features/information. The proposed methodology experiments give the outperform state of the art based on object-based, i.e., image features and visual context can boost the result up to 1.5%.

Quintanilla et al. [24] proposed methods to solve the personalized tag recommendation problem and proposed method to perform the experiment on the large-scale dataset using an end-to-end deep network. To improve the user recommendation, combined the visual encoding and user preference including the user's behaviors and which can be learned using an unsupervised way within the network. To recommend the user-generated keyword/tag, the proposed methodology used the adversarial learning technique. The proposed experiment performs on the publicly available dataset and two large different large-scale, NUS-WIDE and YFCC100M effectively which gives the improved performance significantly on both datasets when compared with the other state-of-the-art baselines [7]. Bharadwaj et al. proposed the methodology to predict the most relevant hashtags for photos/images. In this paper, the personal dataset has been constructed for the recommendations. To classify the hashtag, the dataset includes the image of various classes or categories which have been trained and tested in different ways. To select a reasonable hashtag for the image is a realistic technique, and to do that, machine learning technique is used in the easiest way.

Hachaj and Miazga [17] proposed technique which is a novel voting deep neural network with associative rules mining (VDNN-ARM) algorithm used to recommend the hashtag using multi-label. One of the machine learning technique VDNN-ARM integrate the deep neural network to extract the image features, which are used to categorized the hashtag class according to the image context. Also, the associative rules mining also used to integrate all the hashtags in one single dataset. The proposed methodology used for multi-label classification problem which is performed on HARRISON benchmark dataset. The evaluation parameters are accuracy, precision, and recall. VDNN-ARM methodology gives the outperform state of the art that improve the result. Wang et al. [7] proposed a novel method for hashtag recommendation that resolves the data sparseness problem by exploiting the most relevant twitter post data from other sources of expertise. In addition, based on user's remarkable position, the proposed method includes semantic knowledge which used vocabulary information and contextual attributes based on the technique of word embedding methodology. To take use of diverse hashtag recommendation methods based on distinct attributes, our suggested approach aggregates these methods with learning-to-rank and provides top-ranked hashtags. The experimental findings reveal that the suggested strategy outperforms the present state-of-the-art methods substantially.

Yang et al. [6] proposed technique which is an attention-based multi-modal neural network model (AMNN) to learn the representations of microblog using multi-modal which is used to recommend the appropriate hashtag. The proposed methodology for hashtag recommendation combines hybrid NN technology along with the sequence-to-sequence model to recognise the image's properties and related text. There are two public dataset and self-constructed Instagram dataset on which the proposed method experiment has been performed to achieve the best result that outperforms state of the arts. Moreover, three evaluation parameter are used to compare the result between the three different dataset that is, accuracy, precision, and recall. Ma et al. [8], the proposed method have used the novel co-attention memory network to recommend the hashtag for multi-modal microblogs. Using the co-attention technology integrate the user's past tagging history tag and multi-modal microblog to signify the hashtag. The proposed method demonstrates on the dataset of Twitter which improves the performance by comparing the baseline methodology to solve the multi-class classification problem. Park et al. [9] proposed the hashtag recommendation on real-world images of social media networks and they have introduced the HARRISON dataset. The HARRISON has been constructed from Instagram including 57,383 images with an associated hashtag for individual images. The proposed methodology has been evaluated on the HARRISON dataset to extract the visual features using convolutional neural networks (CNN). Based on the multi-label classifier on neural network integrating the object-based, scene-based model, the multi-model features-based model is evaluated on the HARRISON dataset. To recommend the hashtag on the dataset are required a contextual and wide understanding of the images that are to be conveyed in a form of a hashtag.

2.3 Social Tag Recommendation for Micro-video

See Table 1.

2.4 Social Tag Recommendation for Image

See Table 2.

Table 1 Social tag recommendation for micro-video

S. No	Technique/approach	Class/categories	Dataset	Evaluation parameters	Open issues/critics
2020 [11]	Extracting visual features using Inception-v3 Encoding user representations using a one-hot encoder Encoding hashtag representations Embedding	General	From one of the social Network **Musical.ly** the dataset has been constructed	Precision@10, Recall@10, F1-Mesure@10	Not mentioned
2020 [4]	Multiple modalities fusion: Multi-modal fusion techniques including the parallel and sequential structure learning	General	Using the Public API of **Vine,** micro-video is extracted in form of a dataset including tags	Recall and NDCG	For hashtag recommendation for micro-videos, need to consider external knowledge of hashtag which is used to strengthen the recommendation of hashtag in online manner
2019 [21]	For recommending the hashtag need to integrate the user's behavior, hashtag, and micro-video features using embedding technique, GCN, and multiview learning technique	Micro-video of the dataset includes the three categories of micro-video, i.e., **singing, dance, and sports**	They have constructed their INSVIDEO is a personal dataset comprising 213,847 micro-videos and 6786 users. To construct the dataset, data extracted from Instagram Micro-videos and its associated description extracted for active user and atmost 50 published video micro-video extracted from his/her account	To analyze the effectiveness, we utilized the generally used the evaluation metrics: Recall@N and NDCG@N	In the future, need to decrease the redundant hashtags for micro-videos Integrating user's behavior, hashtags, micro-video micro-videos features, hashtag recommendation will be improved using the attention mechanism and embedding model which is used to focus on the important clues of micro-videos

(continued)

Table 1 (continued)

S. No	Technique/approach	Class/categories	Dataset	Evaluation parameters	Open issues/critics
2020 [22]	Sentiment common space learning Network, self-attentive content common space learning network	Real-word dataset, i.e., general	**Vine**, one of the most popular micro-video streaming social media platforms, was scraped for a micro-video dataset	Recall@K and NDCG@K	In the future, need to integrate the semantic info. of hashtags and micro-video features associated their description for more accurate hashtag recommendation for micro-videos. also need to consider the sentiment info. of micro-video modality and emoji in the textual modality for better results
2019 [11]	To filter the redundant message which is generated from micro-videos that is implemented using graph convolution network (GCN) using an attention mechanism	General	Dataset self-constructed from the Social Media App. Instagram and YFCC100M Publicly available dataset	Accuracy, Precision, and Recall are evaluation metrics	Not mentioned
2016 [13]	To extract the video features using the neural network, NLP technique and skip-gram model of embedding function	29 categories of short video clips of Sports	From **Vine.co**. social media application videos and related hashtags extracted	For better understanding average relevance scores calculated	Integrate the sentiment information and contextual information using the Tag2Vec Space which gives the popular and trending hashtag

(continued)

Table 1 (continued)

S. No	Technique/approach	Class/categories	Dataset	Evaluation parameters	Open issues/critics
2020 [5]	Using the deep learning technique TF-SIM algorithm, features of video extracted also generated hashtag/keywords are arranged according to the relevant to the video	Film and animation, pets and animals, sports, travel and events, gaming, humor, entertainment, news and politics, science and technology were all chosen at random	**YouTube**	Proposed methodology evaluated in three different way: YouTube, Campus volunteers and professional survey website Amazon Mechanical Turk (AMT)	In future, not only the hashtag recommended but title and description of video also need to generate using video semantic technology and NLP technique which are combined so that viewers can find the video of their interests
2010 [25]	Query Construction Significant Tag Extraction SVM classifiers	Class 15 are used in MCG-WEBV	**MCG-WEBV (dataset containing 80,031 representative YouTube videos)**	MAP	To design a more robust query construction method need to extract keyword/tag whether its noisy or irrelevant or not. and to do that need to integrate the video features for better result
2010 [23]	Collaborative Tagging Video Features extraction technique	Arcade games, cars, dances	**YouTube**	Average relevance precision@K	Not mentioned

(continued)

Table 1 (continued)

S. No	Technique/approach	Class/categories	Dataset	Evaluation parameters	Open issues/critics
2019 [12]	Aggregation Layer Bayesian Personalized Ranking (BPR) GNN CF method	General, i.e., real-world dataset	**Publicly available datasets: Tiktok, Kwai, and MovieLens**	Precision Recall NDCG	Micro-videos and associated object in modality are represented using multi-modal knowledge graph Also need to integrate the user's history behavior with objects as micro-videos and trending social networks possibility To perform the recommendation on how social leader's influence

Table 2 Social tag recommendation for image

S. No	Technique/approach	Class/categories	Dataset	Evaluation parameters	Open issues/critics
2017 [3]	To recommend the tag, random walk approach integrating with the neighbor voting approach using tag co-occurrence	General	YFCC100M dataset (from Flickr, 100 million media data collected)	Accuracy@K and Recall@K	In future, need to improve the precision, recall, and accuracy using proposed approach
2017 [15]	Factorization models deep learning Model for image classification object detection Techniques convolutional neural network deep neural network	contains 269,648 images	MS COCO data set NUS-WIDE	F1@K Precision@K Recall@K	In the future, tag recommendations need to consider the users' tagging behaviors and vocabularies of users also integrate the image multi-modality features
2021 [24]	Multi-label classification Encoder–decoder network Adversarial learning	General	YFCC100M and NUS-WIDE	Precision, recall, and accuracy	Apply the proposed methodology for image captioning in the network
2020 [26]	Machine learning techniques such as KNN and multi-label classification and deep learning techniques such as CNN	Animals, art, babies, fashion, fitness, flowers, food, memes, nature, selfie	Harrison dataset (Dataset collected from Instagram as 57,383 images and hashtag associated with each images are an average 4.5%)	Compare various methods for an accurate prediction of hashtags	Not mentioned
2020 [17]	Multi-label Classification hashtag recommendation CNN DNN	57,383 photos from Instagram	Real-world photo dataset, HARRISON	Precision Recall Accuracy	Developing methods that utilize additional information, like geo-positioning information of the image, features of images with multi-modalities and so forth

(continued)

Table 2 (continued)

S. No	Technique/approach	Class/categories	Dataset	Evaluation parameters	Open issues/critics
2021 [27]	Word-embedding-based framework Content-based hashtag recommendation using Probabilistic Latent Semantic Analysis (PLSA) and Latent Dirichlet Allocation (LDA) technology	General	Using public API of Twitter, i.e., Twitter Streaming API dataset has been created with publicly available tweets	Precision@k Recall@k	Not mentioned
2020 [6]	Attention-based neural network framework encoder–decoder architecture GRU network	General	HARRISON NUS-WIDE MM-INS	Evaluate the proposed model on the different dataset using different evaluation metrics Accuracy, Precision, Recall	In future, the proposed methodology of hashtag recommendation are integrate with the characteristics of hashtag
2019 [7]	FP-Rank PageRank Collaborative filtering (CF) Adjacency Matrix Construction	General	YFCC100M	Compare the suggested approaches to five other ways that have been recommender	Conduct experiments on more dataset
2021 [8]	Attention mechanism Embedding LSTM, RNN, CNN Co-attention-based tweet modeling	General	Data set collected from Twitter Twitter's API	Recall Precision F1-Score Hits@3 Hits@5	Not mentioned
2016 [9]	Convolutional neural networks (CNN) VGG-16 VGG-Object VGG-Scene VGG-Object + VGG-Scene Multi-label classifier	57,383 images (Illustration, art, architecture. Ootd, drawing, design, Hollywood, California)	HARRISON dataset	Precision@1, Recall@5, and Accuracy@5	In the future, this HARRISON dataset can be used to improve the recommendation of the hashtag

3 Filtering Methods for Hashtag Recommendation

Hashtag recommendation is an important strategy in online service retrieval and navigation, and it is typically based on integrating the co-occurrence of users, tags, and content and target content information.

3.1 Collaborative Filtering

Collaborative filtering methodology in the domain of hashtag recommendation is used to determine the similarities in the user's history and trending latest hashtag and make recommendations to the user based on a similar preference with other users in which the user may be interested in. Collaborative filtering (CF) is a well-known recommendation technique that has long been studied by researchers [22, 28]. Review publications explain several CF-based approaches [18, 29] and be able to be used to recommend tags based on knowledge like as tagging behavior histories. Several studies extended collaborative filtering by using matrix factorization [9] and a neural network (NN)-based approach [8].

The collaborative filtering technique-based methods are used in integrating the relationship between users' posts, hashtags, and user history in a direct manner. For the personalized hashtag recommendation, need to capture the user-specific information which includes the hashtag semantics and users' behaviors in tagging history. For building a modern hashtag recommender system, a collaborative filtering technique has developed a new emerging technology which includes the multiple domains of the object to discover the relevant object in the potential that defines the fundamental operational methodology of a collaborative filtering-based hashtag recommender system. Existing collaborative filtering techniques can be generally classified into latent-factor model-based techniques and memory-based approaches. For hashtag recommendations, identify the users and objects relations which is used to find out the relevant users and objects from their information. These techniques are often receptive to the sparsity of the user-item interaction matrix since a sparser matrix reduces the likelihood of locating a high-quality neighborhood of users/items. The latent-factor models solve this issue in part by exploiting the relationship matrix's broad form to uncover a smaller set of latent factors that contain all of the items' intrinsic properties including the interests of the users for these aspects.

User Similarity Calculation
When creating a user CF, the selection of a similarity function is a significant design decision.

Pearson Coefficient Correlation
To determine the similarity of two users, this approach calculates the statistical relationship (also known as Pearson's Coefficient) between their shared scores.

Mean calculation

$$\overline{v} = \frac{\sum_{i \in S_u} v_{ui}}{|s_u|} \tag{3.1}$$

To begin, we may apply the aforementioned method to get a user's mean rate.

Pearson Coefficient

$$\omega(a, u) = \frac{\sum_{i \in S_a \cap S_u} (v_{ai} - \overline{v}_a)(v_{ui} - \overline{v}_u)}{\sqrt{\sum_{i \in S_a \cap S_u} (v_{ai} - \overline{v}_a)^2 \sum_{i \in S_a \cap S_u} (v_{ui} - \overline{v}_u)^2}} \tag{3.2}$$

The following is a better approximation to the preceding equation:

$$\omega(a, u) = \frac{\sum v_{ai} v_{ui} - \frac{\sum v_{ai} \sum v_{ui}}{n}}{\sqrt{\sum v_{ai}^2 - \frac{\sum v_{ai}^2}{n}} \sqrt{\sum v_{ui}^2 - \frac{\sum v_{ui}^2}{n}}}$$

The approximation is superior because the value may be obtained in a single pass, minimizing computing time.

Cosine similarity
0–0 matches are ignored by cosine similarity. It is defined as follows:

Cosine Similarity

$$\cos(x, y) = \frac{x.y}{\|x\| \times \|y\|} \tag{3.3}$$

where denotes the dot product and $\|x\|$ denotes the vector x's length A vector's length is defined as

$$\|x\| = \sqrt{\sum_{i=1}^{n} x_i^2}$$

Collaborative filtering is further classified into three types:

Memory-based CF uses the complete user-item database to create tag recommendations. Every user belongs to a group of individuals who have similar interests, and by identifying the user's neighbors, a forecast of the user's preferences on new items may be generated as an aggregate of the neighbors' ratings. This category includes both item-based and user-based CF [16].

Model-based CF trains and models users' preferences based on a collection of ratings using data mining technologies and machine learning algorithms. To predict the test and real-world dataset model-based CF is used.

Hybrid CF predicts the hashtag by mixing memory-based and model-based CF algorithms, or by integrating other recommendation methodologies with collaborative filtering.

3.2 Content-Based Filtering

Content-based approaches are another well-known recommendation strategy. Users and items must be appropriately distinguished by characteristics, like user actions, user and item characteristics, and social connection are some examples. Nevertheless, profiling an individual or object is a difficult process in general. A good user/item profiling strategy should define a user using both explicit and hidden or implicit attributes or aspects [5]. The content-based approach is a domain-specific technique/algorithm, and it focuses more on the evaluation of item properties in order to make predictions. It is proposed that publications and news are recommended while using media such as websites; the furthermost efficient filtering strategy is content-based filtering. In the content-based filtering approach, recommendations are created grounded on information gleaned from user profiles and traits the content of previous items assessed by the user.

In recommender systems, the content-based recommendation is an important technique. The main concept is to predict items which are comparable to what the customer has liked [25]. The primary of a content-based recommender system objective is to compute the degree of correlation among items. There are various approaches for the purpose of modeling items, the utmost familiar of which is the Vector Space Model [12]. The algorithm analyzes the keywords for the item/objects and TF-IDF worked upon to determine the substance. Set k_i as the ith keyword of item di, and wn_i as the weight of k_i for d_i, then the content of d_i may be specified as follows:

$$\text{Items}(d_i) = \{w1_i, w2_i, \ldots\}$$

As said before, a content-based recommendation engine predicts items that are comparable to what the user used to like. As a consequence, a person's choices may be anticipated based on his or her past preferences. Using Content-Based Profile(u) by way of the u's user interest matrix; the specification is:

$$\text{Content\_ Based\_ Profile}(u) = \frac{1}{|N(u)|} \text{sum} \, d \in N(u)\text{Items}(d)$$

$N(u)$ is what you previously admired about the user. After computing the item matrix Content(.) and the information preference vector Content-Based Profile(.) among all users, for any user u and an item d, the similarity in between user and the item is defined as the resemblance between the user and the item. Content_Based_Profile(u) and Content(d):

$$p(u, d) = \text{sim}(\text{Content\_Based\_Profile}(u), \text{Content}(d))$$

Many recommender systems rely heavily on keywords to represent products. However, collecting keywords from an item is also a challenging task, particularly in the media area, because it is extremely hard to retrieve textual keywords from with a video. There are two major approaches to dealing with this type of issue. One option is to let experts tag the things, while the other is to allow consumers tag them.

3.3 Filtering Based on Hybrids Approach

The hybrid filtering approach integrates several recommendation algorithms to improve system optimization and avoid some of the restrictions and issues associated with only a recommendation mechanism. The concept underlying hybrid approaches is that an integration of models will produce more reliable and realistic suggestions than a learning algorithm since one algorithm's disadvantages may be solved by another algorithm. Combining approaches can be accomplished in one of three ways: separately implementing methodologies and combining the results, utilizing a few more content-based filtering in a collaborative approach, utilizing a few collaborative filtering in a collaborative approach, or developing a unified recommendation model that utilizes both approaches.

3.4 Social Popularity Prediction

Prediction of social popularity forecasting has proved to be a significant topic. In [30–33, 49], the internet popularity content and content created by users, such as news and videos from YouTube, has been examined and recommended. Along with the advancement of SNSs, academics and industries are focusing on the Prediction of uploaded images' social popularity [4, 13, 17, 20, 26, 27, 34] and videos [12, 16, 21]. The majority of these popularity predictions are carried out through phases of data-driven feature-based learning, as well as some studies have integrated individuals with dynamic models that represent time fluctuations [1, 33]. Furthermore, multi-modal characteristics such as information in both word and image, as well as social relationships, have been analyzed and integrated with techniques such as to enhance the attention model prediction performance.

3.5 Technique for Tag Recommendation for Images/Videos

The technique for tag recommendation for images and videos is shown in Fig. 3.

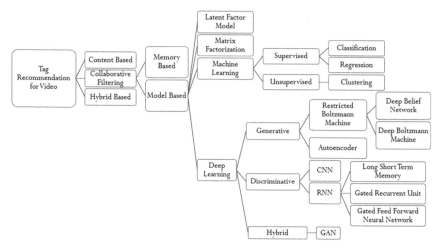

Fig. 3 Technique for tag recommendation for images/videos

4 Evaluation Parameter

To analyze the effectiveness of the suggested technique, metrics for evaluation are utilized. Recommender system efficiency may be analyzed using metrics for evaluation namely precision, recall, and F-measure. Precision and recall are both major measurements for statistics that are used to examine the effectiveness of statistical results. Precision has been used to compute the proportion of related documents to documents chosen at random. Recall is used to compute the proportion of related papers in selected documents to all related documents. In the context of hash tag recommendation, accuracy and recall are examples of definitions [27].

Assume TP_i describes the number of sample documents associated with C_i and they are categorized to C_i as well. FP_i describes the number of sample papers which do not relate to any of the groups C_i are categorized to C_i. FN_i represents the number of sample documents fit in to C_i are categorized to other classifications. As a result, the evaluation parameters are precision and another evaluation metrics recall in categorization C_i are defined as follows:

$$P = \frac{TP_i}{TP_i + FP_i}$$

$$R = \frac{TP_i}{TP_i + FN_i}$$

Mostly, after thoroughly considering accuracy and recall, we offer F-measure, which is defined below.

$$F = \frac{2 \times P \times R}{P + R}$$

We can also calculate the error rate for every method that used the mean absolute error rate (MAE) and root mean squared error (RMSE) making use of the following formulas

$$MAE = \frac{1}{|T|} \sum_{(u,i,r)\in T} |p_{ui} - r|$$

$$RMSE = \sqrt{\frac{1}{|T|} \sum_{(u,i,r)\in T} (p_{ui} - r)^2}$$

where T denotes the total amount of test data, the algorithm's projected rating is presented as p_{ui}, and the actual rate that the user is rating is denoted by r.

5 Conclusion and Future Scope

When users give hashtags regarding their posted images/videos, users prefer to use those that, in their opinion, define the material that they are fascinated in. Because of individuals' own tastes for the picture or video material, as well as their personal hashtag usage patterns or comprehension, they may use vastly diverse hashtags for the same image or video. This discovery suggests that it is critical to understand the intricate relationships between users, hashtags, and images/videos in order to provide appropriate hashtag recommendations. In this study, we will look at several research in which hashtags are suggested. Our research study is classified into two parts: pictures and micro-videos. According to the research, reducing noise has a significant impact in boosting the algorithm's efficiency and accuracy. As well, consisting of user profiles, user's past hashtag history, their interests, and popular trending hashtag and population can aid in the enhancement of recommendation systems. The approach recommends hashtags using computer vision, deep learning, and natural language processing, which will benefit a wide range of social media accounts. We hope that the list of open challenges we identified might serve as a guideline for future investigation efforts and improvements in this vital field.

References

1. Thomee, B., Elizalde, B., Shamma, D. A., Ni, K., Friedland, G., Poland, D., Borth, D., & Li, A. L. J. (2016). YFCC100M: The new data in multimedia research. *Communications of the ACM, 59*(2), 64–73. Association for Computing Machinery. https://doi.org/10.1145/2812802
2. Baltrušaitis, T., Ahuja, C., & Morency, L. -P. (2017). *Multimodal machine learning: A survey and taxonomy.*
3. Shah, R. R., Samanta, A., Gupta, D., Yu, Y., Tang, S., & Zimmermann, R. (2017). PROMPT: Personalized user tag recommendation for social media photos leveraging personal and social

contexts. In *Proceedings—2016 IEEE International Symposium on Multimedia, ISM 2016* (pp. 486–492).

4. Cao, D., Miao, L., Rong, H., Qin, Z., & Nie, L. (2020). Hashtag our stories: Hashtag recommendation for micro-videos via harnessing multiple modalities. *Knowledge-Based Systems, 203.*

5. Zhou, R., Xia, D., Wan, J., & Zhang, S. (2020). An intelligent video tag recommendation method for improving video popularity in mobile computing environment. *IEEE Access, 8,* 6954–6967.

6. Yang, Q., Wu, G., Li, Y., Li, R., Gu, X., Deng, H., & Wu, J. (2020). AMNN: Attention-based multimodal neural network model for hashtag recommendation. *IEEE Transactions on Computational Social Systems, 7*(3), 768–779.

7. Wang, X., Zhang, Y., & Yamasaki, T. (2019). User-aware folk popularity rank: User-popularity-based tag recommendation that can enhance social popularity. In *MM 2019—Proceedings of the 27th ACM International Conference on Multimedia* (pp. 1970–1978).

8. Ma, R., Qiu, X., Zhang, Q., Hu, X., Jiang, Y. G., & Huang, X. (2021). Co-attention memory network for multimodal microblog's hashtag recommendation. *IEEE Transactions on Knowledge and Data Engineering, 33*(2), 388–400.

9. Park, M., Li, H., & Kim, J. (2016). *HARRISON: A benchmark on HAshtag recommendation for real-world images in social networks.*

10. Isinkaye, F. O., Folajimi, Y. O., & Ojokoh, B. A. (2015). Recommendation systems: Principles, methods and evaluation. *Egyptian Informatics Journal, 16*(3), 261–273. Elsevier B.V.

11. Wei, Y., Zhao, Z., Cheng, Z., Zhu, L., Yu, X., & Nie, L. (2019). Personalized hashtag recommendation for micro-videos. In *MM 2019—Proceedings of the 27th ACM International Conference on Multimedia* (pp. 1446–1454).

12. Wei, Y., He, X., Wang, X., Hong, R., Nie, L., & Chua, T. S. (2019). MMGCN: Multi-modal graph convolution network for personalized recommendation of micro-video. In *MM 2019—Proceedings of the 27th ACM International Conference on Multimedia* (pp. 1437–1445).

13. Singh, A., Saini, S., Shah, R., & Narayanan, P. J. (2016, December 18). Learning to hash-tag videos with Tag2Vec. In *ACM International Conference Proceeding Series.*

14. Institute of Electrical and Electronics Engineers, IEEE Computer Society, IEEE Circuits and Systems Society, IEEE Communications Society, & IEEE Signal Processing Society. (n.d.). In *2020 IEEE International Conference on Multimedia and Expo (ICME)*, 06–10 July 2020, London, UK.

15. Nguyen, H. T. H., Wistuba, M., & Schmidt-Thieme, L. (n.d.). *Personalized tag recommendation for images using deep transfer learning.*

16. Zheng, L., Tianlong, Z., Huijian, H., & Caiming, Z. (2020). Personalized tag recommendation based on convolution feature and weighted random walk. *International Journal of Computational Intelligence Systems, 13*(1), 24–35.

17. Hachaj, T., & Miazga, J. (2020). Image hashtag recommendations using a voting deep neural network and associative rules mining approach. *Entropy, 22*(12), 1–13.

18. Association for Computing Machinery. Special Interest Group on Information Retrieval. (2008). In *SIGIR '08: The 31st annual International ACM SIGIR Conference on Research and Development in Information Retrieval*, July 20–24, 2008, Singapore. Association for Computing Machinery.

19. Song, Y., Zhang, L., & Giles, C. L. (2011). Automatic tag recommendation algorithms for social recommender systems. *ACM Transactions on the Web, 5*(1).

20. Belém, F. M., Almeida, J. M., & Gonçalves, M. A. (2017). A survey on tag recommendation methods. *Journal of the Association for Information Science and Technology, 68*(4), 830–844. Wiley.

21. Li, M., Cheng, Z., Gan, T., Yin, J., Liu, M., & Nie, L. (2019). Long-tail hashtag recommendation for micro-videos with graph convolutional network. In *International Conference on Information and Knowledge Management, Proceedings* (pp. 509–518).

22. Yang, C., Wang, X., & Jiang, B. (2020). Sentiment enhanced multi-modal hashtag recommendation for micro-videos. *IEEE Access, 8,* 78252–78264.

23. Toderici, G., Aradhye, H., Pasçca, M., Sbaiz, P. L., & Yagnik, J. (n.d.). *Finding meaning on YouTube: Tag recommendation and category discovery.*

24. Quintanilla, E., Rawat, Y., Sakryukin, A., Shah, M., & Kankanhalli, M. (2021). Adversarial learning for personalized tag recommendation. *IEEE Transactions on Multimedia, 23*, 1083–1094.

25. Rappa, M., & ACM Digital Library. (2010). In *Proceedings of the 19th international conference on World wide web.* ACM.

26. Bharadwaj, P. N., Shivashankaran, T., & N, S. D. (n.d.). *Prediction of hashtags for images.* www.ijert.org

27. Kumar, N., Baskaran, E., Konjengbam, A., & Singh, M. (2021). Hashtag recommendation for short social media texts using word-embeddings and external knowledge. *Knowledge and Information Systems, 63*(1), 175–198.

28. IEEE Computational Intelligence Society, International Neural Network Society, Institute of Electrical and Electronics Engineers, & IEEE World Congress on Computational Intelligence (2016 : Vancouver, B. C.). (n.d.). In *2016 International Joint Conference on Neural Networks (IJCNN), 24–29 July 2016, Vancouver, Canada.*

29. Smys, S., Chen, J. I. Z., & Shakya, S. (2020). Survey on neural network architectures with deep learning. *Journal of Soft Computing Paradigm (JSCP), 2*(03), 186–194.

30. Zhang, C., Yang, Z., He, X., & Deng, L. (2019). *Multimodal intelligence: Representation learning, information fusion, and applications.*

31. Tajbakhsh, M. S., & Bagherzadeh, J. (2016). Microblogging hashtag recommendation system based on semantic TF-IDF: Twitter use case. In *Proceedings—2016 4th International Conference on Future Internet of Things and Cloud Workshops, W-FiCloud 2016* (pp. 252–257). https://doi.org/10.1109/W-FiCloud.2016.59

32. Kim, J., & Shim, K. (2017). General chairs' preface. In *Lecture Notes in Computer Science (including subseries Lecture Notes in Artificial Intelligence and Lecture Notes in Bioinformatics): Vol. 10235 LNAI.* Springer.

33. Zhang, Y., Hu, J., Sano, S., Yamasaki, T., & Aizawa, K. (2017). A tag recommendation system for popularity boosting. In *MM 2017—Proceedings of the 2017 ACM Multimedia Conference* (pp. 1227–1228). https://doi.org/10.1145/3123266.3127913

34. Tonge, A., Caragea, C., & Squicciarini, A. (2018). Privacy-aware tag recommendation for image sharing. In *HT 2018—Proceedings of the 29th ACM Conference on Hypertext and Social Media* (pp. 52–56).

35. Zhang, J., Nie, L., Wang, X., He, X., Huang, X., & Chua, T. S. (2016). Shorter-is-better: Venue category estimation from micro-video. In *MM 2016—Proceedings of the 2016 ACM Multimedia Conference* (pp. 1415–1424).

36. Lops, P., de Gemmis, M., Semeraro, G., Musto, C., & Narducci, F. (2013). Content-based and collaborative techniques for tag recommendation: An empirical evaluation. *Journal of Intelligent Information Systems, 40*(1), 41–61.

37. *Click-Through Rate Prediction with Multi-Modal Hypergraphs.* (n.d.). https://www.tiktok.com/

38. Wang, S., Lo, D., Vasilescu, B., & Serebrenik, A. (2018). EnTagRec ++: An enhanced tag recommendation system for software information sites. *Empirical Software Engineering, 23*(2), 800–832.

39. Krizhevsky, A., Sutskever, I., & Hinton, G. E. (2017). ImageNet classification with deep convolutional neural networks. *Communications of the ACM, 60*(6), 84–90.

40. Yamasaki, T., Hu, J., Sano, S., & Aizawa, K. (2017). *FolkPopularityRank: Tag recommendation for enhancing social popularity using text tags in content sharing services.*

41. Tripathi, M. (2021). Analysis of convolutional neural network based image classification techniques. *Journal of Innovative Image Processing (JIIP), 3*(02), 100–117.

Smart Door Locking System Using IoT—A Security for Railway Engine Pilots

K. Sujatha, N. P. G. Bhavani, U. Jayalatsumi, T. Kavitha, B. Latha, A. Ganesan, and A. Kalaivani

Abstract Many IoT devices are connected together via gateways in smart door locking systems for railway locomotives. The importance of gateways in smart systems cannot be emphasized, yet due to their centralized structure, they are vulnerable to a variety of security concerns, including integrity, certification, and availability. In this article, a blockchain based smart door locking system for railway engine with gateway network to address these security vulnerabilities is discussed. This smart door integrated with the network, guards against the gateway attacks from misuse by terrorists to operate and collapse railway engine without proper authentication. Sensors and actuators are connected and communicated in a dispersed manner in the smart door lock setup for railway engines with limited resources. Because the information is sensitive, one of the issues in this area is data storage and transfer. The fingerprint images are processed using boundary detection algorithms and artificial neural networks (ANN) for feature extraction. With cloud computing technologies, the retrieved feature data may be evaluated and monitored in real time. Identity is one of the thread models for railway engines' IoT-based smart door locking systems.

K. Sujatha (✉)
Department of EEE, Dr. MGR Educational and Research Institute, Chennai, Tamil Nadu, India
e-mail: drksujatha23@gmail.com

N. P. G. Bhavani
Department of ECE, Saveetha School of Engineering, SIMATS, Chennai, Tamil Nadu, India

U. Jayalatsumi
Department of ECE, Dr. MGR Educational and Research Institute, Chennai, Tamil Nadu, India

T. Kavitha
Department of Civil Engineering, Dr. MGR Educational and Research Institute, Chennai, Tamil Nadu, India

B. Latha
Department of Physics, Dr. MGR Educational and Research Institute, Chennai, Tamil Nadu, India

A. Ganesan
Department of EEE, RRASE College of Engineering, Chennai, Tamil Nadu, India

A. Kalaivani
Department of CSE, Saveetha School of Engineering, SIMATS, Chennai, Tamil Nadu, India

© The Author(s), under exclusive license to Springer Nature Singapore Pte Ltd. 2023   263
S. Shakya et al. (eds.), *Sentiment Analysis and Deep Learning*,
Advances in Intelligent Systems and Computing 1432,
https://doi.org/10.1007/978-981-19-5443-6_20

Finally, this paper highlights block chain technology, which has the ability to solve IoT application security issues by utilizing fingerprint recognition for authentication. This system can also be extended for development as a smartphone app. The network using Ethereum blockchain technology is used and was tested for industry standards and security, in terms of response time and accuracy. The proposed system can offer a safe and secure locking of the smart door in railway engines to prevent accidents. Some possible solutions were also suggested to tackle the security and privacy problems in IoT based on block chain to illustrate how block chain helps to IoT.

Keywords Internet of Things · Blockchain · Smart door locking system · Security · Privacy · Gateway · Railway engine

1 Introduction

The Internet of Things (IoT) is one of the most promising technologies to emerge in the preceding decade. Smart devices can detect their environment and communicate with other smart devices on the network. In a talk about supply chains, the Internet of Things [1] was mentioned. The Internet of Things (IoT) is one of the key components of a smart home or city. Low-power embedded systems, being used in conjunction with the twenty-first century; the internet has become so ubiquitous that the number of devices connected to it has surpassed billions, making the world's population a minority. As billions of devices are connected, a vast amount of data and information is created and transferred. As a result, IoT data processing and storage has become a challenging topic. The Internet of Things has a lot of promise for building a variety of applications that will enhance people's lives on a daily basis. Traditional networks and applications become more real time as a result of the Internet of Things, enabling for the deployment of smart devices to be monitored in real time. Data analysis may be done with the aid of associated technology, which speeds up the functioning of the system. By performing also known as edge computing, speeds up processing and calculation [2].

The Internet of Things is used in a variety of applications, including healthcare and earthquake early detection. The creation of a smart IoT-based home system is our primary objective. Each of the eight components that make up block chain has its own set of standards. Block chain's goal is to create a ledger, which is an immutable and distributed historical record. To store, update, and maintain ledgers, a peer network is employed. Every node in the network replicates the ledger. This network's goal is to obtain an agreement on each update's content. This eliminates the need for a centrally duplicated ledger and guarantees that all copies of the ledger are identical. User permissions, authentication, and identity management are all handled by the Membership Services department. The term "smart house" refers to a private residence that transmits and receives data in real-time. It provides automated and intelligent services through various household products such as televisions, lights, and refrigerators. These gadgets are part of a home-based communication system that allows gadgets to communicate with one another and with the rest of the world without the need for human intervention. Users may monitor and manage themselves using a variety of home devices based on their preferences and network settings [3].

However, as a result of this approach, the smart home ecosystem has become overly reliant on gateways. Smart technology relies on centralized networks to connect a variety of devices, which poses a significant security risk. Smart TVs and refrigerators, which are integral parts of smart systems, have previously been hacked to send phishing and spam emails. One example is the hacking of newborn monitoring cameras in a Texas house in order to collect obscene sounds across networks, making them an excellent target for DDoS attacks. Due to cyberattacks on IoT server and gateway systems, security challenges such as data forgery and manipulation, access to illicit devices, and insufficient device management are developing as the IoT era continues. IoT applications make advantage of a variety of smart devices. The great majority of these smart devices are low-resource, low-processing, and low-memory devices. As previously stated, each layer architecture has its own set of security challenges. To make applications realistic enough for end users to security and privacy problems must be addressed [4].

2 Literature Review

Users can freely share a public electronic ledger that functions similarly to a relational database and produces an immutable record of their transactions, each of which is time-stamped and connected to the one before it. Each thread block is a digital deal or record that allows an unlimited or limited number of users to interact with the centralized database. When new data is introduced to the blockchain, it can never be changed or withdrawn, guaranteeing data integrity. Every transaction that has ever taken place in the system is logged on the blockchain and can be validated. Blockchain, from the perspective of a network, is a distributed file system in which members keep copies of files and agree on alterations through consensus. Each block contains a collection of transactions as well as important data including the previous block's time stamp and cryptographic signature (hash), the current block's hash, and other details. The preceding block's hash links the current block to the prior block, and subsequent blocks need the current block's hash, resulting in a chain of blocks. If anything in the block is altered, a hash can be computed and a value found that differs from the one supplied, causing the block to be refused. The authors of this study devised a three-tier architecture termed wireless sensor for safe and efficient data processing [5, 6].

There are two types of block chains: public and private. A public block chain is one that does not require any permission. Anyone may engage in a productive and useful way. They can join in by looking at the block chain or adding to it. This public chain has no one person in command of the network due to its decentralized nature. That is, data on the block chain cannot be changed once it has been validated. This public block chain is advantageous because it allows users to freely add and see data; the database is distributed rather than centralized; it is immutable to prevent data change; and it is safe thanks to the 51% rule [5] demonstrates that the great majority of the network's peers are within its own system. This means that the peer-to-peer network is not properly linked, which could result in relay issues for newly created blocks on the blockchain. The authors show in [6] that when a large number of nodes are controlled by an attacker with substantial computational power (or not), the total computational power in a blockchain with few scans reaches a proportion that is considered excessive in relation to the total computational power. Due to the attacker's ability to cause forks on purpose, the system's integrity may be threatened in this case.

In a selfish mining attack [7], a malevolent mining pool decides to preserve blocks it finds. As a result, a blockchain split is produced. It continues to mine on the private branch until both branches are the same length, at which time it publishes the information. As a result, it's possible that it'll become the longest branch, and other miners will follow suit. After a certain amount of time has passed, the public branch, along with all of the data it contains, may be erased. The miner's ratio between both branches may provide the damaging pool an edge. The authors of [8] described another assault known as the history-revision attack. Then, by combining Proof of Work's hard words, one may create a fork and destructive branch while avoiding the initial branch. Other miners may accept it, converting the history of the blockchain.

It consists of a number of devices that are all managed and monitored via a gateway. This type of network architecture might allow data hacking at home, cause privacy breaches, cause device malfunctions, and put people in risk. When a person is exposed to a smart home network, data collected by devices in a targeted manner may be disclosed. Due to a lack of security standards for smart devices and gadgets, it is difficult to connect many heterogeneous devices. As a result, providing numerous services to consumers is tough. Smart door locking gateway security is critical, and gateway security criteria are necessary for successful operation [7–9]. Smart home networks collect and store a wide range of data, including sensitive data given by residents. Only authorized personnel should have access to this data, which is a crucial part of smart system security. The encryption technology is used and customized with the help of the smart door locking system's essential characteristics. There must be no erroneous transmission when data is transferred and received across configurations. The hash function reduces the risk of tampering with this data and allows for tracking and verification of exactly what data is preserved. In a smart door locking system with network settings, authentication stops attackers from behaving maliciously within a traditional network from the outside. Blockchain is used to verify network members' validity, and it may be checked at any moment to guarantee appropriate smart network architecture [10, 11].

3 Methodology

The railway engine pilot will put his or her fingerprint into a fingerprint scanner, which is connected to the door lock through the Arduino board's microprocessor. The machine searches its database for a match after scanning the print. If a match is detected, the latch will open, unlocking the engine door. When the engine pilot wishes to lock the engine door, he or she does something similar. The latch closes when the fingerprint is correct, locking the engine door behind the pilot. If the erroneous fingerprint is entered, the system sounds a buzzer and shows "Try again" on the LCD display, as well as sending the information to the relevant engine pilot's mobile phone. If the system receives a series of incorrect fingerprints, i.e., if someone other than the concerned engine pilot attempts to open the door repeatedly, the system enters a protected mode, sending an alert and displaying "Panic Mode" on the LCD panel. Unknown people attempted to break into the engine door, according to a message sent to the pilot's cellphone and the railway control room.

Our fingerprint smart door lock system in railway engines is sophisticated, efficient, and secure when compared to a typical protected system. A typical security system consists of locks that open when they come into contact with the correct keys. An accepted and authentic fingerprint is the sole way to access the guarded lock mechanism in our system. The block diagram in Fig. 1 is shown. Locking mechanisms are critical in our daily lives. When it comes to protecting important and sensitive objects, as well as one's privacy, there is no replacement for lock and key. However, the type of technology employed may and will determine how secure something is. The fingerprint door lock system is a biometric lock that unlocks by using the fingerprint interface. Fingerprints are one-of-a-kind and cannot be duplicated, making them more secure and safe. There are a few key differences between various locking methods. Traditional lock and key systems, fingerprint lock systems, password/pin code systems, and biometric lock systems are only a handful of the available security methods. The benefits and drawbacks of each system make it efficient, secure, recognizable, and tough to break. Lower-level headers are not numbered and are styled as run-in headings.

4 Results and Implementation

Engine pilot identification based on classifier output and fusion score level will talk about simulation output for classifiers using MATLAB in this part. All three processes have MATLAB coding that includes information on the identification database and false accept error rates. One fingerprint picture was chosen as the verification image from the database in this identifying procedure. The supplied image is then compared to all of the entries in the database. If the submitted input matches the data base, the user is identified as a valid user who may start the engine, and the maximum matching score value is presented. If the user's input does not match the database, he or she is

Fig. 1 Block diagram for smart door locking systems in railway engine

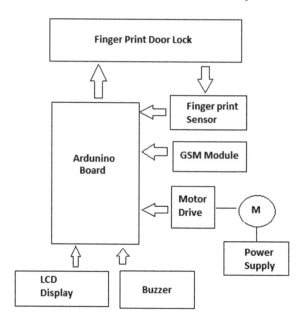

flagged as an unauthorized user, and the minimal matching score value is displayed along with the mistake %. Figures 2 and 3 illustrate the results of the fingerprint authentication simulation as well as the matching score value.

The completely developed system will resemble as shown in Fig. 3 with the following features as a part of advancement in smart door lock system in Railway

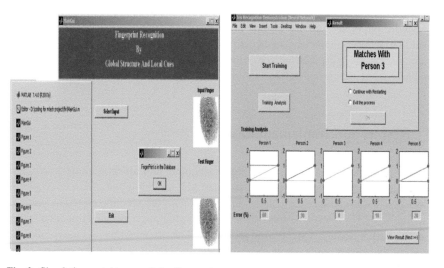

Fig. 2 Simulation matching result for fingerprint

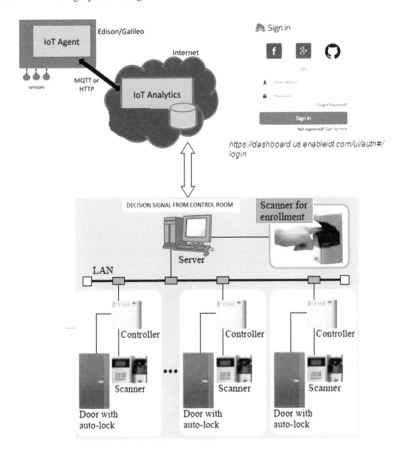

Fig. 3 Blockchain for smart door lock system in railway engines with IoT

Table 1 Performance evaluation

Method	Mean squared error	Pattern match score
Proposed	0.0021	0.9728
Existing	0.0192	0.7628

Engines. The proposed method is more efficient as compared with the existing method as illustrated in Table 1.

5 Conclusion

The fingerprint-based door lock technology may be customized and utilized in a number of different ways. This door locking mechanism is less costly than the current lock systems. Our fingerprint-based lock technology is very accurate and quick to identify fingerprints, allowing for seamless interaction with users and increased security. Despite the existence of several security and privacy standards, they are not applicable to owing limits and the usage of light weight devices to link to smart door locking systems. Then we came upon a few more assault models. This architecture provides the following methods to handle the secrecy, integrity, and authentication challenges of heterogeneous IoT and centralized gateways that make up a smart door locking system. Encryption is used to address secrecy and authentication concerns in smart applications, gateways, and heterogeneous IoT. In addition, block chain technology is utilized to ensure that data kept in gateway is secure. Finally, it is discovered that Block chain security and privacy may be achieved in an IoT-based smart door locking environment. The permission blockchain is used to solve the scalability problem.

Anonymity or privacy can be accomplished by storing the digital asset's fingerprint rather than the digital asset itself. Cross fault tolerance is a technique used for building trustworthy and secure distributed systems. As a consequence of the simulation findings, it is obvious that the suggested multimodal biometric identification is more safe and 97.2% efficient for pilot security automation in railway engines.

References

1. Geneiatakis, D., Kounelis, I., Neisse, R., Nai-Fovino, I., Steri, G., & Baldini, G. (2017). Security and privacy issues for IoT based smart home. In *2017 40th International Convention on Information and Communication Technology, Electronics and Microelectronics (MIPRO)* (pp. 1292–1297). IEEE.
2. Jose, & Malekian, R. (2017). Improving smart home security: Integrating logical sensing into smart home. *IEEE Sensors Journal, 17*(13), 4269–4286.
3. Lin, C., He, D., Kumar, N., Huang, X., Vijayakumar, P., & Choo, K. R. (2019). Home chain: blockchain-based secure mutual authentication system for smart homes. *IEEE Internet of Things Journal, 7*(2), 818–829.
4. Jain, A. K., Feng, J., & Nandakumar, K. (2010). Matching fingerprints. *IEEE Computer, 43*(2), 36–44.
5. Mary Lourde, R., & Khosla, D. (2010). Fingerprint identification in biometric security systems. *International Journal of Computer and Electrical Engineering, 2*(5).
6. Rathore, S., Kwon, B. W., Park, J. H. (2019). BlockSecIoTNet: Blockchain-based decentralized security architecture for IoT network. *Journal of Network and Computer Applications, 143*, 167–177.
7. Huang, X., Yu, R., Kang, J., Xia, Z., & Zhang, Y. (2018). Software defined networking for energy harvesting internet of things. *IEEE Internet of Things Journal, 5*(3), 1389–1399.
8. Jain, A. K., Ross, A., & Prabhakar, S. (2004). An introduction to biometric recognition. *IEEE Transactions on Circuits and Systems for Video Technology, Special Issue on Image and Video Based Biometrics, 14*(1).
9. Sharma, P. K., Rathore, S., Jeong, Y. S., & Park, J. H. (2018). Soft Edge Net: SDN based energy-efficient distributed network architecture for edge computing. *IEEE Communications Magazine, 56*(12), 104–111.
10. Suma, V. (2021). Wearable IoT based distributed framework for ubiquitous computing. *Journal of Ubiquitous Computing and Communication Technologies (UCCT), 3*(1), 23–32.

11. Pandian, M. D. (2019). Enhanced network performance and mobility management of IoT multi networks. *Journal of Trends in Computer Science and Smart Technology (TCSST)*, *1*(2), 95–105.

AdaSmooth: An Adaptive Learning Rate Method Based on Effective Ratio

Jun Lu

Abstract It is well known that we need to choose the hyper-parameters in Momentum, AdaGrad, AdaDelta, and other alternative stochastic optimizers. While in many cases, the hyper-parameters are tuned tediously based on experience becoming more of an art than science. We introduce a novel per-dimension learning rate method for stochastic gradient optimization called AdaSmooth. The method is insensitive to hyper-parameters thus it requires no manual tuning of the hyper-parameters like Momentum, AdaGrad, and AdaDelta methods. We show promising results compared to other methods on different convolutional neural networks, multi-layer perceptron, and alternative machine learning tasks. Empirical results demonstrate that AdaSmooth works well in practice and compares favourably to other stochastic optimization methods in neural networks.

Keywords Stochastic optimization · Adaptive learning rate · Stochastic gradient descent

1 Introduction

Over the years, stochastic gradient-based optimization models have become a standard method in a variety of domains of science and engineering including automatic speech recognition processing and computer vision [5, 6, 11]. Deep neural networks (DNNs) and stochastic gradient descent (SGD) are essential in training stochastic loss/objective functions. However, when a new deep neural network is created for a given (new) task, some hyper-parameters related to the network training must be selected heuristically. A new network is typically trained from scratch and evaluated over and over again for each possible and feasible combination of such hyper-parameters. While great progress has been made on both hardwares [e.g. graphical processing units (GPUs)] and softwares (e.g. cuDNN) to speed up the training time

J. Lu (✉)
Trexquant, New York, USA
e-mail: jun.lu.locky@gmail.com

© The Author(s), under exclusive license to Springer Nature Singapore Pte Ltd. 2023   273
S. Shakya et al. (eds.), *Sentiment Analysis and Deep Learning*,
Advances in Intelligent Systems and Computing 1432,
https://doi.org/10.1007/978-981-19-5443-6_21

of a single structure of a DNN, the exploration of a vast range of potential structures or combinations remains very slow necessitating the use of a stochastic optimizer that is hyper-parameters insensitive.

1.1 Gradient Descent

Gradient descent (GD) is one of the most widely used algorithms to optimize objective functions and it is by far the most often used method to perform machine learning optimization tasks. And this is especially true when it comes to neural network optimization models. In general, the neural networks or machine learning models find the set of parameters $x \in \mathbb{R}^d$ (a.k.a., weights) in order to optimize an objective/loss function $L(x)$. The gradient descent then finds a sequence of parameters

$$x_1, x_2, \ldots, x_T, \tag{1}$$

such that the objective function $L(x_T)$ achieves the optimal minimum value when $T \to \infty$. At each iteration t, a step Δx_t is applied to modify the parameters. Suppose the parameters at the tth iteration are denoted as x_t, the update rule then becomes

$$x_{t+1} = x_t + \Delta x_t. \tag{2}$$

The simplest method of stochastic gradient descent is the vanilla update: the parameter moves in the opposite direction of the gradient which finds the steepest descent direction since the gradients are orthogonal to level surfaces (a.k.a., level curves, see Lemma 16.4 in Lu [16]):

$$\Delta x_t = -\eta g_t = -\eta \frac{\partial L(x_t)}{\partial x_t} = -\eta \nabla L(x_t), \tag{3}$$

where the positive value η is known as the learning rate whose value depends on specific problems, and $g_t = \frac{\partial L(x^t)}{\partial x^t} \in \mathbb{R}^d$ represents the gradient of the parameters over the loss function $L(\cdot)$. The learning rate η determines how large of a step to take in the direction of the negative gradient in order to reach a (local) minimum. While the local estimate of the direction can be produced by iteratively following the negative gradient of a single sample or a batch of samples, which is known as the stochastic gradient descent (SGD) [22]. The objective function in the stochastic gradient descent framework is stochastic and is composed of a sum of subfunctions that are evaluated at different subsamples of the data.

Gradient descent produces a monotonic improvement for a modest step size at every iteration. As a result, it always converges, albeit to a local minimum. The vanilla GD approach, on the other hand, is typically sluggish, as it can take an exponential amount of time when the curvature condition is poor. While choosing higher than this rate may cause the procedure to diverge in terms of the objective function. For many

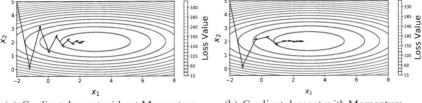

(a) Gradient descent without Momentum.   (b) Gradient descent with Momentum.

Fig. 1 Two-dimensional convex function $L(x) = 2(x_1 - 3)^2 + 20(x_2 - 2)^2 + 5$ and $\frac{\partial L(x)}{\partial x} = (4x_1 - 12, 8x_2 - 16)^\top$. Starting point to descent is $(-2, 5)^\top$. **Without momentum**: A higher learning rate may result in larger parameter updates in dimension across the valley (direction of x_2) which could lead to oscillations back as forth across the valley. **With momentum**: Though the gradients along the valley (i.e. in the direction of x_1) are much smaller than the gradients across the valley (i.e. in the direction of x_2), they are usually going in the same direction, and thus the momentum term builds up to speed up movement, reduce oscillations, and propel us to barrel through narrow valleys, small humps, and (local) minima

problems, determining an appropriate learning rate (either global or per-dimension) becomes more of an art than a science problem. Though the previous work has been done to alleviate the need for selecting a good global learning rate (e.g. [4, 32]), they are still sensitive to other hyper-parameters.

The main contribution of this paper is to propose a novel stochastic optimization algorithm which is insensitive to different choices of hyper-parameters resulting in adaptive learning rates, and naturally performs a type of annealing called the step size annealing. We propose the AdaSmooth algorithm to both increase optimization efficiency and out-of-sample accuracy. While previous works propose somewhat algorithms that are insensitive to the global learning rate (e.g. [32]), the methods are still sensitive to hyper-parameters that influence per-dimension learning rates. The proposed AdaSmooth approach, a method for efficient stochastic optimization that only has the requirement on the calculation of first-order gradients and (accumulated) past update steps with little memory demand, allows flexible and adaptive per-dimension learning rates. Meanwhile, the method is memory efficient and easy to implement. Our method is designed to combine the advantages of the following methods: AdaGrad, which performs well with sparse gradients [4], RMSProp, which is suitable for both non-stationary and online settings [7], and AdaDelta, which is less sensitive in global learning rate [32].

2 Related Work

There are several variants of gradient descent methods to use heuristics to estimate an appropriate learning rate at each iteration of the progress. These methods either attempt to slow down the learning near a local minima or to accelerate learning when suitable [10, 15, 23, 32].

2.1 Momentum

Learning can be quite slow if the loss surface is not spherical since the learning rate must be kept low to avoid divergence along the steep curvature directions [21, 24, 30]. When suitable, the version of SGD with Momentum (which may be applied to mini-batch or full batch learning) tries to exploit the previous steps to speed up learning such that it has superior rate of convergence on deep neural networks. The primary idea behind the Momentum method is to accelerate the learning along dimensions where the gradients consistently point in the same directions; and to slow down the rate along dimensions where the gradients' signs are changing consistently. Figure 1a demonstrates a series of vanilla GD updates where we may observe that the update along dimension x_1 is consistent; and the zigzag pattern of movement along dimension x_2. The method of GD/SGD with Momentum uses an exponential decay to keep track of past parameter adjustments. Formally, the update method has the following step:

$$\Delta x_t = \rho \Delta x_{t-1} - \eta \frac{\partial L(x_t)}{\partial x_t}, \tag{4}$$

in which case the algorithm remembers the most recent update Δx_{t-1} and multiplies it by a parameter ρ (called the *momentum parameter*) to add it to the current update $-\eta \frac{\partial L(x_t)}{\partial x_t}$. In other words, the amount we alter the parameter is proportional to the negative gradient plus the previous weight change; the extra *momentum factor* acts as both an accelerator and a smoother. The momentum parameter ρ acts as a *decay constant* such that Δx_1 may have an influence on Δx_{100}; though its effect is decayed by this decay constant. The momentum parameter ρ is usually set at 0.9 by default in practice. The added momentum factor simulates the concept of inertia in physics. It means that, in each iteration, the update mechanism is not only related to the gradient descent, which refers to the dynamic term, but also maintains a component that is related to the direction of last update iteration, which refers to the momentum.

The momentum works extremely better in ravine-shaped loss curves. A ravine is an area where the surface curves in one dimension are substantially steeper than in the other (see the loss curve in Fig. 1, i.e. a long narrow valley). In deep neural networks, ravines are quite common near local minima; however, vanilla SGD/GD has trouble navigating them. Vanilla SGD/GD will tend to oscillate over the narrow ravine directions, as shown by the toy example in Fig. 1a, since the negative gradient will point down one of the steep sides rather than along the ravine towards the optimum. Momentum, on the other hand, aids in the acceleration of update towards the correct direction (see the updates in Fig. 1b by GD with Momentum).

2.2 AdaGrad

The learning rate annealing process alters a single global learning rate that affects all dimensions of the parameter [15, 28]. Duchi et al. [4] propose a method called AdaGrad where the learning rate is adjusted on a per-dimension basis. The learning rate for each parameter is determined by the history of gradient updates of that parameter in a way such that parameters with a scarce history of updates are updated faster via a larger learning rate. To put it another way, parameters that have not been updated much in the past are more likely to have higher learning rates in the current iteration. Let $\alpha \odot \beta$ denote the element-wise vector multiplication between vectors α and β, then the AdaGrad has the following update step formally:

$$\Delta x_t = -\frac{\eta}{\sqrt{\sum_{\tau=1}^{t} g_\tau^2 + \epsilon}} \odot g_t, \tag{5}$$

where η is the global learning rate shared by all dimensions, ϵ is a smoothing term to better condition the division, g_τ^2 indicates the element-wise square $g_\tau \odot g_\tau$, and the denominator computes the l_2 norm of sum of all previous squared gradients in a per-dimension fashion. Though the global learning rate η is shared by all dimensions, The l_2 norm of accumulated gradient magnitudes controls the adaptive learning rate of each dimension. Since this dynamic learning rate develops with the inverse of the accumulated gradient magnitudes, smaller gradient magnitude has larger learning rates, and larger absolute value of gradients has smaller learning rates in turn. Therefore, the accumulated gradient in the denominator has the same effects as the learning rate annealing [15].

On the one hand, the AdaGrad method partly eliminates the need to tune the learning rate manually since this is controlled by the accumulated gradient in the denominator. The continual accumulation of the squared gradients in the denominator, on the other hand, is AdaGrad's fundamental flaw since every added term is positive such that the accumulated sum keeps growing or even exploding to infinity during the training steps. As a result, the per-dimension learning rates shrink and subsequently decrease during training, becoming infinitesimally small before the loss is decently small and thus terminating training. Furthermore, because the magnitudes of gradients are factored out in AdaGrad, this method can be sensitive to parameter and gradient initializations. The per-dimension learning rates will be low for the remainder of training if the initial magnitudes of the gradients are high or even infinitesimally massive though this can be partially mitigated by increasing the global learning rate making the AdaGrad technique more sensitive to learning rate selection. Moreover, since AdaGrad considers that the parameter with fewer updates should use a higher learning rate; while the one with more movement should use a lower learning rate making it solely consider information from squared gradients, or the gradients' absolute value. As a result, AdaGrad does not incorporate information for the entire migration (i.e. the sum of updates; not the sum of absolute updates).

To be more succinct, AdaGrad has the following main drawbacks: (1) *Global learning rate*: the need for a manually selected global learning rate; (2) *Learning rate decay*: the continual decay of per-dimensional learning rates throughout training; (3) *Not enough information*: considering only the magnitude of gradients.

2.3   AdaDelta

Before going to our proposed method, we here discuss the details of the AdaDelta method that sheds lights on our algorithm [7, 32]. AdaDelta is a straight extension of AdaGrad that addresses the main weaknesses of AdaGrad. The original idea of AdaDelta is straightforward: it limits the window of cumulative prior gradients from t steps (i.e. current time step) to some fixed size $w < t$. However, the AdaDelta presented in Zeiler [32] implements this accumulation as an exponentially decaying average of the squared gradients since storing all w previous squared gradients is memory inefficient. This is akin to the concept of momentum term.

AdaDelta: Form 1 (RMSProp) First, we will go through the exact form of the original AdaGrad (we here call it AdaGradWin for clarity). Suppose the running average is $E[g^2]_t$ at step t, then we obtain the decaying average by:

$$E[g^2]_t = \rho E[g^2]_{t-1} + (1 - \rho)g_t^2, \tag{6}$$

where ρ is a decay constant similar to that used in the momentum method and again g_t^2 represents the element-wise square $g_t \odot g_t$. Since the denominator of Eq. (5) is just the root mean squared (RMS) error criterion of the gradients, we can replace it with the RMS criterion short-hand. Denote $\text{RMS}[g]_t = \sqrt{E[g^2]_t + \epsilon}$, where again a constant ϵ is added to better condition the denominator. Then, the final update step can be calculated:

$$\Delta x_t = -\frac{\eta}{\text{RMS}[g]_t} \odot g_t, \tag{7}$$

where again $\odot$ is the element-wise vector multiplication.

As previously stated, the form in Eq. (6) is derived from the exponential moving average (EMA). $1 - \rho$ is also known as the smoothing constant (SC) in the original form of EMA where the SC can be approximately represented by $\frac{2}{N+1}$ and the period N can be thought of as the number of past steps to do the moving average calculation [14]:

$$SC = 1 - \rho = \frac{2}{N+1}. \tag{8}$$

The decay constant ρ, the smoothing constant (SC), and the period number N are all linked in the above Eq. (8) constant. It can be shown that if $\rho = 0.9$, then $N = 19$. That is, roughly speaking, $E[g^2]_t$ at step t is approximately equal to the (simple) moving average of 19 previous squared gradients and the present one (i.e. moving average of 20 squared gradients totally). Though this relationship is not discussed in Zeiler [32], it is important to decide the lower bound of the decay constant ρ. Typically, a time period of $N = 3$ or 7 is thought to be a relatively small frame making the lower bound of decay constant $\rho = 0.5$ or 0.75; when $N \to \infty$, the decay constant ρ approaches 1.

However, we can find that the AdaGradWin still only considers the absolute value of gradients and a fixed number of past squared gradients is not flexible which can cause a small learning rate near (local) minima as we will discuss in the sequel.

RMSProp The AdaGradWin is actually the same as the RMSProp method developed independently by Geoff Hinton in [7] both of which are motivated by the need to address AdaGrad's radically diminishing per-dimension learning rates. In Hinton et al. [7], the authors suggest $\rho = 0.9$ as the default value.

AdaDelta: Form 2 The author also shows that the units of the step size shown above do not match (so as the vanilla SGD, the SGD with Momentum, and the AdaGrad methods) in Zeiler [32]. To overcome this difficulty, the author considers rearranging Hessian to identify the quantities involved from the correctness of Newton's method. Though calculating Hessian or approximating the Hessian matrix is a computation-expensive task, its curvature information is useful for optimization since the units in Newton's method are well-matched. The update step in Newton's method can be described as follows if given the Hessian matrix $\boldsymbol{H}$ [1, 2]:

$$\Delta \boldsymbol{x}_t \propto -\boldsymbol{H}^{-1} \boldsymbol{g}_t \propto \frac{\frac{\partial L(x_t)}{\partial x_t}}{\frac{\partial^2 L(x_t)}{\partial x^2}} \quad \longrightarrow \quad \frac{1}{\frac{\partial^2 L(x_t)}{\partial x_t^2}} = \frac{\Delta \boldsymbol{x}_t}{\frac{\partial L(x_t)}{\partial x_t}}, \tag{9}$$

i.e. the right-hand side term of the above equation can approximate the units of the Hessian matrix. Because the RMSProp update in Eq. (7) already includes RMS$[\boldsymbol{g}]_t$ in the denominator, i.e. the units of the gradients. Putting another unit of the order of $\Delta \boldsymbol{x}_t$ in the numerator can match the same order as Newton's method. To do this, define another exponentially decaying average of the update steps:

$$\mathrm{RMS}[\Delta \boldsymbol{x}]_t = \sqrt{E[\Delta \boldsymbol{x}^2]_t} = \sqrt{\rho E[\Delta \boldsymbol{x}^2]_{t-1} + (1-\rho)\Delta \boldsymbol{x}_t^2}. \tag{10}$$

Since $\Delta \boldsymbol{x}_t$ for the current iteration is not known, RMS$[\Delta \boldsymbol{x}]_{t-1}$ can be used as a surrogate for RMS$[\Delta \boldsymbol{x}]_t$ since the curvature can be assumed to be locally smoothed. Therefore, we can replace the computationally expensive $\boldsymbol{H}^{-1}$ with an estimate of $1/\frac{\partial^2 L(x_t)}{\partial x_t^2}$. This is an approximation to the diagonal Hessian using simply RMS measures of $\boldsymbol{g}$ and $\Delta \boldsymbol{x}$, and it yields an update step with matching units:

$$\frac{\Delta \boldsymbol{x}_t}{\frac{\partial L(x_t)}{\partial x_t}} \sim \frac{\mathrm{RMS}[\Delta \boldsymbol{x}]_{t-1}}{\mathrm{RMS}[\boldsymbol{g}]_t} \quad \longrightarrow \quad \Delta \boldsymbol{x}_t = -\frac{\mathrm{RMS}[\Delta \boldsymbol{x}]_{t-1}}{\mathrm{RMS}[\boldsymbol{g}]_t} \odot \boldsymbol{g}_t. \tag{11}$$

The concept of AdaDelta from the second-order method eliminates the need for the annoying choosing of learning rate. Similarly, in Kingma and Ba [10], an exponentially decaying average is incorporated into the gradient information such that convergence in online convex setting is improved. Meanwhile, the diminishing problem was also attacked in second-order methods [26]. The idea of averaging gradient or its alternative is not unique, previously presented in Moulines and Bach [19], Polyak-Ruppert averaging has been shown to improve the convergence of vanilla GD/SGD methods [20, 25]. Other stochastic optimization methods, including the natural Newton method, AdaMax, Nadam, all set the step size per-dimension by estimating curvature from first-order information [3, 10, 12]. The LAMB adopts layerwise normalization due to layerwise adaptivity [31]; powerpropagation applies weight-parameterization for neural networks that leads to inherently sparse models [27]. And we shall not go into the details.

3   AdaSmooth Method

We now discuss the form of *effective ratio (ER, a.k.a., efficiency ratio)* based on past updates in the stochastic optimization steps and how it can be applied to accomplish adaptive learning rates per-dimension via the flexible smoothing constant, hence the name AdaSmooth. The idea presented in the paper is derived from the concept of AdaDelta algorithms [32] in order to improve two main drawbacks of the method: (1) *Not enough information*: consider only the magnitude of the gradients rather than the total movement in each dimension; (2) *Sensitive to hyper-parameters*: the need of manual selecting for hyper-parameters.

3.1   Effective Ratio (ER)

Kauffman suggested in [8, 9] to replace the smoothing constant in the EMA formula with a constant based on the *effective ratio*. Previously, the ER has been found to produce promising performance for financial forecasting with classic quantitative strategies [14] where the effective ratio of the portfolio closing price is calculated to decide the upward/downward trends of the asset. This indicator, defined within a range from -1.0 to $+1.0$ where the larger magnitude indicates a larger upward or downward trend, is originally designed to assess the *strength of a trend*; Lu and Yi [17] recently demonstrate that the effective ratio can also be used to reduce overestimation/underestimation in time series prediction. Given a series $\{h_1, h_2, \ldots, h_T\}$ and the window size M, the ER is calculated using the following formula:

$$e_t = \frac{s_t}{n_t} = \frac{h_t - h_{t-M}}{\sum_{i=0}^{M-1} |h_{t-i} - h_{t-1-i}|} = \frac{\text{Total move for a period}}{\text{Sum of absolute move for each bar}},$$

(12)

where e_t denotes the ER of the sequence at time t. At a strong trend (i.e. the input sequence is moving either up or down consistently, in a certain direction), the ER will gravitate 1 in absolute value; if there is no directed movement, the value will be a bit more than 0.

However, we want to calculate the ER of the moving direction in the update methods for each parameter instead of calculating the ER of assets' closing price of asset. And in the gradient descent methods, we are primarily concerned with how much each parameter deviates from its initial value in each period, whether positively or negatively. So here, we only consider the magnitude of the ER making it ranges from 0 to 1. To be more specific, the ER for the parameters is obtained as follows:

$$e_t = \frac{s_t}{n_t} = \frac{|x_t - x_{t-M}|}{\sum_{i=0}^{M-1} |x_{t-i} - x_{t-1-i}|} = \frac{|\sum_{i=0}^{M-1} \Delta x_{t-1-i}|}{\sum_{i=0}^{M-1} |\Delta x_{t-1-i}|}, \tag{13}$$

where $e_t \in \mathbb{R}^d$ whose ith element $e_{t,i}$ is in the range of [0, 1] for all $i \in [1, 2, \ldots, d]$. A larger value of $e_{t,i}$ indicates the descent method in the i-th dimension is moving in a certain direction; while a smaller value approaching 0 means the parameter in the i-th dimension is moving in a pattern that is interleaved by positive and negative movement, i.e. in a zigzag pattern. The M is chosen as the batch index for each epoch in practice and in all of our experiments. That is, $M = 1$ if the training procedure is in the first batch of each epoch; and $M = M_{\max}$ (where $M_{\max}$ is the maximal number of batches per epoch) if the training is in the last batch of the epoch. As a result, the value of $e_{t,i}$ reflects how the i-th parameter has moved in the most recent epoch. Alternatively, and more aggressively, the window size can range from 1 to the total number of batches (for all epochs) seen during the whole training progress. The advantage of using the adaptive window size M rather than a fixed one is that we do not need to keep the past $M + 1$ steps $\{x_{t-M}, x_{t-M+1}, \ldots, x_t\}$ in memory to calculate the signal and noise vectors $\{s_t, n_t\}$ in Eq. (13) due to the accumulated fashion.

3.2 AdaSmooth

If the absolute effective ratio of each parameter is small (i.e. closing to 0), the movement and update in this dimension are in a zigzag pattern. The proposed AdaSmooth method tends to use a long period average as the scaling constant to slow down the pace in that dimension. While the absolute ER per-dimension is large (tend to 1), the movement in that dimension is moving in a certain direction (not in a zigzag way), and the learning actually is happening and the descent is moving in a correct direction where the learning rate should be assigned to a relatively large value for that dimension. Thus, the AdaSmooth tends to choose a small period which yields a small compensation in the denominator; this is because the gradients in the closer periods are smaller than ancient ones when it is near the (local) minima. To see this, a toy example is shown in Fig. 2, where the gradient descent method is moving in a certain direction, and the gradient in the near periods is small; if we choose a larger period to compensate for the denominator, the descent will be slower due to the large factored denominator. In short, we want a smaller period to calculate the exponential

Fig. 2 Toy example of how the effective ratio works for AdaSmooth. The steps tend to update a large step when it is far from the (local) minima; and a relatively small update when it is close to the (local) minima

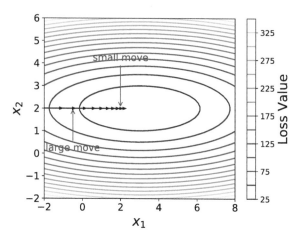

average of the squared gradients in Eq. (6) if the update is moving in a certain direction. However, the period for the EMA should be adapted larger accordingly when the parameter is updated in a zigzag fashion.

The obtained value of ER is used in the EMA formula. What we want to do now is to set the time period N discussed in Eq. (8) to be a smaller value when the effective ratio is closer to 1 in absolute value; or a larger value when the effective ratio moves towards 0. When N is small, SC is known as a "*fast SC*"; otherwise, SC is known as a "*slow SC*".

For example, let the small time period be $N_1 = 3$, and the large time period be $N_2 = 199$. The smoothing ratio for the fast movement must be as for EMA with period N_1 ("fast SC" $= \frac{2}{N_1+1} = 0.5$), and for the period of no trend EMA period must be equal to N_2 ("slow SC" $= \frac{2}{N_2+1} = 0.01$). Thus, the new changing smoothing constant is introduced, we call it the "*scaled smoothing constant*" (SSC), and denote it by a vector $c_t \in \mathbb{R}^d$:

$$c_t = (\text{fast SC} - \text{slow SC}) \times e_t + \text{slow SC}. \tag{14}$$

By Eq. (8), from the small and large time period, we can define the *fast decay constant* $\rho_1 = 1 - \frac{2}{N_1+1}$, and the *slow decay constant* $\rho_2 = 1 - \frac{2}{N_2+1}$. Then, the SSC vector can be obtained by:

$$c_t = (\rho_2 - \rho_1) \times e_t + (1 - \rho_2), \tag{15}$$

where the smaller the value of e_t, the smaller that of c_t. For the obtained smoothing constant to have a more effective influence on the averaging period, we follow Kaufman [8, 9] to square it. The end formula is then obtained by:

$$E[g^2]_t = c_t^2 \odot g_t^2 + \left(1 - c_t^2\right) \odot E[g^2]_{t-1}. \tag{16}$$

or after rearrangement:

$$E[g^2]_t = E[g^2]_{t-1} + c_t^2 \odot (g_t^2 - E[g^2]_{t-1}). \tag{17}$$

We notice that $N_1 = 3$ is a small period to calculate the average (i.e. $\rho_1 = 1 - \frac{2}{N_1+1} = 0.5$) such that the EMA sequence will be noisy if N_1 is less than 3. Therefore, the minimal value of ρ_1 in practice is set to be larger than 0.5 by default. While $N_2 = 199$ is a large period to compute the average (i.e. $\rho_2 = 1 - \frac{2}{N_2+1} = 0.99$) such that the EMA sequence almost depends only on the previous value leading to the default value of ρ_2 no larger than 0.99. Empirical study shows that the AdaSmooth update is not sensitive to these hyper-parameter choices. We also carefully notice that when $\rho_1 = \rho_2$, the proposed AdaSmooth method reduces to the RMSProp algorithm with decay constant $\rho = 1 - (1 - \rho_2)^2$ since we have squared it in Eq. (16). After developing the AdaSmooth method, we realize the main idea behind the method is similar to that of SGD with Momentum, both of them try to speed up (i.e. compensate less in the denominator) the learning along dimensions where the gradient consistently points in the same direction; and consider to slow the movement (i.e. compensate more in the denominator) along dimensions where the signs of the gradients continue to change.

Empirical evidence shows the ER used in simple moving average with a fixed windows size w can also reflect the trend of the series/movement in quantitative strategies [14]. However, this again needs to store w previous squared gradients in the AdaSmooth case, making it inefficient and we shall not adopt this extension.

3.3 AdaSmoothDelta

Notice the effective ratio can also be applied in the same setting of the AdaDelta method:

$$\Delta x_t = -\frac{\sqrt{E[\Delta x^2]_t}}{\sqrt{E[g^2]_t + \epsilon}} \odot g_t, \tag{18}$$

where

$$E[g^2]_t = c_t^2 \odot g_t^2 + \left(1 - c_t^2\right) \odot E[g^2]_{t-1}, \tag{19}$$

and

$$E[\Delta x^2]_t = (1 - c_t^2) \odot \Delta x_t^2 + c_t^2 \odot E[\Delta x^2]_{t-1}. \tag{20}$$

The different lies in that the $E[\Delta x^2]_t$ chooses a larger period when the effective ratio is small. This is particularly true in the sense that $E[\Delta x^2]_t$ is in the numerator while $E[g^2]_t$ appears in the denominator of Eq. (18). Thus, their compensations should go towards different directions. Or even, we can apply a fixed decay constant for $E[\Delta x^2]_t$ alternatively as follows:

$$E[\Delta x^2]_t = (1 - \rho_2)\Delta x_t^2 + \rho_2 E[\Delta x^2]_{t-1}, \tag{21}$$

The AdaSmoothDelta algorithm introduced above can further help to eliminate the need for a manually specified global learning rate, which is set to $\eta = 1$ from the Hessian context. However, due to the adaptive smoothing constants in Eqs. (19) and (20), the $E[g^2]_t$ and $E[\Delta x^2]_t$ are less locally smooth making so that it is less insensitive to the global learning rate than the AdaDelta method. As a result, AdaSmoothDelta prefers a smaller global learning rate, e.g. $\eta = 0.5$. The complete procedure for computing AdaSmooth is then formulated in Algorithm 1.

Algorithm 1 Computing AdaSmooth at iteration t for the proposed AdaSmooth algorithm. All operations on vectors are performed element-wise. Good default settings for the tested tasks are $\rho_1 = 0.5$, $\rho_2 = 0.99$, $\epsilon = 1e - 6$, $\eta = 0.001$; see Sect. 3.2 or Eq. (8) for a detailed discussion on the explanation of the decay constants' default values. Empirical study in Sect. 4 shows that the AdaSmooth algorithm is not sensitive to the hypermarater ρ_2, while $\rho_1 = 0.5$ is relatively a lower bound in this setting. The AdaSmoothDelta iteration can be calculated in a similar way

---

1: **Input**: Select initial parameter x_1 to descent, constant ϵ for better condition;
2: **Input**: Select global learning rate η, by default we set $\eta = 0.001$;
3: **Input**: Select parameters to calculate ER: fast decay constant ρ_1, slow decay constant ρ_2;
4: **Input**: Assert $\rho_2 > \rho_1$, by default $\rho_1 = 0.5$, $\rho_2 = 0.99$;
5: **for** $t = 1 : T$ **do**
6:     Compute the gradient vector $g_t = \nabla L(x_t)$;
7:     Compute effective ratio (ER) $e_t = \frac{|x_t - x_{t-M}|}{\sum_{i=0}^{M-1} |\Delta x_{t-1-i}|}$;
8:     Compute smoothing $c_t = (\rho_2 - \rho_1) \times e_t + (1 - \rho_2)$;
9:     Compute normalization term: $E[g^2]_t = c_t^2 \odot g_t^2 + (1 - c_t^2) \odot E[g^2]_{t-1}$;
10:    Compute step to update $\Delta x_t = -\frac{\eta}{\sqrt{E[g^2]_t + \epsilon}} \odot g_t$;
11:    Apply the update $x_t = x_{t-1} + \Delta x_t$;
12: **end for**
13: **Return**: Output the resulting parameters x_t, and the loss $L(x_t)$.

---

4 Experiments

To evaluate the strategy and demonstrate the main advantages of the proposed AdaSmooth method, we conduct experiments with different machine learning tasks on different data sets including Census Income data set[1] and MNIST data set (a real handwritten digit classification task) [13].[2] In all experiments, the same procedure for parameter initialization is adopted when training with different stochastic opti-

---

[1] The Census Income data set has 48842 number of samples totally where 70% of them are used as training set in our case: https://archive.ics.uci.edu/ml/datasets/Census+Income.

[2] The MNIST data set has a training set of 60,000 examples, and a test set of 10,000 examples.

mization models. We compare the results with respect to the generalization and the convergence speed. In a wide range of scenarios across various models, AdaSmooth improves optimization rates, and leads to out-of-sample performances that are as good or better than existing stochastic optimization algorithms in terms of loss and accuracy.

4.1 Experiment: Multi-layer Perceptron

Multi-layer neural networks (a.k.a., multi-layer perceptrons or MLP) are effective tools for optimizing machine learning tasks that detect intrinsic linear and non-linear features behind the model's inputs and outputs. In our experiment, we employ the simplest MLP structure: an input layer, a hidden layer, and an output layer. We notice that rectified linear unit (Relu) outperforms hyperbolic tangent unit (TanH), Sigmoid, and other nonlinear function in practice. Thus, we use Relu function as the default nonlinear unit in our structures. Since the dropout technique has become a standard tool for neural network training [29], by default, we adopt a 50% dropout noise to the network architecture during training to prevent overfitting. To be more specific, suppose the detailed architecture for each fully connected layer is described by $F(\langle num\ outputs \rangle : \langle activation\ function \rangle)$; and for a dropout layer is described by $DP(\langle rate \rangle)$. Then, the network structure we use can be described as follows:

$$F(128:\text{Relu}) \cdot DP(0.5) \cdot F \text{ (num of classes:Softmax)}. \qquad (22)$$

All the optimization methods are trained with mini-batches of 64 images per batch for 60 (on MNIST) or 200 (on Census Income) epochs through the training set, respectively. We set the hyper-parameter of epsilon $\epsilon = 10^{-6}$ to better condition. The global learning rates for all models are set to $\eta = 0.001$ in all experiments if not especially mentioned. While a relatively large global learning rate (with $\eta = 0.01$) is used for the AdaGrad method due to its continual accumulation of the l_2 norm in the denominator; the global learning rate for the AdaDelta algorithm is selected to be 1 as suggested by the original paper in Zeiler [32] and for the proposed AdaSmoothDelta algorithm is set to 0.5 as mentioned in Sect. 3.3 above while we will show that the learning rate for AdaSmoothDelta does not alter the end result significantly in the sequel. Figure 3a, b shows the comparison among the results of SGD with Momentum, RMSProp, AdaGrad, AdaDelta, and the proposed AdaSmooth/AdaSmoothDelta methods in optimizing the training set losses for the Census Income and MNIST data sets, respectively. The SGD with Momentum optimizer performs the worst in this experiment. AdaSmooth works slightly better than AdaGrad and RMSProp in the MNIST case and performs much better than the letters in the Census Income data set. The proposed AdaSmooth optimizer shows fast rate of convergence from the initial epochs while continuing to decrease the training losses in both the two data sets. We here show two sets of slow decay constant for AdaSmooth, i.e. "$\rho_2 = 0.9$" and "$\rho_2 = 0.95$". Because we have squared the SSC in

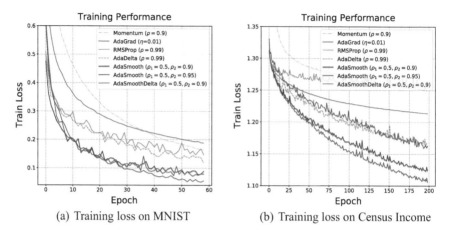

(a) Training loss on MNIST   (b) Training loss on Census Income

Fig. 3 **MLP**: demonstration of various optimization methods for 60 epochs on the MNIST data set; and for 200 epochs on the Census Income data set with MLP structure

Table 1 **MLP**: best in-sample evaluation in training accuracy (%)

Method	MNIST (%)	Census (%)
Momentum ($\rho = 0.9$)	98.64	85.65
AdaGrad ($\eta = 0.01$)	98.55	86.02
RMSProp ($\rho = 0.99$)	99.15	85.90
AdaDelta ($\rho = 0.99$)	99.15	86.89
AdaSmooth (0.5, 0.9)	**99.34**	**86.94**
AdaSmooth (0.5, 0.95)	**99.45**	**87.10**
AdaSmDel. (0.5, 0.9)	**99.60**	86.86

AdaSmDel is short for AdaSmoothDelta. The parameters in AdaSmooth and AdaSmDel are ρ_1 and ρ_2, respectively [see Eq. (15)]

Eq. (16), the AdaSmooth reduces to RMSProp with $\rho = 0.99$ when $\rho_1 = \rho_2 = 0.9$ (so as the AdaDelta and AdaSmoothDelta case). The AdaSmooth results perform better in all scenarios; however, there is no essential difference between the different AdaSmooth outcomes for various hyper-parameter settings in the MLP model. The best training set accuracy is given in Table 1 for different algorithms, indicating the superiority of the proposed AdaSmooth algorithm. While we notice the best test set accuracy for different algorithms are fairly close in this experiment; we then report only the best ones in Table 2 for the first 5 epochs. The proposed AdaSmooth approach converges slightly faster than other optimization methods when it comes to the test accuracy in all scenarios.

Table 2 **MLP**: best held-out evaluation in test accuracy with the first 5 epochs

Method	MNIST (%)	Census (%)
Momentum ($\rho = 0.9$)	94.38	83.13
AdaGrad ($\eta = 0.01$)	96.21	84.40
RMSProp ($\rho = 0.99$)	97.14	84.43
AdaDelta ($\rho = 0.99$)	97.06	84.41
AdaSmooth (0.5, 0.9)	97.26	84.46
AdaSmooth (0.5, 0.95)	97.34	84.48
AdaSmDel. (0.5, 0.9)	97.24	**84.51**

AdaSmDel is short for AdaSmoothDelta. The parameters in AdaSmooth and AdaSmDel are ρ_1 and ρ_2, respectively

4.2 Experiment: Convolutional Neural Networks

Convolutional neural networks (CNNs) are known as powerful models with non-convex objective functions. CNNs with several layers of convolution, pooling, and nonlinear units have mostly demonstrated remarkable success in computer vision tasks, e.g. face identification, traffic sign detection, speech recognition [5, 6], and medical picture segmentation [11]; which is partly from the local connectivity of the convolutional layers, and the rotational and shift invariance due to the pooling layers. To demonstrate the main advantages and evaluate the strategy of the proposed AdaSmooth/AdaSmoothDelta methods, a real handwritten digit classification task, MNIST is used. For comparison with Zeiler [32]'s method, we train with Relu non-linearities and 2 convolutional layers in the front of the structure, followed by two fully connected layers. Again, the dropout with 50% noise is adopted in the network to prevent from overfitting.

To make it more specific, suppose the architecture for each convolutional layer is described by $C(\langle kernel\ size \rangle : \langle num\ outputs \rangle : \langle activation\ function \rangle)$; for each fully connected layer is described by $F(\langle num\ outputs \rangle : \langle activation\ function \rangle)$; for a max pooling layer is described by $MP(\langle kernel\ size \rangle : \langle stride\ number \rangle)$; and for a dropout layer is described by $DP(\langle rate \rangle)$. Then, the convolutional neural network structure we adopt can be presented as follows:

$$C(5{:}10{:}\text{Relu}) \ \cdot \ MP(2{:}2) \ \cdot \ C(5{:}20{:}\text{Relu}) \ \cdot \ MP(2{:}2)$$
$$- \ DP(0.5) \ \cdot \ F(50{:}\text{Relu}) \ \cdot \ DP(0.5) \ \cdot \ F(10{:}\text{Softmax}). \tag{23}$$

All descent methods are trained on mini-batches of 64 images per batch for 50 epochs through the training set. Setting the hyper-parameters to $\epsilon = 1e - 6$ and $\rho_1 = 0.5$, $\rho_2 = 0.9$ or 0.99, i.e. the number of periods chosen finds the exponential moving average are between 3 and 19, or between 3 and 199 iterations as discussed in Sects. 2.3 and 3. The AdaSmooth with $\rho_2 = 0.9$ or 0.99 is a wide range of upper bound on the decay constant from this context; and we shall see the AdaSmooth results are not sensitive to these choices. If not specially described, a small learning

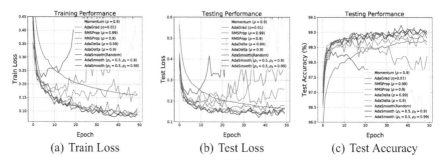

(a) Train Loss   (b) Test Loss   (c) Test Accuracy

Fig. 4 CNN: loss curves of various optimization methods on MNIST data set for 50 epochs with CNN. Though not significantly better than RMSProp or SGD with momentum, AdaSmooth converges slightly faster and obtains better out-of-sample accuracy

rate $\eta = 0.001$ for the convolutional networks is used in our experiments when applying stochastic descent. While again, the learning rate for the AdaDelta method is selected to be 1 as suggested by the original paper [32] and for the AdaSmoothDelta method is selected to be 0.5 as mentioned in Sect. 3.3.

In Fig. 4a, we compare results of SGD with Momentum, RMSProp, AdaGrad AdaDelta, and the proposed AdaSmooth/AdaSmoothDelta methods in optimizing the training set loss (negative log likelihood). The RMSProp ($\rho = 0.9$) does the worst in this case, whereas tuning the decay constant to $\rho = 0.99$ can significantly improve performance making the RMSProp sensitive to the hyper-parameter (hence the learning rates per-dimension). Further, though the training loss of AdaDelta ($\rho = 0.9$) decreases fastest in the first 3 epochs, the performance becomes poor at the end of the training, with an average loss larger than 0.15; while the overall performance of AdaDelta ($\rho = 0.99$) works better than the former, however, its overall accuracy is still worse than AdaSmooth. This also reveals the same drawback of AdaGrad for the AdaDelta method, i.e. they are sensitive to the choices of hyper-parameters.

In order to evaluate whether the AdaSmooth actually can find the compensation we want, we also explore a random selection for the number of periods/iterations N, termed as *AdaSmooth(Random)* in the sequel, which selects the decay constant between ρ_1 and ρ_2 randomly during different batches. The result shows the AdaSmooth(Random) cannot find the adaptive learning rates per-dimension in the right way showing our method actually finds the correct choices between the lower and upper bound of the decay constant.

The analysis of different methods in test loss and test accuracy is consistent with that of training loss as shown in Fig. 4b, c. We further save the effective ratios in magnitude as stated in Eq. (13) during each epoch; the distributions of them are shown in Fig. 5. In the first few epochs, the ERs in magnitude have more weights in large values, while in the last few epochs, the probability decreases into a stationary distribution; in other words, the ER distributions in the last few epochs are similar. In Fig. 6, we put a threshold on the ER values, the percentages of ERs larger than the threshold

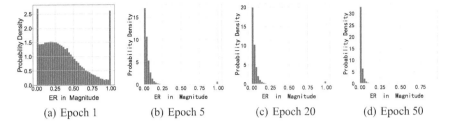

Fig. 5 CNN: the distribution of ERs for all parameters in different epochs when training MNIST data for AdaSmooth ($\rho_1 = 0.5$, $\rho_2 = 0.99$) model

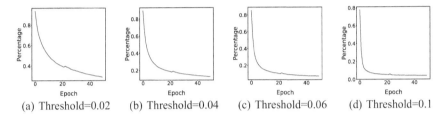

Fig. 6 CNN: percentage of ERs when they are larger than a threshold for different epochs in AdaSmooth ($\rho_1 = 0.5$, $\rho_2 = 0.99$) model

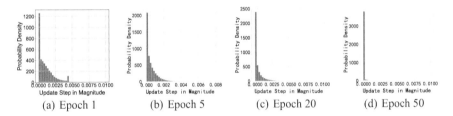

Fig. 7 CNN: the distribution of step Δx_t in magnitude for different epochs in AdaSmooth ($\rho_1 = 0.5$, $\rho_2 = 0.99$) model

are plotted; when the training is in progress, the ERs tend to approach smaller values indicating the algorithm goes into stationary points, the (local) minima. While Fig. 7 shows the absolute values of update step Δx_t for different epochs; when the training is in the later stage, the parameters are closer to the (local) minima, indicating a smaller update step and favouring a relatively smaller period (resulting from both the step and the direction of the movement) of exponential moving average in Eq. (16), which in turn compensates relatively less for the learning rate per-dimension and makes the update faster. In Table 3, we further show test accuracy for AdaSmooth and AdaSmoothDelta with various hyper-parameters after 10 epochs in which case different choices of the parameters do not significantly alter performance.

Table 3 CNN: best held-out evaluation in test accuracy for AdaSmooth and AdaSmoothDelta with various hyper-parameters after 10 epochs

Method	MNIST (%)
AdaGrad ($\eta = 0.01$)	96.82
AdaGrad ($\eta = 0.001$)	89.11
RMSProp ($\rho = 0.99$)	97.82
RMSProp ($\rho = 0.9$)	97.88
AdaDelta ($\rho = 0.99$)	97.83
AdaDelta ($\rho = 0.9$)	98.20
AdaSmooth ($\rho_1 = 0.5$, $\rho_2 = 0.9$)	98.13
AdaSmooth ($\rho_1 = 0.5$, $\rho_2 = 0.95$)	98.12
AdaSmooth ($\rho_1 = 0.5$, $\rho_2 = 0.99$)	98.12
AdaSmoothDelta ($\rho_1 = 0.5$, $\rho_2 = 0.9$, $\eta = 0.5$)	98.86
AdaSmoothDelta ($\rho_1 = 0.5$, $\rho_2 = 0.95$, $\eta = 0.5$)	98.91
AdaSmoothDelta ($\rho_1 = 0.5$, $\rho_2 = 0.99$, $\eta = 0.5$)	98.78
AdaSmoothDelta ($\rho_1 = 0.5$, $\rho_2 = 0.99$, $\eta = 0.6$)	98.66
AdaSmoothDelta ($\rho_1 = 0.5$, $\rho_2 = 0.99$, $\eta = 0.7$)	98.66
AdaSmoothDelta ($\rho_1 = 0.5$, $\rho_2 = 0.99$, $\eta = 0.8$)	98.58

The parameters ρ_1, ρ_2, η of AdaSmooth and AdaSmoothDelta are described in Algorithm 1

4.3   Experiment: Logistic Regression

Logistic regression is an analytic technique for multivariate modelling of categorical dependent variables with a well-studied convex loss function, allowing it to be used to compare different optimizers without worrying about going into local minima [18]. To evaluate, again, the learning rates are set to $\eta = 0.001$ in all experiments if not mentioned explicitly. And a relatively large learning rate (with $\eta = 0.01$) is applied for the AdaGrad method due to its accumulation of decaying factors in the denominator; the global learning rate for the AdaDelta method is set to 1 as suggested by the original paper [32] and for the proposed AdaSmoothDelta algorithm is selected to be $\eta = 0.5$ as mentioned in Sect. 3.3. The loss curves for each algorithm of the training processes are shown in Fig. 8a, b, where we compare vanilla SGD, SGD with Momentum, AdaDelta, RMSProp, and the proposed AdaSmooth/AdaSmoothDelta models in optimizing the training set losses for both Census Income and MNIST data sets, respectively. The unaltered SGD method performs the poorest in the logistic regression task. However, the AdaSmooth algorithm works slightly better than RMSProp in the MNIST case and performs much better than the latter in the Census Income experiment. AdaSmooth matches the rapid convergence of AdaDelta (in which case AdaSmooth converges slightly slower than AdaDelta in the first few epochs), while AdaSmooth continues to reduce the training loss, converging to the best performance in these models. As aforementioned, when $\rho_1 = \rho_2 = 0.9$ are selected for the AdaSmooth model, the proposed AdaSmooth model reduces to the

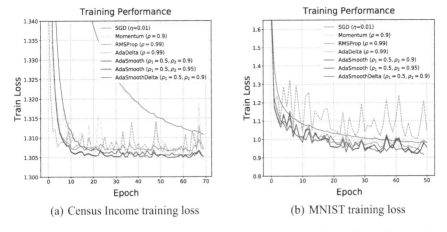

(a) Census Income training loss　　　(b) MNIST training loss

Fig. 8 Logistic regression: loss curves of various optimization methods on Census Income data set and MNIST data set for 70 and 50 epochs, respectively, with logistic regression. Notice that the training curve of AdaSmooth ($\rho_1 = 0.5$, $\rho_2 = 0.9$) is close to ($\rho_1 = 0.5$, $\rho_2 = 0.95$) in this case for both data sets

Table 4　Logistic regression: best in-sample evaluation in training accuracy

Method	MNIST (%)	Census (%)
SGD ($\eta = 0.01$)	93.29	84.84
Momentum ($\rho = 0.9$)	93.39	84.94
RMSProp ($\rho = 0.99$)	93.70	84.94
AdaDelta ($\rho = 0.99$)	93.48	84.94
AdaSmooth ($\rho_1 = 0.5$, $\rho_2 = 0.9$)	**93.74**	84.92
AdaSmooth ($\rho_1 = 0.5$, $\rho_2 = 0.95$)	93.71	84.94
AdaSmoothDelta ($\rho_1 = 0.5$, $\rho_2 = 0.9$, $\eta = 0.5$)	93.66	**84.97**

The parameters ρ_1, ρ_2, η of AdaSmooth and AdaSmoothDelta are described in Algorithm 1

RMSProp with $\rho = 0.99$ (similarly for the AdaSmoothDelta and AdaDelta case). In all experiments, the results of the proposed AdaSmooth model perform better while the difference between various hyper-parameters for AdaSmooth is not significant; there are almost no differences in AdaSmooth results with different hyper-parameter settings, indicating its insensitivity to hyper-parameters. In Table 4, we compare the training set accuracy of various algorithms; AdaSmooth works best in the MNIST case, while the superiority in Census Income data is not significant. Similar results as the MLP scenario can be observed in the test accuracy and we shall not give the details for simplicity.

5 Conclusion

The aim of this paper is to solve the hyper-parameter tuning in the gradient-based optimization methods for machine learning problems with large data sets and high-dimensional parameter spaces. We propose a simple and computationally efficient algorithm that requires little memory and is easy to implement for gradient-based optimization of stochastic objective functions. Overall, we show that AdaSmooth is a versatile algorithm that scales to large-scale high-dimensional machine learning problems and AdaSmoothDelta is an algorithm insensitive to both the global learning rate and the hyper-parameters with caution to set the global learning rate smaller than 1.

References

1. Becker, S., & LeCun, Y. (1988). Improving the convergence of back-propagation learning with second-order methods.
2. Dauphin, Y. N., Pascanu, R., Gulcehre, C., Cho, K., Ganguli, S., & Bengio, Y. (2014). Identifying and attacking the saddle point problem in high-dimensional non-convex optimization. In *Advances in neural information processing systems* (Vol. 27).
3. Dozat, T. (2016). *Incorporating Nesterov momentum into Adam.*
4. Duchi, J., Hazan, E., & Singer, Y. (2011). Adaptive subgradient methods for online learning and stochastic optimization. *Journal of Machine Learning Research, 12*(7).
5. Graves, A., Mohamed, A.-R., & Hinton, G. (2013). Speech recognition with deep recurrent neural networks. In *2013 IEEE International Conference on Acoustics, Speech and Signal Processing* (pp. 6645–6649). IEEE.
6. Hinton, G., Deng, L., Yu, D., Dahl, G. E., Mohamed, A.-R., Jaitly, N., Senior, A., Vanhoucke, V., Nguyen, P., Sainath, T. N., et al. (2012). Deep neural networks for acoustic modeling in speech recognition: The shared views of four research groups. *IEEE Signal Processing Magazine, 29*(6), 82–97.
7. Hinton, G., Srivastava, N., & Swersky, K. (2012). Neural networks for machine learning lecture 6a overview of mini-batch gradient descent. *Neural Networks for Machine Learning, 14*(8), 2.
8. Kaufman, P. J. (1995). *Smarter trading.*
9. Kaufman, P. J. (2013). *Trading systems and methods, + website* (Vol. 591). Wiley.
10. Kingma, D. P., & Ba, J. (2014). Adam: A method for stochastic optimization. arXiv:1412.6980
11. Krizhevsky, A., Sutskever, I., & Hinton, G. E. (2012). Imagenet classification with deep convolutional neural networks. In *Advances in neural information processing systems* (Vol. 25)
12. Le Roux, N., & Fitzgibbon, A. W. (2010). A fast natural newton method. In *ICML.*
13. LeCun, Y. (1998). The MNIST database of handwritten digits. http://yann.lecun.com/exdb/mnist/.
14. Lu, J. (2022a). Exploring classic quantitative strategies. arXiv:2202.11309
15. Lu, J. (2022b). Gradient descent, stochastic optimization, and other tales. arXiv:2205.00832
16. Lu, J. (2022c). Matrix decomposition and applications. arXiv:2201.00145
17. Lu, J., & Yi, S. (2022). Reducing overestimating and underestimating volatility via the augmented blending-arch model. *Applied Economics and Finance, 9*(2), 48–59.
18. Menard, S. (2002). *Applied logistic regression analysis* (Vol. 106). Sage.
19. Moulines, E., & Bach, F. (2011). Non-asymptotic analysis of stochastic approximation algorithms for machine learning. In *Advances in neural information processing systems* (Vol. 24).
20. Polyak, B. T., & Juditsky, A. B. (1992). Acceleration of stochastic approximation by averaging. *SIAM Journal on Control and Optimization, 30*(4), 838–855.

21. Qian, N. (1999). On the momentum term in gradient descent learning algorithms. *Neural Networks, 12*(1), 145–151.
22. Robbins, H., & Monro, S. (1951). A stochastic approximation method. In *The annals of mathematical statistics* (pp. 400–407).
23. Ruder, S. (2016). An overview of gradient descent optimization algorithms. arXiv:1609.04747
24. Rumelhart, D. E., Hinton, G. E., & Williams, R. J. (1986). Learning representations by back-propagating errors. *Nature, 323*(6088), 533–536.
25. Ruppert, D. (1988). *Efficient estimations from a slowly convergent Robbins-Monro process.* Technical report, Cornell University Operations Research and Industrial Engineering.
26. Schaul, T., Zhang, S., & LeCun, Y. (2013). No more pesky learning rates. In *International Conference on Machine Learning* (pp. 343–351). PMLR.
27. Schwarz, J., Jayakumar, S., Pascanu, R., Latham, P. E., & Teh, Y. (2021). Powerpropagation: A sparsity inducing weight reparameterisation. In *Advances in neural information processing systems* (Vol. 34, pp. 28889–28903).
28. Smith, L. N. (2017). Cyclical learning rates for training neural networks. In *2017 IEEE Winter Conference on Applications of Computer Vision (WACV)* (pp. 464–472). IEEE.
29. Srivastava, N., Hinton, G., Krizhevsky, A., Sutskever, I., & Salakhutdinov, R. (2014). Dropout: A simple way to prevent neural networks from overfitting. *The Journal of Machine Learning Research, 15*(1), 1929–1958.
30. Sutskever, I., Martens, J., Dahl, G., & Hinton, G. (2013). On the importance of initialization and momentum in deep learning. In *International Conference on Machine Learning* (pp. 1139–1147). PMLR.
31. You, Y., Li, J., Reddi, S., Hseu, J., Kumar, S., Bhojanapalli, S., Song, X., Demmel, J., Keutzer, K., & Hsieh, C.-J. (2019). Large batch optimization for deep learning: Training Bert in 76 minutes. arXiv:1904.00962
32. Zeiler, M. D. (2012). Adadelta: An adaptive learning rate method. arXiv:1212.5701

Designing and Implementing a Distributed Database for Microservices Cloud-Based Online Travel Portal

Biman Barua, Md Whaiduzzaman, M. Mesbahuddin Sarker, M. Shamim Kaiser, and Alistair Barros

Abstract Designing and implementing a distributed database for a microservices cloud-based online travel portal has been proven essential for handling a large volume of the database, load balancing of an application, and applying for global access with live data. Microservices become more popular as the applications become more complex and distributed. The main goal of microservices is to build an application by splitting the application into small services from large business components which can be deployed and run independently. In this research, we show the distributed databases design and distribution using different types of fragmentation techniques, data allocation, and data integration using relational algebra, union and join. We also illustrate how distributors, agents, and customers from all different countries can be managed in a database table from an individual country by setting up a database on a nearby site instead of searching from a whole global database. This research will help design and implement distributed databases for cloud-based online travel portals, including microservices applications, to save from a complete system failure, huge data handling, and load balancing of an application.

Keywords Distributed database · Online travel agent · Microservices · Cloud computing

B. Barua (✉)
Department of CSE, BGMEA University of Fashion and Technology, Nishatnagar, Turag, Dhaka 1230, Bangladesh
e-mail: biman@buft.edu.bd

B. Barua · Md Whaiduzzaman · M. Mesbahuddin Sarker · M. Shamim Kaiser
Institute of Information Technology, Jahangirnagar University, Dhaka 1342, Bangladesh

Md Whaiduzzaman · A. Barros
School of Information Systems, Queensland University of Technology, Brisbane, Australia

© The Author(s), under exclusive license to Springer Nature Singapore Pte Ltd. 2023
S. Shakya et al. (eds.), *Sentiment Analysis and Deep Learning*,
Advances in Intelligent Systems and Computing 1432,
https://doi.org/10.1007/978-981-19-5443-6_22

1 Introduction

As digital technology and e-commerce are proliferating, it has become essential to design and develop applications for the growing online travel industry to smooth operation and handle a huge volume of data globally. The online airline reservation system has already become very popular for all people, and the airline reservation is the only system that updates the fare and flight information so rapidly. To design an online airline reservation system, a large number of modules have to be designed. Moreover, handling a large volume of data retrieved from multiple vendors is critical where it has to search and display the real-time fare from different vendors using global distribution system (GDS) and low-cost carrier (LCC) airlines. To access the data from different locations and applications using application programming interface (API), the decentralization of data is essential, and it becomes more critical when operations are run from different countries.

When ticket issues are from any location, the customer's data are stored in a centralized database, whereas the same data need not be accessed from other locations. In the case of accessing any local area-wise data, it has to be searched from the whole database, which takes much time to execute the query. The execution time of any query is increased, and on the other hand, the storage capacity is also increased.

To overcome the above situation, the integration of distributed databases is essential. In a distributed system, the data can be stored area and region-wise; in this system, all client's data will be stored in different regional locations, and whenever any user executes a query, it will search to the local database; if available, it will be displayed; otherwise, it will go to a centralized database.

2 Literature Review

Designing and implementing a distributed database on a cloud-based online travel portal is a critical job. Most of the researchers developed systems to convert manual travel booking systems to computerized systems. Ceri et al. [1] defined how the airline's database is a part of a global database design method; they defined the gathering of requirements from the disbursed records and programs and a way to construct a distribution of a schema. Some described the current status of online tourism. The behavior of retailers in the online tourism distribution system [2] and the future of online travel agents (OTA). A few researchers designed the application for online travel to make it easier, flexible, evaluate the right documentation, and on time [3], by budget development costs to improve productivity [4] and quality of the travel agents. Jauhari et al. [5] designed the application for tourism agents and tourism agent providers, where they included education about tourist places. Sekarhati et al. [6] worked on data analysis, data maturity, and management using HESA and discussed IATA theory and innovation, OTA theory, and different frameworks. Lee et al. [7] researched developing, operating, and maintaining a travel agency website with how a travel agency could understand consumer behavior. Suma [8] has described a model of a hybrid deep fuzzy hashing algorithm to retrieve information in a distributed cloud environment, ensuring a high retrieval efficiency and accuracy than a conventional system. In this model, he showed 97.6% efficiency in information retrieval. Kumar [9] described and reviewed in his paper how the task can be scheduled on the ubiquitous cloud; he also described the hybridized meta-heuristic methods for improving the privacy of services. Wiese proposed a framework for intelligent fragmentation and data replication [10] for a cloud-based database environment where described cloud storage is able to execute semantically-guided flexible query mechanisms. Barua and Whaiduzzaman [11] described how a web-based application can be hosted at cloud and used as Software as a Service (SaaS).

Most researchers designed the online travel agent using a centralized database, cloud data retrieval, and data privacy. Some are proposed methodologies for increasing the query execution time using the algorithm. Hence, there is a gap in designing an online travel portal using a distributed database, where the database will be distributed to a nearer server location to access the information by searching the global database.

3 Problem Definition

The existing system of online travel agent (OTA) uses a centralized database where all data is stored in a single database server. If the server receives a large volume of data requests, the server may become hung or slow. There is no option for load balancing, which is also a good technique to smooth the operation of the database. Moreover, in an existing centralized database, all agents send requests to the same database, and customers worldwide issue tickets. For example, if an agent uses a cloud online travel portal from Bangladesh and the database server location is in the USA, all customer's data from Bangladesh is stored in the USA server. At the same time, if an agent uses from India, all customer data from India is also stored in the USA server.

If the database server is nearer to the application, the query execution time and data accessibility will be faster than in a centralized database.

3.1 Why Distributed Database?

A distributed database (DDB) is a set of multiple, logically interrelated databases disbursed over a computer network [12]. It is a collection of databases stored on multiple servers connected through a TCP/IP protocol network. The distributed database system is also a collection of distributed databases (DDB) and distributed database management systems (DDBMS).

3.2 DDB Systems Types

The distributed database is mainly categorized into two types:

Homogeneous Distributed System: In the types of homogeneous systems, the data is distributed but uses identical databases and operating systems. This type of database is easy to manage and design [12].

Heterogeneous Distributed System: This type of distributed database [13] uses a different database, schemas, and different data models.

3.3 Advantages of Distributed Database System

The management of DDDB is easy as it manages various levels of transparency, including distributional transparency, fragmentation transparency, transaction transparency, replication transparency, etc.

- The reliability of distributed databases is much better than of centralized databases. If any failure occurs, the total system becomes halted, but in distributed systems, if any portion of the system is down, the full system will remain active.
- As DDB is stored near the site, the accessibility speed of data is faster than the centralized system.
- It is easy to integrate or add a new database or new site to an existing system without any interruption of the existing system.
- In a distributed database system, the data can be organized in an organized fashion according to the department, which is easier to access the data hierarchy.

3.4 Limitations of Relational Database

Though there are a lot of benefits of using RDBMS, there are some limitations also, mostly

- **Costly**: The users or the company need to purchase the software and tools to achieve the full benefits of RDBMS. It is also costly to set up and maintain the relational database system. It is difficult to implement the RDBMS system with a minimum budget. Needs expert DBA to work as system admin.
- **Maintenance Problem**: we use RDBMS for a large volume of data. So it becomes a critical job to maintain the DB administration. We need to engage a highly expert BDA.
- **Physical Storage**: As the database is stored in a collection of rows and columns and every set of data needs separate storage, it needs a lot of physical memory. The requirement for storage is increased when the storage of application data becomes high.

- **Complex Structure**: The relational data is stored in rows and columns, and most applications need to access data from multiple tables, similar to storing data in different tables.
- **Performance Issue**: The relational database becomes slower because it depends on multiple tables. Also, it becomes more complex when a large volume of tables and data is needed in the system. It also becomes slow query response time and can be a failure if there is a large volume of users' login and access data.

3.5 Major Characteristics of RDBMS

Support ACID Properties: The BDMS should be supported by (atomicity, consistency, isolation, and durability) (ACID). It ensures that the transaction in a database like reading, writing, deleting, and updating are up to date. The ACID transaction also ensures data reliability and integrity.

Concurrent Data Access: In an application, there is a chance that multiple users access the same data concurrently at the same time; in this case, the DMBS should handle the concurrent data handling without any problem.

Data Integrity: Data integrity is one of the essential characteristics of a DBMS. It ensures the quality and reliability of the database, and it also ensures more protection of a database so that no unauthorized person is able to access the database.

Backup and Recovery: In today's business scenarios handling data is a challenging task when handling a large volume of data. There are chances of data failure in a partial or full database; by losing essential data, a user may fall into great trouble, so backup and recovery are great options to save the user from the failure of the database. If any issues exist, the database can be restored in a previous stage.

3.6 Why Microservice Architecture?

It is an emerging architecture and a development methodology where a large single application can be fragmented into a smaller autonomous service, each smaller services will execute on its own process and interact with lightweight mechanism. Figure 1 represents the difference between Monolithic and Microservices architecture. The services can be build using different business logic and can be written [14] using verities programming languages also can implement different data storage technologies. Microservices are getting popularity as industries are building larger and complex applications for their need which is becoming critical in handling and slower. A large number of peoples are currently moving forward to redesign their traditional monolithic application into a single independent services. To reduce the complexity of code, alteration of code without effecting the whole code and deployed

Monolithic Architecture Microservices Architecture

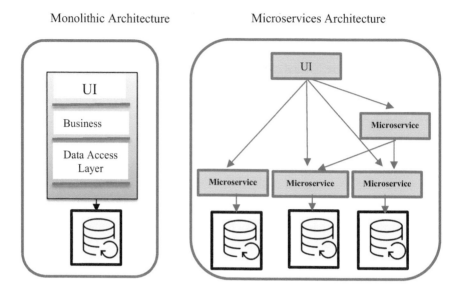

Fig. 1 Monolithic versus microservices architecture

program again, avoiding slow down the application most of the developers are moving to Microservices architecture.

Microservices can be build and implemented using a different horde of frameworks, tools, and versions like Java, C++, Node JS, and Microsoft.Net, etc.

4 Methodology

In integrating and distributing databases to online travel agents, as a large volume of data is handled for different geographic locations, it was critical to distribute the database. In a centralized database, all data is stored on a single database server; if an agent needs the client's data, needed to search the whole data without searching specific countries' data. To overcome this issue, it was better to distribute data to different regions so that if any agent searches customers' data, it could be searched for their nearest database without searching the whole database.

To design and distribute the database needed to study the whole reservation process and what objects related to the system's design. It found multiple objects like airlines, airports, customers, agents, etc. After finalizing the objects, the attributes were found. The primary data was collected from local agents, and secondary data was collected from online, newspapers, etc.

4.1 System Design

The design phase of the online travel agent contains different levels, in this back-end part of the system contains table designs for different geographic locations. As the system is used in different geographical locations, data will be accessed country or region-wise. Consider if any agents are using the system from Bangladesh; all low-cost carriers (LCC) that are not connected to GDS can be stored in a database that resides in Bangladesh regions. If data is decentralized and distributed region-wise, data access speed will be better. Moreover, it is unnecessary to search the whole database for local users' or customers' information. It will also keep the system free from unwanted query requests.

4.2 Data Collection Methods

We used Google Forms to collect data from local agents using some questionnaires and secondary data collected from google search and the company website.

4.3 Database Design

A database is a collection of integrated and interrelated tables. The design concept of database distribution depends on the architecture of the system [12]. In a tightly integrated distributed database system the design process follows a top-down approach [15]. In a multi-database distributed system the design process follows a bottom-down approach which is integrated with the system existing database.

4.3.1 Top-Down Approach

In this type of system, the database is designed from scratch which starts from the requirement analysis [16]. The requirement analysis for any company is used to find out the company situation, barriers, or constraints with any problem [17]. At the time of designing any distributed database need to follow two types: conceptual design and view design. Figure 2 represents the top-down approach of a distributed database. The conceptual view deals with entity normalization and relationship, whereas the view design deals with the interface of any application.

After the above two steps, the next step is distribution design. The distribution design includes fragmentation and allocation.

Fig. 2 Top-down approach
of distributed database

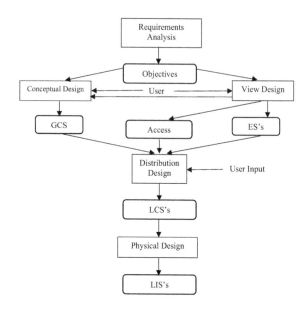

4.3.2 Bottom-Up Approach

The bottom-up approach is used whenever the design is required to integrate with any existing database. In this phase first, we need to identify the elements and then group the items into a dataset. In the bottom-up approach, the process starts from the initial stage with lower-level elements like attributes and functional dependencies and then forward to an acceptable logical data model [13].

4.3.3 Distribution of Data Across Different Regions and Cloud

In the current competitive market and globalization, it has become challenging to deliver service to its customer for delivering faster and securely at shortest possible time. Also, deploying application and maintaining from different location and serving global customer it is more challenging. To access data from anywhere at any devices cloud is an another paradigm of on demand delivery based services [18]. So, it is essential to integrate and develop new applications where applications and data will be hosted and distributed in nearer location to its customer. In this case, data needs to be replicated in a manner where it can access with low latency. In Fig. 3 represents the data distribution and load balancing. In this scenario, the client will automatically forward to next available local cluster if any region failed. In this case, need to have a load balancer as front-end in an application.

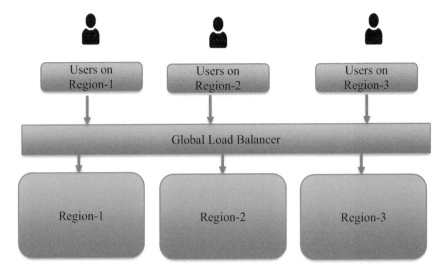

Fig. 3 Data distribution and load balancing

4.3.4 Components of the Distributed System

To design the online airlines agent (OTA) the main consideration is as follows

- Need to partition the database into different fragments.
- The fragmented data to be replicated in different servers.
- The allocation of fragments and replicas in an appropriate manner.

Designing a distributed database mainly requires two things: fragmentation and allocation of any data. Fragmentation is a system that divides main data into different pieces of data which are called fragments [1]. On the other hand, allocation is the system of mapping individual fragments to different sites.

Horizontal Fragmentation: Horizontal fragmentation means taking rows from a table and placing rows in different locations or sites. In our online travel portal, the horizontal fragmentation of the agents management table is enumerated below, as agents are connected from different geographical locations or countries, the data from different countries can be stored in their nearest node so that it can be accessed faster.

In relational algebra, we can create the horizontal fragmentation on table (T) may be represented as shown in Table 1.

$$\sigma p(T)$$

here

σ is relational algebra operator for selection.

Table 1 Complete agent table of global data storage

ID	Agent code	Company name	Mobile	Email	Password	Login ID	Address	Country	Agent class ID
1	TS001	Trip Sheba	018..	info@tripsheba.com	Pass1	trips	Dhaka	BD	1
2	TS002	Travel BD	019..	info@tbd.com	Pass2	Tbd	CTG	IN	2
3	TS003	Flight Expert	016	info@flightexpert.com	pass3	flighte	DKA	BD	2
4	TS04	Yatra	910	info@yatra.com	pass4	yatra	MUM	IN	1

p is the condition which satisfied by a horizontal fragment.

In the below table, a union operator is used on the fragment to design the table T. Having all the rows of a table T is called complete horizontal fragment [19].

To make the distribution of data from the AgentManagement table, it can be broken into different fragments as shown in Tables 2 and 3.

$$\text{AgentMan } 1 = \sigma \text{Country} = \text{BD AgentManagement}$$

$$\text{AgentMan } 2 = \sigma \text{Country} = \text{IN AgentManagement}$$

Now, we can consider the fragment T_1 of Country = BD.

In the same way, the fragment T_2 of country = IN.

Now, if we like to access the whole information of the AgentManagement table we can use as

$$T = T_1 \cup T_2 \cup T_3 \ldots T_N$$

Therefore, by distributing and storing the different fragments to different sites or locations we can speed up the process of accessing data to online travel portals. In this system, setting up server location to the nearest site of the system can be made faster. Moreover, as data is stored on different servers so if any database server is down the whole system will not be stopped.

Vertical Fragmentation: Vertical fragmentation is a process of dividing the table vertically by partitioning the columns. In this type of partitioning or fragmentation, some attributes are stored on tables and some attributes are stored on different tables. The main objective of the process is most of the time all attributes are not necessary for all the time. We can call specific attributes for specific tasks, which is most important to make the application more efficient. We have to create the fragmentation in such a way, so that whenever we need all the attributes we can rebuild the table using natural JOIN. In this case, we may use another attribute Tuple_ID at each fragmented part.

$$\pi a_1, a_2, \ldots, a_n(T)$$

here

π is relational algebra operator

$a_1, a_2, \ldots, a_n$ are the attributes of T

T is the table (relation).

In our AgentManagement table, we can consider T_1 as shown in Table 4.

Where we have added an extra attribute Tuple_ID, on the second fragmented part after vertically dividing the table (T_2) as shown in Table 5.

To access the all attributes, we can use both fragments T_1 and T_2 as bellow

$$\pi \text{ AgentManagement}(T_1 \infty T_2)$$

Table 2 Horizontal fragmentation part-1

ID	Agent code	Company name	Mobile	Email	Password	Login ID	Address	Country	Agent class ID
1	TS001	Trip Sheba	018..	info@tripsheba.com	Pass1	trips	Dhaka	BD	1
3	TS003	Flight Expert	016	info@flightexpert.com	pass3	flighte	DKA	BD	2

Table 3 Horizontal fragmentation part-2

ID	Agent code	Company name	Mobile	Email	Password	Login ID	Address	Country	Agent class ID
2	TS02	Travel BD	019..	info@tbd.com	Pass2	Tbd	CTG	IN	2
4	TS04	Yatra	910	info@yatra.com	pass4	yatra	MUM	IN	1

Table 4 Vertical fragmentation part-1 with extended column

ID	Agent code	Company name	Mobile	Email	Tuple_ID
1	TS001	Trip Sheba	018..	info@tripsheba.com	1
2	TS002	Travel BD	019..	info@tbd.com	2
3	TS003	Flight Expert	016	info@fe.com	3
4	TS04	Yatra	910	info@yatra.com	4

Table 5 Vertical fragmentation part-2

Password	Login ID	Address	Country	Agent class ID	Tuple_ID
Pass1	trips	Dhaka	BD	1	1
Pass2	tbd	CTG	IN	2	2
pass3	flighte	DKA	BD	2	3
pass4	yatra	MUM	IN	1	4

A. Database Design Using ER Diagram

The entity-relationship diagram (ERD) of a cloud online travel portal represents the relationship of all entities in a database. An entity is an object which is an element of data [20]. All entities will have attributes that define properties. The ER diagram of the cloud online travel portal shows the all visual components of a database table with relationships between airports, airlines, customers, travel agents, tickets, agent class, fare details, payments, booking, users, etc.

To design the ER diagram of the system, first need to identify the entity and its attributes, the next step needs to be to design the entity-relationship diagram, so that it can be cleared to design the database based on ERD.

In our research, the entity and attributes as shown in Table 6.

Table 6 Entity and attributes

Entity	Attributes
Airlines	ID, AirlineName, AirlineCode, Status, Logo
Airport	AirportID, AirportName, City, Country, Status
FareDetails	FareID, DepartTime, AriveTime, Duration, SegmentCount, AddFareID
Ticket	TicketID, TicketNo, IssueDate, ValidatingAirLine, ServiceFee, Status, Remarks
AgentManagement	AgentCode, AgentName, ContactNo, LoginID, Password, ContactPerson, Address, Country, AgentClassID
AgentClass	ID, ClassName, Details, Status, Remarks
Bookings	BookID, Title, BookType, BookDate, Description
Payment	PayID, PayCustID, PayDate, PayAmount, PayDescription
Customer	CustID, CustName, Email, CustUserName, CustPass, CustAddress

The details of travel agencies are stored on the AgentManagement table; airline information is stored on the airlines table the same way every entity has an individual table. Figures 4, 5, 6, and 7 represents the four parts of full database diagram. Also, every table has an individual primary key and unique key. A primary key is unique key in a column [21] or columns which can be uniquely identified each row in a table.

Hence, a large number of tables are needed to design the full system. Figure 8 represents the whole database design and with the relationship of all tables.

Fig. 4 Database design part-1

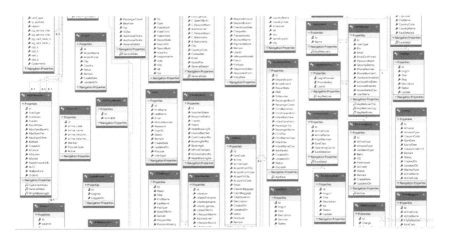

Fig. 5 Database design part-2

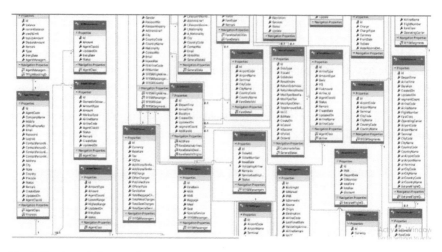

Fig. 6 Database design part-3

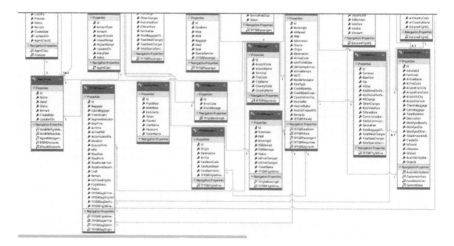

Fig. 7 Database design part-4

4.4 Data Analysis Tools

After analyzing the objects and attributes we designed the scratch data, tables and ER diagrams were designed to analyze the flow of data.

To design the database, we used the most popular relational database management system (RDBMS) of Microsoft SQL Server as it was more user-friendly [22].

We designed the database to create a relationship. Finally, we used to query, subquery, and join an application that was designed in another Microsoft product Visual Studio.Net MVC to access data from different locations as the system was

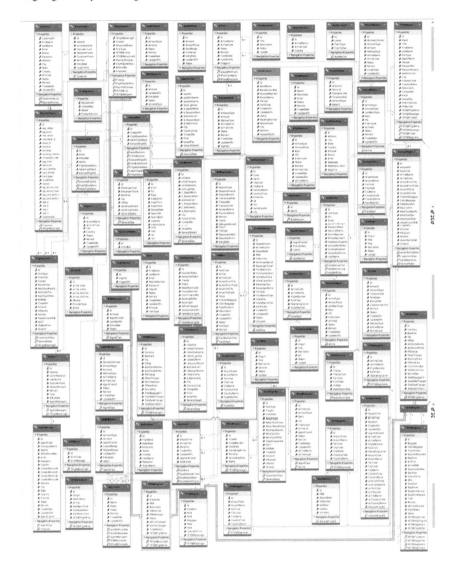

Fig. 8 Database design full

designed for distribution. In the future, a resilient distributed database (RDD) can be used [18]. A single RDD, we can divide into multiple logical units of partition [23, 24]. The partition can be processed and stored in the different machines of a cluster.

5 Results and Discussions

The solution and methodology that we have experimented with in our research, implemented in the first phase, we have designed and implemented a centralized database using the most popular Microsoft RDBMS SQL Server 2016 version and interface designed with another popular product of Microsoft Visual Studio.Net MVC framework. The application was hosted on the cloud and tested for different errors, validation, and functionalities. The application was designed for using multiple partners with different access rights like airline distributors, agents, and direct customers. The integrated modules were Distributor, B2B, and B2C. As we designed the application for cloud-based use as Software as a Service (SaaS). So, parties could sign-up with their information and use the service as they go. As the number of users is increasing rapidly and data handling is becoming a challenging task, in this second phase, the database was converted to distribute databases where the fragmented databases were placed on different servers nearer to the site.

In this system, all agents from different countries can locate and store their customers' data in a nearby location instead of searching from the whole database. If any part of the database server is down, it will not affect the whole system. After hosting the site and database in a different location, it is found at ping if data is used from a global database located in the USA, the ping time takes 313 MS, whereas if it is located at the nearest site, like Bangladesh, it takes 1-2 MS.

6 Conclusion

Our main goal was to design and implement distributed databases for cloud-based online travel agencies with microservices, which will help split the application into different servers, load the balance of the application, provide faster retrieval of data access, and decentralize the database. If any part of the database server is down in this system, the overall system will be available. Adding integration of replication techniques on data, down of any site still will run the site by accessing replicated data from other sites. We have also discussed different design approaches of top-down and bottom-up approaches, data integration, horizontal, vertical, and mixed fragmentation.

Acknowledgements This research is partly supported through the Australian Research Council Discovery Project: DP190100314.

References

1. Ceri, S., Pernici, B., & Wiederhold, G. (1987). Distributed database design methodologies. *Proceedings of the IEEE, 75*(5), 533–546.
2. Marzo-Navarro, M., Berne-Manero, C., Gómez-Campillo, M., & Pedraja-Iglesias, M. (2019). Strengths of online travel agencies from the perspective of the digital tourist. In *Predicting trends and building strategies for consumer engagement in retail environments* (pp. 187–210). IGI Global.
3. Soegoto, E. S., & Fadillah, R. (2018, August). Design and development of ticket reservation information system in travel business. *IOP Conference Series: Materials Science and Engineering, 407*(1), 012026).
4. https://www.geeksforgeeks.org/fragmentation-in-distributed-dbms/
5. Jauhari, A. F. D., et al. (2021). Design and implementation of travel agent in the face of the COVID-19 pandemic. In *E3S Web of Conferences* (Vol. 328). EDP Sciences.
6. Sekarhati, D. K. S., Nefiratika, A., Hidayanto, A. N., & Budi, N. F. A. (2019, August). Online travel agency (OTA) data maturity assessment: Case study PT Solusi Awan Indonesia-"Flylist". In *2019 International Conference on Information Management and Technology (ICIMTech)* (Vol. 1, pp. 492–497). IEEE.
7. Lee, J. J. Y., Sung, H. H., Defranco, A. L., & Arnold, R. A. (2005). Developing, operating, and maintaining a travel agency website: Attending to e-consumers and internet marketing issues. *Journal of Travel & Tourism Marketing, 17*(2–3), 205–223.
8. Suma, V. (2020). A novel information retrieval system for distributed cloud using hybrid deep fuzzy hashing algorithm. *JITDW, 2*(03), 151–160.
9. Kumar, D. (2019). Review on task scheduling in ubiquitous clouds. *Journal of ISMAC, 1*(01), 72–80.
10. Wiese, L. (2014). Clustering-based fragmentation and data replication for flexible query answering in distributed databases. *Journal of Cloud Computing, 3*(1), 1–15.
11. Barua, B., & Whaiduzzaman, M. (2019, July). A methodological framework on development the garment payroll system (GPS) as SaaS. In *2019 1st International Conference on Advances in Information Technology (ICAIT)* (pp. 431–435). IEEE.
12. Özsu, M. T., & Valduriez, P. (2011). *Principles of distributed database systems*. Springer Science & Business Media.
13. Ozsu, M. T., & Valduriez, P. (1991). Distributed database systems: Where are we now? *Computer, 24*(8), 68–78. https://doi.org/10.1109/2.84879
14. Aderaldo, C. M., Mendonça, N. C., Pahl, C., & Jamshidi, P. (2017, May). Benchmark requirements for microservices architecture research. In *2017 IEEE/ACM 1st International Workshop on Establishing the Community-Wide Infrastructure for Architecture-Based Software Engineering (ECASE)* (pp. 8–13). IEEE.
15. Kung, H.-J., Kung, L. A., & Gardiner, A. (2012). Comparing top-down with bottom-up approaches: Teaching data modeling. In *Proceedings of the Information Systems Educators Conference* (Vol. 2167).
16. Tomar, P., & Megha. (2014). An overview of distributed databases. *International Journal of Information and Computation Technology, 4*(2), 207–214. ISSN 0974-2239.
17. Gadicha, A. B., et al. (2012). Top-down approach process built on conceptual design to physical design using LIS, GCS schema. *International Journal of Engineering Sciences & Emerging Technologies, 3*, 90–96.
18. Whaiduzzaman, M., Mahi, M. J. N., Barros, A., Khalil, M. I., Fidge, C., & Buyya, R. (2021). BFIM: Performance measurement of a blockchain based hierarchical tree layered fog-IoT microservice architecture. *IEEE Access, 9*, 106655–106674.
19. Sese Tuperekiye, E., & Zuokemefa Enebraye, P. Framework for client-server distributed database system for an integrated payroll system.
20. Chaki, P. K., Sazal, M. M. H., Barua, B., Hossain, M. S., & Mohammad, K. S. (2019, February). An approach of teachers' quality improvement by analyzing teaching evaluations data. In *2019*

Second International Conference on Advanced Computational and Communication Paradigms (ICACCP) (pp. 1–5). IEEE.

21. Barua, B. (2016). M-commerce in Bangladesh-status, potential and constraints. *International Journal of Information Engineering and Electronic Business, 8*(6), 22.

22. Chaki, P. K., Barua, B., Sazal, M. M. H., & Anirban, S. (2020, May). PMM: A model for Bangla parts-of-speech tagging using sentence map. In *International Conference on Information, Communication and Computing Technology* (pp. 181–194). Springer.

23. Whaiduzzaman, M., Barros, A., Shovon, A. R., Hossain, M. R., & Fidge, C. (2021, September). A resilient fog-IoT framework for seamless microservice execution. In *2021 IEEE International Conference on Services Computing (SCC)* (pp. 213–221). IEEE.

24. Hossen, R., Whaiduzzaman, Md., Uddin, M. N., Jahidul Islam, Md., Faruqui, N., Barros, A., Sookhak, M., & Julkar Nayeen Mahi, Md. (2021). BDPS: An efficient spark-based big data processing scheme for cloud fog-IoT orchestration. *Information, 12*(12), 517.

A Comparative Study of a New Customized Bert for Sentiment Analysis

Fatima-ezzahra Lagrari and Youssfi ElKettani

Abstract This paper presents a new customized Bert method based sentiment analysis classification. The blank space removal method, stop word removal, and stemming methods were used in the data pre-processing step. Then tokens are extracted in tokenization step. Standard lion algorithm is used to optimize the weight and bias of the proposed Bert algorithm. Many comparisons were done in order to analyze the performance of the new proposed Bert model. Results obtained showed that the proposed model outperforms other classical models in term of accuracy and performance other metrics.

Keywords Bert · Twitter sentiment analysis · Deep learning · Lion algorithm · Optimization

1 Introduction

The is a huge increase today in Internet with the growth of social medias, e-commerce websites, and platforms of exchanging ideas. This increase leads to generate a huge amount of datasets. Hence, the usefulness of sentiment analysis classification.

The topic of sentiment analysis or opinion mining is one of the natural language processing (NLP) domain that attracts big attention of researchers, industry and government around the world [1]. It is a mechanism that allows us to determine the emotions in a text document which help us to define either the opinion, perception or the expressed sentiment. As specifies by Liu, an opinion consists of a target, an object or a feature of an entity, and a mood (positive, negative, or neutral) [2]. This sentiment can contain author and time of the expressed opinion [3].

Emotions can be categorized into different classes depending on the objective of the classification. Lagrari and Elkettani [4] and Rachiraju and Revanth [5] classified tweets into positive, negative and neutral, although Kalchbrenner et al. [6] classified movie reviews to positive, somewhat positive, neutral, negative, somewhat negative.

F. Lagrari (✉) · Y. ElKettani
Department of Mathematics, Ibn Tofail University of Science, Kenitra, Morocco
e-mail: Fatima.ezzahra.lagrari@gmail.com

Several sentiment analysis techniques have been developed. The classical techniques are categorized into two classes: lexicon-based and machine learning techniques. The lexicon-based approach involves calculating orientation for a document from the semantic orientation of words or phrases in the document [7].

The text classification approach or machine learning approach involves building classifiers from labeled instances of texts or sentences [8].

Traditional sentiment analysis approaches have achieved good accuracy according to literature however the step of pre-processing data is time consuming and complex add to that the huge data generated nowadays needs robust knowledge and methods to analyze and manage it that's why deep learning methods was born to fulfill this need and to facilitate the learning phase without the need of the feature engineering step. Researches indicates that deep learning methods outperforms classical methods in sentiment analysis [9].

The major contribution of this research work is that:

- Standard lion algorithm is used to modify the maximum sequence length of Bert encoder [10].
- Standard lion algorithm is used to finetune the Bert weight and bias.

2 Proposed Sentiment Analysis Framework

2.1 Proposed Model

A new twitter sentiment analysis classification model was proposed in this work as in Fig. 1. We have applied stop word removal method, stemming and blank space removal in the pre-processing step as in Fig. 2. Only meaningful tokens were kept and classified by the optimized Bert model which classifies twitter dataset into positive, negative, and neutral.

2.2 Pre-processing and Tokenization Steps

2.2.1 Pre-processing Step

The pre-processing step is an important step that must be applied to raw data before the embedding step in order to standardize text and make it easier to use [11]. In this work, tree methods were used is order to pre-proceed the data: stop words removal, stemming, and blank space removal.

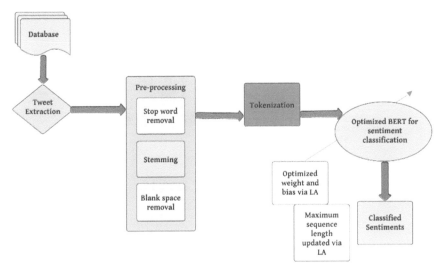

Fig. 1 Proposed sentiment classification model

2.2.2 Tokenization Step

The tokenization is the mechanism to split a text document into a bag of words or terms that is called tokens. These words can be separated with whitespace and will be classified via proposed optimized BERT framework [12].

3 Proposed Customized Bert with Lion Algorithm

Bidirectional Encoder Representations from Transformers (BERT) is a method developed by Devlin et al. [9]. The BERT has tree important steps:

Input layer, BERT encoder and output layer.

The proposed customized Bert is a multi-layer bidirectional transformer encoder with 12 transfer blocks with maximum sequence length of 512 tokens.

The output layer from the encoder is the representations of the sequence and it can be a hidden state vector or the "hidden state vector's time-step sequence".

In this work, standard LA was used to predict the best sequence token (hidden state vector) is used.

Standard LA is used to modify the maximum sequence count of Bert encoder to make it applicable for big datasets.

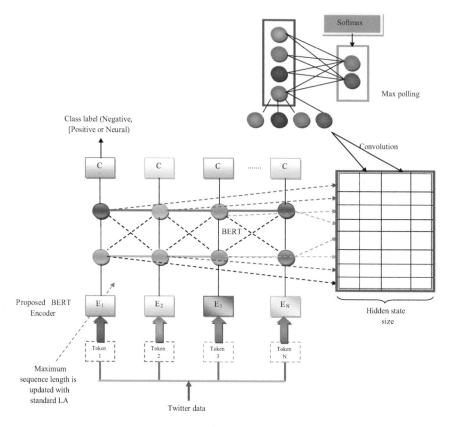

Fig. 2 Proposed customized Bert model

4 Material and Methods

The proposed model was developed in Python. A dataset called TWCS (Customer support on twitter) is used. TWCS data set is a large modern corpus of tweets. the objective of this dataset is to study modern customer care practices, conversations and their impact. In addition, the proposed approach was compared with other conventional classification methods as Instance-Based with k-nearest neighbors (IB-K) [13], Naïve Bayes (NB) [14], sequential minimal optimization (SMO) [13], Bayesian Network (BN) [15], jRip [13], j48 [13], PART [13], CNN [16] and BERT [9]. Different measures have been evaluated like precision, accuracy, specificity, sensitivity.

4.1 Analysis in TWCS Dataset: Performance and Errors

Figures 3, 4 and 5 shows performance of accuracy, precision, sensitivity and balanced accuracy of the optimized Bert compared to standard Bert model, instance-based with k-nearest neighbors (I-BK), naïve Bayes (NB), sequential minimal optimization (SMO), Bayesian network (BN), Jrip, J48, PART, and CNN.

In this work, the key parameter of the research, i.e., the accuracy of the classifier. The proposed method shows the highest value. An accuracy of 89.6% is achieved using our proposed approach which is better than accuracy's other conventional models at every variation of TP.

The proposed method shows the high value of measures for each variation. An accuracy of 89.9 is achieved for the optimized Bert compared to other models. Sensitivity shows the high value of our proposed method too at every variation. The same thing for the specificity and balanced accuracy.

Here, we can see that the presented work shows highest performance comparing to other models in every variation of TP in terms of negative predictive value, positive

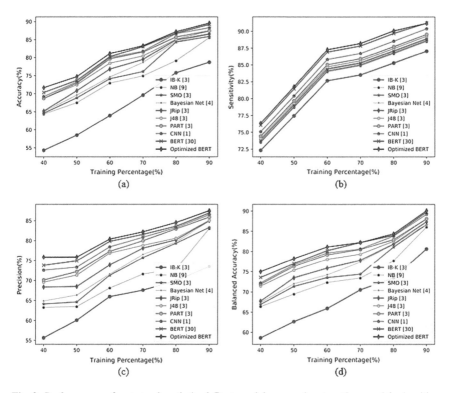

Fig. 3 Performance of proposed optimized Bert model comparing to other models (positive measures)

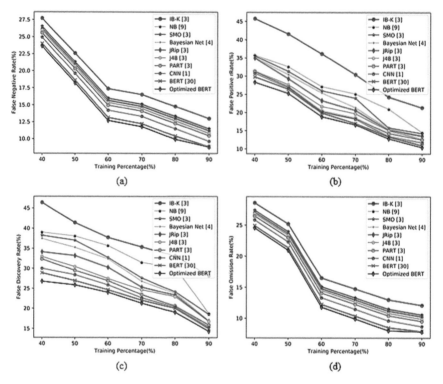

Fig. 4 Performance of proposed optimized Bert model comparing to other models (negative measures)

predictive value and Matthews correlation coefficient, a value of 91.8% is achieved at TP = 90 which is better than other values of the comparing models.

From Table 1, the proposed method shows the lowest value error performance (TP = 70).

5 Conclusion

In this paper, we have presented a customized Bert model for sentiment analysis in twitter datasets. Standard LA was used to tune the weight and the bias of Bert and to update the maximum sequence length of encoder. The accuracy of the proposed model is compared to 9 other conventional models, results shows that the accuracy of the proposed optimized Bert overcomes the accuracy of the other conventional models. The accuracy of the proposed model for TWCS dataset at TP = 90, the maximal accuracy of 89.6 is recorded by the presented work, and it is 13.3, 8.49, 7.46, 6.98, 67.71, 4.72, 31.12, 1.67, and 70.75% greater than the other mentioned conventional approaches.

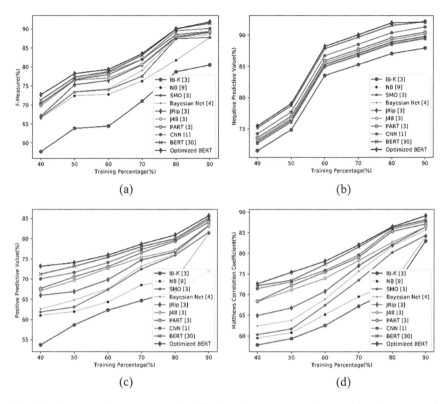

Fig. 5 Performance of proposed optimized Bert model comparing to other models (other measures)

Table 1 TWCS error performance

Approaches	TP = 40	TP = 50	TP = 60	TP = 70	TP = 80	TP = 90
IB_K [13]	3.38×10^1	3.12×10^1	2.98×10^1	2.52×10^1	2.11×10^1	1.89×10^1
NB [14]	2.16×10^1	2.06×10^1	1.99×10^1	1.95×10^1	1.76×10^1	1.18×10^1
SMO [13]	2.13×10^1	1.91×10^1	1.86×10^1	1.82×10^1	1.22×10^1	1.15×10^1
Bayesian Net [15]	2.08×10^1	1.81×10^1	1.80×10^1	1.55×10^1	1.20×10^1	1.09×10^1
jRip [13]	2.06×10^1	1.67×10^1	1.56×10^1	1.46×10^1	1.16×10^1	1.08×10^1
j48 [13]	1.63×10^1	1.48×10^1	1.39×10^1	1.36×10^1	1.14×10^1	9.92
PART [13]	1.62×10^1	1.42×10^1	1.25×10^1	1.23×10^1	1.0×10^1	9.87
CNN [16]	1.56×10^1	1.39×10^1	1.22×10^1	1.21×10^1	9.77 × 10¹	9.01
BERT [9]	1.43×10^1	1.34×10^1	1.19×10^1	1.07×10^1	9.63×10^1	8.07
Optimized BERT	1.26×10^1	1.21×10^1	1.09×10^1	1.04×10^1	9.18×10^1	7.68

References

1. Lagrari, F., & Elkettani, Y. (2021). Traditional and deep learning approaches for sentiment analysis: A survey. *Advances in Science, Technology and Engineering Systems Journal, 6*(4), 1–7.
2. Liu, B. (2012). Sentiment analysis and opinion mining. *Synthesis Lectures on Human Language Technologies, 5*(1), 1–167. https://doi.org/10.2200/S00416ED1V01Y201204HLT016
3. Habimana, O., Li, Y., Li, R., Gu, X., & Yu, G. (2020). Sentiment analysis using deep learning approaches: An overview. *Science China Information Sciences, 63*(1), 1–36.
4. Lagrari, F., & Elkettani, Y. (2020). Customized BERT with convolution model: A new heuristic enabled encoder for Twitter sentiment analysis.
5. Rachiraju, S. C., & Revanth, M. (2020). Feature extraction and classification of movie reviews using advanced machine learning models. In *2020 4th International Conference on Intelligent Computing and Control Systems (ICICCS)* (pp. 814–817).
6. Kalchbrenner, N., Grefenstette, E., & Blunsom, P. (2014). A convolutional neural network for modelling sentences. In *Proceedings of the 52nd Annual Meeting of the Association for Computational Linguistics* (Vol. 1: Long Papers, pp. 655–665), Baltimore, Maryland. https://doi.org/10.3115/v1/P14-1062
7. Turney, P. D. (2001). Thumbs up or thumbs down? Semantic orientation applied to unsupervised classification of reviews. In *Proceedings of the 40th Annual Meeting on Association for Computational Linguistics—ACL'02* (p. 417), Philadelphia, Pennsylvania. https://doi.org/10.3115/1073083.1073153
8. Pang, B., Lee, L., & Vaithyanathan, S. (2002). Thumbs up? Sentiment classification using machine learning techniques. In *Proceedings of the ACL-02 Conference on Empirical Methods in Natural Language Processing—EMNLP'02* (Vol. 10, pp. 79–86). https://doi.org/10.3115/1118693.1118704
9. Devlin, J., Chang, M.-W., Lee, K., & Toutanova, K. (2018). Bert: Pre-training of deep bidirectional transformers for language understanding. ArXiv Preprint ArXiv:1810.04805
10. Boothalingam, R. (2018). Optimization using lion algorithm: A biological inspiration from lion's social behavior. *Evolutionary Intelligence, 11*(1–2), 31–52.
11. Lagrari, F., & Elkettani, Y. (2020). Customized BERT with convolution model: A new heuristic enabled encoder for Twitter sentiment analysis. *International Journal of Advanced Computer Science and Applications, 11*(10), 9.
12. Axelbrooke, S. *Customer Support on Twitter.* kaggle.com/datasets/thoughvector/customer-support-on-twitter.
13. Iqbal, F., Hashmi, J. M., Fung, B. C. M., Batool, R., Khattak, A. M., Aleem, S., & Patric. (2019). A hybrid framework for sentiment analysis using genetic algorithm based feature reduction. *IEEE Access, 7.*
14. Sailunaz, K., & Alhajj, R. (2019). Emotion and sentiment analysis from Twitter text. *Journal of Computational Science, 36.*
15. Ruz, G. A., Henríquez, P. A., & Mascareño, A. (2020). Sentiment analysis of Twitter data during critical events through Bayesian networks classifiers. *Future Generation Computer Systems, 106,* 92–104.
16. Jianqiang, Z., Xiaolin, G., & Xuejun, Z. (2018). Deep convolution neural networks for Twitter sentiment analysis. *IEEE Access, 6.*

Crops Recommendation System Model Using Weather Attributes, Soil Properties, and Crops Prices

Sudarshan Reddy Palle and Shital A. Raut

Abstract Agriculture is a respectful and predominant occupation in India. Approximately, 60% population works in this industry contributing to 18% GDP of the country. Weather and soil conditions are key factors to be considered for selecting a crop to be farmed. Even after growing a crop successfully, it should provide profitable returns to the farmer. Traditionally, farmers choose a crop to be harvested from past experiences and some wise men advices. But there are chances that predictions may go wrong assuming incorrect past experiences since human brains are good at analyzing rather than remembering. There were many articles and research works which tried to recommend crops based on either weather or soil properties. But a methodology which includes both of them along with consideration of crop prices will be of more useful to farmers. So, a proper system using weather attributes, taking soil inputs, and analyzing profit earned could help farmers in taking a right decision of crops to be farmed. Using this model, we propose a system that can recommend highly profitable crops to farmers. For predicting weather attributes like temperature and rainfall, we can use past data of a region applied to ARIMA model. Logistic regression can be used for classification of data based on weather and soil conditions. And for predicting profit that can be attained, we can use investments and returns data of past years applied to ARIMA model.

Keywords Data mining · Crop recommendation · Weather attributes prediction · Profit prediction · Autoregressive integrated moving averages (ARIMA) · Logistic regression

S. R. Palle (✉) · S. A. Raut
Department of Computer Science and Engineering, Visvesvaraya National Institute of Technology, Nagpur, Maharashtra, India
e-mail: pallesudarshanreddy1@gmail.com

S. A. Raut
e-mail: saraut@cse.vnit.ac.in

© The Author(s), under exclusive license to Springer Nature Singapore Pte Ltd. 2023
S. Shakya et al. (eds.), *Sentiment Analysis and Deep Learning*,
Advances in Intelligent Systems and Computing 1432,
https://doi.org/10.1007/978-981-19-5443-6_24

323

1 Introduction

Data mining refers to the extraction of useful information and patterns from data. With the advancements in data storage and machine learning fields, data mining had got a huge hype since the knowledge extracted from this information can be used to improve businesses. Data mining [1] is widely applied in medical, stock market analysis, banking, finance, retail, agricultural, and many other sectors to extract knowledge and thus making better business decisions. Agriculture is a respectful and an ancient occupation in India. The key attributes for agriculture are weather and soil properties. Expected profit for a particular crop can be calculated by predicting investment and return values associated to it. But traditional way of assumptions for crop selection may go wrong since assumptions are made on past experiences, and human brains are not good at storing past data compared to machines. Also there are many problems for farmers like sometimes weather conditions will not be favorable, sometimes fertilizers chosen may be disproportionate in chemical compositions because of misunderstanding soil properties, and finally, returns may not be as good as expected. So, considering all these factors, a proper mechanism or methodology is required that best recommends a set of favorable crops to farmers which are optimally profitable.

There are some techniques or methodologies which recommend one profitable crop based on either farmer's input of weather or soil properties. But a methodology that includes weather attributes prediction could be of more beneficial. So, in this paper, we tried to provide a methodology which could predict weather attributes based on a location selected and recommend highly profitable crops to the farmers according to their region. Farmers need to input mentioned soil properties, their choice of crops, and pick their location using map. Based on these input parameters, the model tries to output best profitable crops and gives comments about farmer's choices and expected profits. From these suggestions and comments about his choices, farmer will decide which crops to be harvested.

2 Literature Survey

The paper [2] tries to give recommendation of crops based on soil properties. The properties such as soil depth, texture, pH, color, permeability, drainage, water holding, and erosion are considered for the crops groundnut, millet, paddy, sugarcane, pulses, vegetables, etc. The model is trained with data collected from Madurai soil testing lab and some general online sources of crop data. "Ensembling" technique is used which included models like CHAID, random tree, KNN, and Naïve Bayes algorithms for classification of data. Once the training is done, voting mechanism has been used to get best accuracy model out of all and recommend the crop.

The paper [3] tries to recommend the crops based on soil and weather properties. The attributes considered are N, P, K, pH, rainfall, temperature, and humidity. The training dataset is taken from Kaggle Website. The farmer needs to enter all these input values manually, and based on these values, best crop will be recommended by the trained model. Models like decision trees, Naïve Bayes, support vector machine, logistic regression, random trees, and XGboost were applied on training dataset. Out of applying all models, XGBoost has given highest accuracy score of 99.31%. This paper also illustrated related works from different papers, given comparative study with accuracy measures.

The paper [4] proposes a model with precision agriculture for small and marginal scale farmers to affect a degree of control over variability. The author gives a detailed explanation of precision agriculture and its practices in foreign countries. The methodology is carried out on small-scale farms in Kerala where average land holdings are less comparatively so that it can be enhanced easily to large farms. Sensors were used at panchayat levels to get dynamic attributes like weather conditions, soil conditions, and field monitoring. SMS and email communication were used to give notifications to farmers for actions to be taken at defined intervals of time like cyclone alert, pest attack, etc.

The paper [5] proposes a system for crop yield prediction based on past historical data with attributes like rainfall, temperature, and soil conditions. Datasets collected were applied on random forest algorithm, and accuracy is proportional to crop yield profit. They also applied various methods like density diagram, box plots upon the data attributes to analyze and understand their behavior.

The paper [6] describes the importance of crop selection to be yield and proposes a model for the same. They discussed about various machine learning algorithms with their relation at each level of model. They used the historical information of crop yield time, production cost, and yield rate which were influenced by parameters like weather, soil type, water density, etc., and predicts best yield crop and proposes that.

Each of the articles listed above makes use of one or more weather and soil properties. Papers [2] and [4] use soil and weather properties, respectively. Both are used in papers [3] and [5]. In paper [6], cost parameters are combined with yield time and factors that influence yield rate. There were no models which are considering all of the factors impacting to farming in a single work which could had been more beneficial. So, a methodology that combines weather, soil conditions, and even crop prices will be more real-time usable. Also, we should not provide a single recommendation every time because farmers may not always be able to harvest the crop suggested. As a result, giving two or more crops as well as the details of necessary funds will be quite beneficial. Also, it is not always the case there were only given number of profitable crops every time. The farmer might have interest toward some particular crops. So giving some comments or information about his preferences would be also beneficial in selecting crop. We attempted to provide a model for such a real-world application which could be an advancement or extension of above mentioned works in this paper.

3 Methodology

Block Diagram for Proposed System
The proposed system tries to create a methodology for recommendation of crops using available and assumed data. It tries to recommend crops based on weather attributes, classification of crop labels, and rearranging crops according to profits. The entire methodology can be split into six components as shown in Fig. 1.

1. Collection and creation of datasets.
2. Data cleaning and preprocessing.
3. Weather attributes generation model creation.
4. Classification model creation.
5. Rearranging crops based on profits.
6. Comments on user's preferences.

3.1 Collection and Creation of Datasets

The model requires following datasets:

- Training dataset for classification model
- State-wise weather datasets
- Crops prices datasets

3.1.1 Training Dataset

Training dataset is obtained from Kaggle Website [7]. It consists of N, P, K, pH, rainfall, humidity, and temperature as input parameters and crop name as output

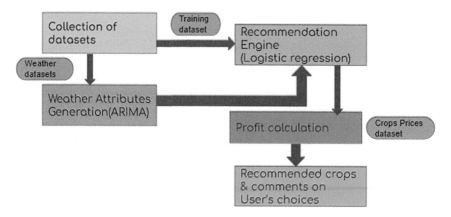

Fig. 1 Block diagram for proposed system

label. The dataset consists of twenty two crops such as rice, maize, pigeon peas, grapes, orange etc., for different values of input parameters.

3.1.2 State-Wise Weather Datasets

The state-wise weather datasets [8] are required for weather attributes prediction. It consists of temperature and rainfall values of each district of a state for past twenty years. We generated such patterned data for all states and districts in year-wise using meteomatics Website. We will have one dataset for each state.

3.1.3 Crops Prices Datasets

Crops prices dataset consists of district name, crop name, investment, and return values of each crop for past twenty years. Some values are available for past data, and remaining values are generated out of pattern basis to support the model.

3.1.4 User Interface for Input

We have Google Maps API [9] for map implementation. The farmer needs to select soil type and enter soil properties like nitrogen (N), phosphorous (P), and potassium (K) which are key attributes in deciding soil contribution in crop harvest. Using the pointer on map needs to select his location so that district name and state name could be extracted which are required in generating weather attributes and crop finances involved as in Fig. 2.

3.2 Data Cleaning and Preprocessing

The data collected will not always be in ready to use format. There might be missing values for attributes, and some attributes may not be required. So, proper analysis is to be done about data acquired and make it processed for proper use with model. For example, the state-wise weather datasets collected does not contain district name initially but are with latitude and longitude values. But, the input that comes from user is in form of district name itself. So, it will be burden for model to dynamically attain latitude and longitude values from district name. Instead data transformation can be done in which we can just create a column with district name initially, assign value once with district name for corresponding latitude and longitude and use when required. During weather attributes prediction, we need to first identify state name and district name for taking out dataframe to be predicted. After identification of dataframe whichever we will operate on data reduction can be applied to pick particular weather attributes and provide to forecasting model to predict upcoming values.

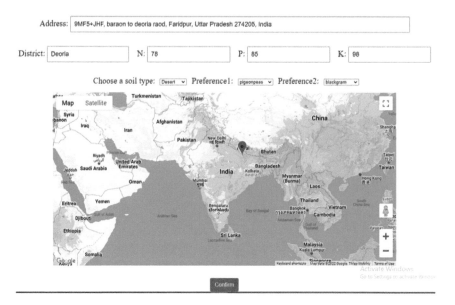

Fig. 2 User interface page

This type of preprocessing activities will help in faster response and high performance of model in real time.

3.3 Weather Attributes Generation Model Creation

As we have seen, weather conditions have a greater impact on farming. So, understanding which factors will impact at most and using them in recommending crops will be of great use. In this model, we had considered temperature and rainfall attributes and predicted upcoming values based on ARIMA model. The dataset that needs to be picked for weather attributes generation will be dynamically selected based on user's location choice from state-wise weather datasets.

ARIMA [10] stands for auto regressive integrated moving averages method. It is a statistical model for analyzing and forecasting time series data. It is constructed from three components AR, I, and MA. AR stands for autoregression is a model that uses the dependent relationship between observation and some number of lagged observations. I stand for integrated which uses differentiation of raw observations to make time series stationary. MA stands for moving averages, which uses dependent relationship between observation and residual error from moving average model that was applied to lagged observations.

ARIMA library function is provided by "statsmodels.tsa.arima.model" package in Python.

Syntactical usage of ARIMA is

1. *arimamodel = ARIMA(dataset, order = (p, d, q))*
2. *model = arimamodel.fit( )*
3. *output = model.forecast( )*

where *p* denotes value for AR, *d* denotes value for *I*, and *q* denotes value for MA.

In line 1, for ARIMA, we need to feed the dataset (the dataset with only considered attribute like temperature or rainfall list of values) upon which we need to forecast the values along with *p, d, q* values.

Line 2 returns a reference for the fitted data to an object.

In line 3, we are forecasting the upcoming value.

The values *p, d, q* should be identified with manual inspection, and they should be considered based on least root mean square errors.

3.3.1   Identifying *I* Term

Here, the dataset that is been used is a historical data of rainfall or temperature of a particular district. So, for predicting the upcoming value, we need to understand the past data clearly and need to difference it to make stationary since regressive methods are well suited on stationary or nearly stationary data. So, we need to identify differencing parameter for every dataset and should be noted, and this becomes d term in the equation.

3.3.2   Identifying AR Terms

For finding number of AR terms that are to be included, we need to inspect the partial autocorrelation (PACF) plot. Partial autocorrelation is said to be the relation between the series and its lag, excluding contributions from intermediate lags. So, it says the purest relation between lag and series.

Mathematically, PACF can be written as

$$Y_t = \alpha_0 + \alpha_1.Y_{t-1} + \alpha_2.Y_{t-2} + \alpha_p.T_{t-p}$$

The count of terms that need to be considered is taken from PACF plot that we obtain by performing first order differencing and observing lags whichever have crossed significance level and that become *p* value in the equation.

3.3.3   Identifying MA Terms

As like in AR, in MA, the Y_t depends on lagged forecast errors. Mathematically, it can be written as

$$Y_t = \varepsilon_t + \phi_1.\varepsilon_{t-1} + \phi_2.\varepsilon_{t-2} + \cdots + \phi_q.\varepsilon_{t-q}$$

For identifying number of MA terms to be considered, ACF plot is required with first order differencing the given dataset and take conservatively the number of terms whichever have crossed significance level so as to eliminate over-fitting and under-fitting issues. This becomes q value in the equation.

3.3.4 Observation

After several observations with different districts datasets and different values of p, q, d, it was observed that at $p = 2$, $d = 1$, and $q = 1$, the model is working good for both temperature and rainfall with least root mean square error.

The following are test observations for temperature predictions using ARIMA model.

From Fig. 3, blue line represents actual values, and red line represents predicted values. X-axis represents numbers of observation, and Y-axis represents values of observation. These values are obtained by performing testing operation on ARIMA with temperature dataset. Since we have twenty records for every value prediction, we have fed 13 values as training data and predicted next 7 values with forecast method. Here, "predicted" is the value predicted by ARIMA model, and "expected" is the actual value in dataset.

Above is an example for temperature value prediction for a district. We can see that the RMSE value turned out be approximately. Generally in the case of temperature, a difference of 2 degrees can be accepted as a good prediction. As the root mean square error goes lower, we can say that value generated will be more precise and accurate for temperature and rainfall. To achieve this lower root mean square, we need to manually try with different combinations of p, d, q values.

The expected rainfall can also be predicted by performing similar kind of training and forecasting in which the rainfall column for past years needs to be fed to ARIMA with appropriate p, d, q values. This can be done for other weather attributes by changing p, q, d values for each parameter based on pattern the attribute values are being followed.

3.4 Classification Model Creation

There are many models available like above observed papers which try to give one best output label based on user's input. But it is not a good idea of recommending only one output every time since there could be more than one crop which has same optimality for given input. Also the farmer may not be able to afford the harvesting recommended crop all times or might not be habituated to farm that recommended crop. So, giving more than one, i.e., three or more recommendation of crops for any given input along with details like investment, return, and profit could be better. It gives an opportunity to user to choose what to harvest based on those recommendations.

```
predicted=24.883, expected=25.700000
predicted=24.399, expected=25.000000
predicted=24.602, expected=22.000000
predicted=25.100, expected=24.000000
predicted=25.583, expected=23.800000
predicted=24.074, expected=21.500000
predicted=23.206, expected=27.300000
Test RMSE: 2.254
```

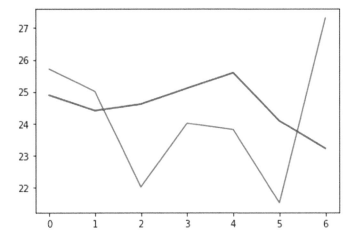

```
Expected temperature: 24.410662907930767
```

Fig. 3 Observations of ARIMA predictions

So, from above observation, we can understand that we need to implement a model which could give closeness or relation of each crop label for a given input. From those closeness values, pick top three or more labels and identify finances associated with them. For this operation, we can use logistic regression with one versus rest multiclass model. This model considers each label as an independent label compared to all other labels and finds probability of that label for any given input. In this way, we can identify probability of each label for any given input fed to the model that got trained with training dataset.

3.4.1 Proposed Model

As per the above requirement, we need to create multiple binary classifiers for each label and find belongingness of input to each label. In training dataset, we have twenty two crops as final labels. So, we need to create twenty two binary classification models every time considering positive class as selected label and negative class

Fig. 4 Multi-logistic
regression

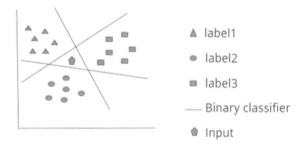

as all remaining labels. Whenever a new input is fed, it will be compared to each classifier, probability with each label is attained and then sorted in decreasing order of probabilities. From this sorted list, we can take out as many suggestions as required to output.

As in Fig. 4, each polygon represents a label of each crop, and each line represents a binary classifier considering each label as positive class and all remaining as negative class. So, whenever a new input from user is received, the model internally creates as many models as labels and finds probabilities with each crop label.

A normal linear regression model can be applied whenever the final labels are proportional to input fed by user and output is binary. But in our requirement, the input fed is multi-input, outputs are multi-labeled, and even the training data cannot be proportional to labels. So, logistic regression with multi-classification is required.

Logistic function or sigmoid function is an activation function that is used to add non-linear attributes in machine learning and decides which attribute to output and which to not. Here as we have discussed, our input attributes are non-linear with multi-directional and multi-valued attributes. So, logistic function will serve better for an activation function, and logistic regression will be good as classification algorithm since we require multi-outputs.

Logistic function or sigmoid function can be represented as

$$p = 1/\left(1 + e^{-(\beta_0 + \beta_1 x)}\right)$$

where p is the probability of positive class for given input.

This function helps in identifying the probability of each class or label for given input values.

Using the above probability function, we need a cost function for each label at given input. So a cost function for this requirement using logistic regression is

$$-\log[L(\theta)] = -\sum y * \log\left[\sigma\left(\Theta^T.x^i\right)\right] + (1 - y) * \log\left(1 - \sigma\left(\Theta^T.x^i\right)\right)$$

where y represents probability of positive class, and $\log\left[\sigma\left(\Theta^T.x^i\right)\right]$ represents probability of feature x belonging to that particular class. Negative sign represents it is a loss function, and we need to minimize it to attain maximum profit. This kind of regression is to be done for as many labels present and store probabilities for each class.

The above proposed model can be achieved by one vs rest multi-class logistic regression [11, 12] model by importing logistic regression library from "sklearn.linear_model" library.

```
X_train, X_test, y_train, y_test =
train_test_split(featureset, targetset, test_size
= 0.3, random_state = 42)
scaler = StandardScaler()
X_std = scaler.fit_transform(X_train)
Xtest_std = scaler.fit_transform(X_test)
clf = LogisticRegression(random_state = 0,
multi_class = 'ovr')
model = Clf.fit(X_std, y_train)
y_pred = model.predict(Xtest_std)
score = accuracy_score(y_test, y_pred) * 100
print("Accuracy score of logistic regression
model:", score)
```

```
Accuracy score of logistic regression model:
94.24242424242424
```

The above description depict the training of model with one vs rest logistic regression and show accuracy of model after training. We have taken 70% data for training and remaining 30% for testing the model. We need to standardize our data to make it suitable for training. StandardScaler() is a function from "sklearn" library which provides us the required functionality. Once we train the model, we can test our data with that model to check its accuracy. We can see it is producing 94% accuracy for the given dataset.

Now, when a new input is fed, it will be calculated as follows.

3.4.2 Pseudo Code for Classification

Probabilities = model.predict_probability(new_input)
dict = {}
for cropname in labels:

dict.add(cropname, probabilities[cropname])

dict.sort() // according to probabilities
for i = 0 to 3 in dict:

print cropname

In the above code portion, new input is the input fed by user. Probabilities of each crop label against given input are calculated. We need to arrange them in decreasing order of probabilities and pick top required labels.

```
[[102, 45, 36, 33.418647854419504, 157.12060635486708,
7.1417935159818935, 7.7]]
cotton
orange
grapes
```

So, for given input, cotton, orange, and grapes are top three recommended crops. Based on *i* value in above code, we can even recommend more than three crops.

3.5 Rearranging Crops Based on Profits

Till this step, we were able to figure out which were best suitable crops to be harvested based on weather and soil conditions. Now, we need to rearrange recommendations based on the profits associated with each crop. So, we have collected datasets of investment and return values of each crop for each state for past twenty years. The dataset consists of district name, crop name, investment, return, year as attributes. So, when we know state name, district name, and crop name, we can get expected investment and return which consequently gives expected profit associated with any crop. Again, we can use ARIMA model for predicting investment and return values based on past years' data since it is also a time series forecasting series. Here, we are getting our input values of state name and district name from location selected by user from maps on user interface which we provide as landing page for application. So, we can calculate profit for the crops, and we received from the previous step based on the district name and state name and arrange crops in decreasing order of profits.

3.5.1 Pseudo Code for Rearrange

```
dict = {}
for crop in weather_crops:

    inv = [list of investment[crop] for past twenty years for selected district]
    ret = [list of return[crop] for past twenty years for selected district]
    arimamodel = ARIMA(inv, order = (p, d, q))
    model = arimamodel.fit()

investment = model.forecast()
```

```
    arimamodel = ARIMA(dataset, order = (p, d, q))
    model = arimamodel.fit()

return = model.forecast(ret)

    profit = return-investment
    dict.add (crop,profit)

dict.sort() // based on profit
```

```
Expected investment for mango is: 148892.49457418337
Expected return for mango is: 314106.30546091526
Expected investment for coffee is: 26753.23843392141
Expected return for coffee is: 101993.10670854531
Expected investment for mothbeans is:
4876.6976868420315
Expected return for mothbeans is: 15438.539499269225
Profits associated with each crop are:
[('mango', 165213.8108867319), ('coffee',
75239.8682746239), ('mothbeans', 10,561.841812427194)]
```

From the above description, we have depicted investment, return, and profits for each crop received from the previous step and are arranged in decreasing order of profits. These output values can be shown to user as final output screen.

3.6 Comments on User's Preferences

We have taken preference choices from user as input so that we can verify if they are feasible to harvest for the available weather and soil conditions. We will strongly recommend the user's choice if it is in top five of recommendations generated, slightly recommend if in top six to ten and oppose recommendation if in more than tenth recommendation. These recommendations are done based on weather and soil conditions, and profits associated are shown to user in all cases.

3.6.1 Pseudo Code for Preferences

```
for crop in preferences:

    Expected investment for crop is: investment[crop]
    Expected return for crop is: return[crop]
    if crop.position in recommended_crops <=5

        good to be grown

    else if crop.position in recommended crop >5 and <=10
```

Table 1 Results of machine learning models used

Function	Model	Metrics	Value
Weather attributes generation	ARIMA	RMSE	2.254
Classification	Multi-logistic regression	Percentage	94.24

slightly good to be grown

else

not good to be grown

```
Expected investment for grapes is: 5247.153693107961
Expected return for grapes is: 155468.85085564727

Expected investment for mango is: 148892.49457418337
Expected return for mango is: 314106.30546091526
grapes is good to be grown
mango is slightly good to be grown
```

4 Results and Discussion

Using ARIMA model, the model is acquiring small root mean square error which accounts to good efficiency of predictions of weather and crop prices. Logistic regression is producing an output of 94.24 for given training dataset from Table 1.

There are models which tried to recommend crops based on weather inputs or soil inputs and some using both. But we need a model which combines all of them and provide a model which gives highly profitable and feasible crops. This model tries to overcome or improvise those limitations using existing models by combining all of them. We tried to include weather attributes data, recommending more than one crop using classification algorithms, and predicting expected crops prices from past years trend. This paper tries to give a model for such a real-time useful project by showing up all required activities using highly realistic patterned data.

5 Conclusion

India is an agriculture-based country, and farmers and villages are backbone for it. Helping the farmers can make economically stronger India and promotes pride among other nations of world by exporting food and farm products to other countries. In this paper, we tried to propose a methodology that could make farming profitable and reduced risk by using attributes like weather, crop prices, and analyzing soil conditions. The earlier works tried to propose or recommend one crop always which is not an efficient way of recommendations. They have used classification algorithms in a way that produces one best output every time. But in our solution, we tried to understand the nature and behavior of every crop for any given input. Then, based on this knowledge, we had tried to suggest top number of recommendations in which the number is configurable by us. Also, comments produced on user's preferences will give an idea about how profitable or feasible to grow that crop.

6 Future Work

- Include real-time data for weather, crop prices.
- Include more attributes for weather and soil conditions to make predictions much more accurate.
- Create a Website or android app for easy access to user.

References

1. Han, J., Kamber, M., & Pei, J. *Data mining concepts and techniques* (3rd ed.). Morgan Kaufmann Publications.
2. Pudumalar, S., Ramanujam, E., Harine Rajashree, R., Kavya, C., Kiruthika, T., & Nisha, J. (2017). Crop recommendation system for precision agriculture. In *2016 IEEE Eighth International Conference on Advanced Computing (ICoAC)*, June 19, 2017. https://ieeexplore.ieee.org/stamp/stamp.jsp?tp=&arnumber=7951740
3. Gosai, D., Raval, C., Nayak, R., Jayswal, H., & Patel, A. Crop recommendation system using machine learning. *International Journal of Scientific Research in Computer Science, Engineering and Information Technology*. ISSN: 2456-3307. https://ijsrcseit.com/paper/CSEIT2173129.pdf
4. Babu, S. (2013). A software model for precision agriculture for small and marginal farmers. In *International Centre for Free and Open Source Software (ICFOSS)*, October 15, 2013. https://ieeexplore.ieee.org/stamp/stamp.jsp?arnumber=6629944
5. Jeevan Nagendra Kumar, Y., Spandana, V., Vaishnavi, V. S., Neha, K., & Devi, V. G. R. R. (2020). Supervised machine learning approach for crop yield prediction in agriculture sector. In *Proceedings of the Fifth International Conference on Communication and Electronics System (ICCES 2020)*, July 10, 2020. https://ieeexplore.ieee.org/stamp/stamp.jsp?tp=&arnumber=9137868
6. Kumar, R., Singh, M. P., Kumar, P., & Singh, J. P. (2015). Crop selection method to maximize crop yield rate using machine learning technique. In *2015 International Conference on Smart Technologies and Management for Computing, Communication, Controls, Energy and Materials (ICSTM)*, August 27, 2015. https://ieeexplore.ieee.org/stamp/stamp.jsp?tp=&arnumber=7225403&tag=1
7. Kaggle Datasets. https://www.kaggle.com/datasets/atharvaingle/crop-recommendation-dataset

8. Meteomatics Weather API. https://www.meteomatics.com/en/weather-api/
9. Google Maps Platform. https://developers.google.com/maps
10. Prabhakaran, S. ARIMA model for time series forecasting in Python, machine learning plus community. https://www.machinelearningplus.com/time-series/arima-model-time-series-forecasting-python/
11. Brownlee, J. One vs rest and one vs one multiclass classification model, machine learning mastery community. https://machinelearningmastery.com/one-vs-rest-and-one-vs-one-for-multi-class-classification/
12. One vs rest classifier, scikit-learn community. https://scikit-learn.org/stable/modules/generated/sklearn.multiclass.OneVsRestClassifier.html

Development of Machine Learning Approaches for Autism Detection Using EEG Data: A Comparative Study

Anoop Kumar and Anupam Agrawal

Abstract Autism is a kind of disorder that impacts the human brain due to which people face some difficulties in interaction and communication with others. The detection and identification of autism at an early stage are a difficult task for researchers. The preprocessing techniques include discrete wavelet transformation (DWT), standard deviation, and mean. DWT helps in preprocessing of the EEG signals, which reduces the noise and decomposes the signals into EEG subbands. In this work, we have discussed various classifiers like SVM, Naïve Bayes, KNN, random forest, decision tree, LSTM, and ANN. These classifiers help to classify the EEG signals into autistic and non-autistic based on the features extracted. When evaluated on the actual dataset obtained from King Abdul-Aziz University(KAU), Saudi Arabia, the techniques produced up to 99.9% encouraging results. This dataset contains 17 subjects, in which there are 4 normal and 13 autistic subjects. We are using the SMOTE technique for data augmentation, which has helped us to improve the performance.

Keywords Autism spectrum disorder · Machine learning · DWT · Classifier · Electroencephalogram

1 Introduction

Autism is a type of mental illness. People suffer greatly as a result of this illness. They have issues with communication and social contact. Using electroencephalogram (EEG) signals, researchers are attempting to create a computer-assisted model [1]. EEG is recommended because it is less expensive than other testing and is more accessible to clinicians. Because electroencephalogram signals are intricate, non-linear, and non-stationary, we cannot employ Fourier processing techniques directly.

A. Kumar (✉) · A. Agrawal
ITMR Lab, Indian Institute of Information Technology, Allahabad, Uttar Pradesh, India
e-mail: mit2020123@iiita.ac.in

As a result, we favor the discrete wavelet transformation (DWT) approach, which can split EEG signals into a semi-analytical temporal domain and detect signal changes.

Our research aims to discover novel approaches for assisting in the diagnosis of ASD using DWT and a combination of classifiers. SVM, Naive Bayes, LSTM, ANN, AdaBoost, random forest, decision tree, and KNN [2] are some examples. The properties of electroencephalogram signals are discovered using statistical approaches (mean and standard deviation). The classifiers are then applied to the retrieved features of the EEG sub-bands. We attained the highest accuracy, up to 99.90%, using ANN for observable features. The rest of the paper is organized as follows: The literature survey is discussed in Sect. 2. The dataset used in this investigation is described in Sect. 3. Section 4 describes the methods for processing and feature extraction. Results are discussed in Sect. 5. Section 6 provides the paper's conclusion as well as information on its future scope.

2   Literature Survey

Many computer-assisted diagnoses have been investigated in the previous years. The research reported by Sheikhani et al. [3] in which they recorded a dataset with 21 electrodes and choose two earlobes. There were two distinct groups; ASD was assigned to 10 boys and 1 female, while non-ASD was assigned to 7 boys and 3 girls. They choose a method (STFT) to extract the features and then applied a KNN classifier to get an accuracy of 84.2%. In their further work [4], they improved the technique and also used a larger dataset, where they achieved an accuracy of 96.4%.

In the work presented by Ibrahim et al. [1], they have worked on different datasets and used different EEG channels and different placements of the electrode. Their work includes DWT of the 6th level with entropy and a KNN classifier. The classifier has achieved an accuracy of 94.6% using KAU data.

In one of the studies by Alhaddad et al. [5], they collected a dataset of 12 subjects, of which 8 (5 male, 3 female) had ASD symptoms, and 4 (all of them male) had no symptoms of ASD. The dataset had 16 channels with the right earlobe as a reference. After that, they used a band-pass filter with a frequency range of (0.1–60) Hz to filter the data. The feature extraction techniques they used were FFT and fisher linear discriminate (FLD). They achieved an accuracy of 90%.

3   Experimental Data

The methods are tested on an EEG dataset provided by KAU University. The raw EEG data contains four normal participants (males, ages 9–16) and 13 autistic participants (male and female, ages 9–16) contributed to this dataset. The data was recorded with a 16-channel EEG [6] at a sampling rate 256 Hz in a relaxed state. The EEG data is presented in .dat files.

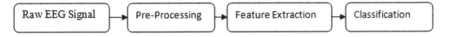

Fig. 1 Flowchart for the proposed approach

Fig. 2 Autistic subject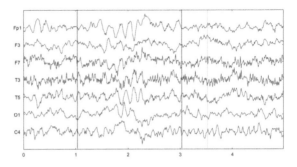

These are the channels (F3, F7, FP1, T3, T5, O1, and C4) [7] which are used in this study.

4 Research Methodology

The following three basic processes illustrated in Fig. 1 are used to create a model for diagnosis:

1. Preprocessing of EEG signals.
2. Extraction of features.
3. Classification.

The raw electroencephalogram signals are subjected to preprocessing to eliminate any noise in the brain patterns. To extract characteristics from the EEG segment, we used the discrete wavelet transformation (DWT), mean, and standard deviation. Many classifiers are applied, which are KNN, linear SVM, Gaussian process, decision tree, random forest, AdaBoost, Gaussian Naive Bayes, LSTM, and ANN for classification.

4.1 Preprocessing

We have used EEGLAB [8] for preprocessing the data and re-sampled [9] data from 256 to 128 Hz. After re-sampling, the data is passed through a FIR filter with a bandwidth of 0.5–60 Hz. The data is re-referenced using an average referencing technique; after this, we merge the files into a single EDF file.

Figures 2 and 3 represent the channels for autistic and non-autistic subjects.

Fig. 3 Non-autistic subject

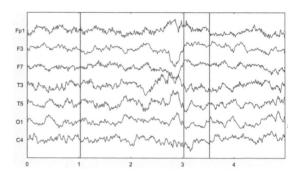

4.2   Feature Extraction

Wavelet techniques are mostly used for signal processing in the case of time-frequency decomposition. The DWT is quite capable of making small changes in electroencephalogram signals [10] by presentation of the signals in terms of approximate and detailed coefficients in the multi-scale time-frequency domain with the help of the Daubechies 4 wavelet [11]. We have used the 6th level of decomposition in this study, which is known for electroencephalogram analysis, we have used these [12] (D1, D2, D3, D4, D5, D6, and A6) coefficients for feature extraction in EEG sub-bands [13]. Here, we have worked on DWT, mean, and standard deviation techniques to extract the features from EEG segments.

4.3   SMOTE

SMOTE is also known as the synthetic minority oversampling technique [14]. It is very helpful to balance the class distribution by increasing the minority class at random. This technique is widely used for oversampling approaches to overcome the imbalance problem. In this study, we find that the shape of the data before SMOTE is (176,898), and after applying SMOTE, it is (268,498), which somehow helps to improve the accuracy.

4.4   Classifiers

KNN KNN is one of machine learning's most basic but crucial categorization algorithms. Face recognition and text mining are just a few of the applications it finds in the supervised learning domain. It is also called the lazy learning algorithm. It is used for classification purposes. The k value represents the nearest neighbors [15], and we will choose k equal to 5 for all tests.

SVM SVMs, or support vector machines, are supervised machine learning algorithms [16]. This approach is well-known for data analysis and is also used to solve classification and regression problems.

Gaussian Process Gaussian is known as a stochastic process (which contains most of the random variables that are indexed by time or space) such that every finite stock of the random variables contains the multivariate variables which are normally distributed [17], i.e., that every finitely distributed combination should be normally distributed.

Decision Tree Decision tree is known for the tree type of structure which represents a tree like model that can represent decisions and their possible outcomes [18], which can include the event outcome, less costly resources, and also helpful for the utility. It is one of the most important algorithms that can be used for conditional and control statements.

Random Forest Random forest is a well-known machine learning ensemble technique that is commonly used for classification and regression issues [18]. All operations are operated by the construction of the multitude of decision tree at the time of training and giving the output by using the mean or average which helps in prediction of the classes of the individuals.

AdaBoost AdaBoost is commonly known as the adaptive boosting method which is a machine learning algorithm. These techniques are mostly used to improve the performance of other algorithms by just combining them together. By doing this, it improves the accuracy and performance of the algorithms. So if an algorithm is performing poorly then that algorithm is combined with the weighted sum which gives the final new and a better output result [19]. This booster is known as adaptive because the poor performer learners which are misclassified are tweaked to the instances by previously misclassified classifiers. There could be cases where it will be high strung to the over-fitting issue than the other machine learning algorithms. So we can say that as long as the accuracy of each one is little bit better at the end, the overall performance will be better.

Gaussian Naive Bayes If we extend the Naïve bayes to the real valued attributes or Gaussian distribution, then it will be known as Naïve Bayes [17]. We may use the normal distribution to calculate the distribution of data because we only need to calculate the mean and standard deviation of the data.

Neural Network Neural networks are multilayer perceptrons which work on forward and backward propagation method. The data is processed through multiple layers of the network [20]. It also has dense layers, which aid in the formation of a connected neural network by connecting all of the separate layers. We apply the activation function and bias while processing the neural network.

LSTM LSTM networks are a sort of recurrent neural network [21] that can learn order dependence in sequence prediction issues. This is a necessary characteristic in complicated problem fields such as machine translation, speech recognition, and others.

Table 1 Confusion matrix

Predicted class		
Actual class	ASD	Non-ASD
ASD	True positive	False negative
Non-ASD	False positive	True negative

The performance of a classifier is generally measured with the help of confusion matrix that is given in Table 1.

Apart from this, we also follow a formula (given in Eq. 1) that is used for the calculation of accuracy.

$$\text{Accuracy}(\%) = \frac{\text{TP} + \text{TN}}{\text{TP} + \text{TN} + \text{FP} + \text{FN}} \tag{1}$$

where

TP True positive
TN True negative
FP False positive
FN False negative

this is how the accuracy of the classifiers is calculated.

5 Results and Analysis

In this work, we have taken both the autistic and normal electroencephalogram signals from the KAU data. Generally, if the spectrum has a low frequency, then it will have more power density, otherwise it will have less power density. As a result, if there is a high frequency, it will be considered noise in an electroencephalogram.

We have worked on many methods, their results are given in Table 3, and detailed information has been given in Sect. 4, which tells us about the methods used in this work. All of these methods have been tested using KAU datasets to ensure that they provide a fair comparison. Figure 4 shows the ROC plot for the 7 classifiers without SMOTE and Fig. 5 with SMOTE. We have plotted the graphs in Figs. 6 and 7 for LSTM and ANN, which show the comparison between accuracy and epochs during testing period. Our method has achieved an accuracy of up to 99.90% by using ANN technique (with SMOTE).

The experiments are performed in different scenarios. The KNN, SVM, Gaussian process, decision tree, random forest, AdaBoost, Gaussian Naive Bayes classifier, LSTM, and ANN were employed directly with the DWT output with statistical features (mean and standard deviation).

Fig. 4 ROC plot without
SMOTE

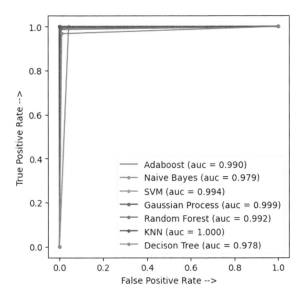

Fig. 5 ROC plot with
SMOTE

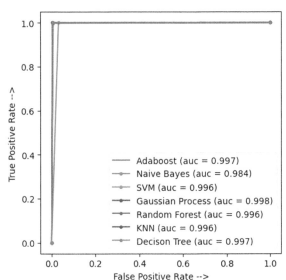

All the models, including ANN and LSTM, are trained over the training and testing data. The dataset was split into (60 and 40) ratios, which means that 60% of the data is used for training purposes, and 40% of the data is used for testing purposes.

ROC curves are used to represent the performance of the classifiers using true positive and false positive rates. We have shown ROC plots for seven different classifiers, as well as separate training versus accuracy plots for LSTM and ANN.

Fig. 6 Accuracy versus epochs plot for LSTM

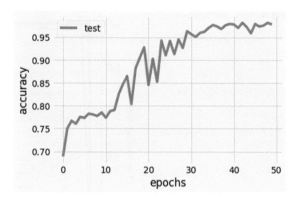

Fig. 7 Accuracy versus epochs plot for ANN

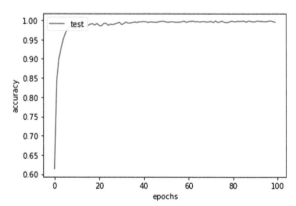

The accuracies of various classification models are shown in Table 2 without SMOTE and in Table 3 with SMOTE, through ANN we got the maximum accuracy for the given classification problem.

In Table 4, we have shown the comparison between the work done previously and made comparisons with their work, and found that the ANN outperforms all other methods and achieves a promising accuracy of 99.90% by applying SMOTE.

The models are executed over many iterations before they produce the results. The term epochs tell us about the number of passes the model has made over the training dataset. Almost each model uses more than one epoch for the training purpose. The data is grouped into batches, and they are passed to the model before it gets update.

6 Conclusion and Future Scope

The methods used in this study help us to detect the autistic or non-autistic nature of a person using electroencephalogram data. This data was provided by the KAU University. We have used DWT, mean, and standard deviation techniques for feature

Table 2 Results without SMOTE

Classifiers	Accuracy (%)
SVM	99.42
Gaussian process	99.85
Decision tree	98.26
Random forest	99.42
AdaBoost	99.13
Naive Bayes	96.67
KNN	99.85
LSTM	97.10
ANN	99.85

Table 3 Results with SMOTE

Classifiers	Accuracy (%)
SVM	99.62
Gaussian process	98.81
Decision tree	99.44
Random forest	99.53
AdaBoost	99.72
Naive Bayes	98.41
KNN	99.34
LSTM	98.60
ANN	99.90

Table 4 Summary of all the comparisons made on KAU data

Author	Feature extraction	Classifiers	Dataset	Accuracy (%) (with SMOTE)
Djemal et al. [20]	DWT	ANN	KAU	99.7
Anas et al. [22]	FFT	RFLDA	KAU	92
Tawhid et al. [23]	STFT	CNN	KAU	99.15
Alhaddad et al. [7]	FFT	FLDA	KAU	90
Our work	DWT, mean, std. deviation	SVM, G. process, D. tree, random forest, AdaBoost, Naive Bayes, KNN, LSTM, ANN	KAU	99.62, 98.81, 99.44, 99.53, 99.72, 98.41, 99.34, 98.60, 99.90

extraction on the EEG data. We have used SMOTE and adaptive techniques in this study due to which accuracy improves. There are many classifiers used in this comparative analysis, but ANN has achieved better accuracy, up to 99.90%. This is the best accuracy that we have achieved by using the SMOTE technique.

Acknowledgements KAU Brain Computer Interface (BCI) Group provided their own autism dataset, which we highly appreciate.

References

1. Ibrahim, S., Djemal, R., & Alsuwailem, A. (2018). Electroencephalography (EEG) signal processing for epilepsy and autism spectrum disorder diagnosis. *Biocybernetics and Biomedical Engineering, 38*(1), 16–26.
2. Acharya, U. R., Yanti, R., Swapna, G., Sree, V. S., Martis, R. J., & Suri, J. S. (2013). Automated diagnosis of epileptic electroencephalogram using independent component analysis and discrete wavelet transform for different electroencephalogram durations. *Proceedings of the Institution of Mechanical Engineers, Part H, Journal of Engineering in Medicine.*
3. Sheikhani, A., Behnam, H., Mohammade, M. R., Noroozian, M., & Golabi, P. (2008). Connectivity analysis of quantitative electroencephalogram background activity in autism disorders with short time Fourier transform and coherence values. In *Proceedings of the 1st International Congress on Image and Signal Processing (CISP'08)* (pp. 207–212). IEEE.
4. Sheikhani, A., Behnam, H., Mohammadi, M. R., Noroozian, M., & Mohamamadi, M. (2012). Detection of abnormalities for diagnosing of children with autism disorders using of quantitative electroencephalography analysis. *Journal of Medical Systems, 36*(2), 957–963.
5. Alhaddad, M. J., Kamel, M. I., Malibary, H. M., Alsaggaf, E. A., Thabit, K., Dahlwi, F., et al. (2012). Diagnosis autism by Fisher linear discriminant analysis FLDA via EEG. *International Journal o Bio-Science Bio-Technology, 4,* 45–54.
6. Sheikhani, A., Behnam, H., Mohammadi, M. R., Noroozian, M., & Mohamamadi, M. (2012). Detection of abnormalities for diagnosing of children with autism disorders using of quantitative electroencephalography analysis. *Journal of Medical Systems, 36*(2), 957–963.
7. Alhaddad, M. J., Kamel, M. I., Malibary, H. M., Alsaggaf, E. A., Thabit, K., Dahlwi, F., & Hadi, A. A. (2012). Diagnosis autism by Fisher linear discriminant analysis FLDA via EEG. *International Journal of Bio-Science and Bio-Technology, 4*(2), 45–54.
8. Delorme, A., & Makeig, S. (2004). EEGLAB an open-source toolbox for analysis of single-trial EEG dynamics. *Journal of Neuroscience Methods, 134,* 9–21.
9. Bosl, W. J., Tager-Flusberg, H., & Nelson, C. A. (2018). EEG analytics for early detection of autism spectrum disorder. A data-driven approach. *Scientific Reports, 8*(1). https://doi.org/10.1038/s41598-018-24318-x
10. Alsuwailem, A., Djemal, R., & Ibrahim, S. (2018). Electroencephalography (EEG) signal processing for epilepsy and autism spectrum disorder diagnosis. *Biocybernetics and Biomedical Engineering, 38*(1), 16–26. https://doi.org/10.1016/j.bbe.2017.08.006.
11. Faust, O., Acharya, U. R., Adeli, H., & Adeli, A. (2015). Wavelet-based EEG processing for computer-aided seizure detection and epilepsy diagnosis. *Seizure, 26*(March), 56–64.
12. Tawhid, M. N. A., Siuly, S., & Wang, H. (2020). Diagnosis of autism spectrum disorder from EEG using a time-frequency spectrogram. The Institution of Engineering and Technology, Current Trends in Cognitive Science and Brain Computing Research and Applications. https://doi.org/10.1049/el.2020.2646
13. Ghosh-Dastidar, S., Adeli, H., & Dadmehr, N. (2007). Mixed-band wavelet-chaos-neural network methodology for epilepsy and epileptic seizure detection. *IEEE Transactions on Biomedical Engineering, 54*(9), 1545–1551.

14. Seo, J.-H., & Kim, Y.-H. (2018). Machine-learning approach to optimize SMOTE ratio in class imbalance dataset for intrusion detection. *Computational Intelligence and Neuroscience*, 11, Article ID 9704672. https://doi.org/10.1155/2018/9704672

15. Weinberger, K. Q., & Saul, L. K. (2019). Distance metric learning for large margin nearest neighbor classification. *Journal of Machine Learning Research, 10*(February), 207–244.

16. Burges, C. J. C. (1998). A tutorial on support vector machines for pattern recognition. *Data Mining and Knowledge Discovery, 2,* 121–167.

17. Jahromi, A. H., & Taheri, M. (2017). A non-parametric mixture of Gaussian Naive Bayes classifiers based on local independent features. In *Artificial Intelligence and Signal Processing Conference (AISP)* (pp. 209–212). https://doi.org/10.1109/AISP.2017.8324083

18. Ray, S. (2019). A quick review of machine learning algorithms. In *International Conference on Machine Learning, Big Data, Cloud and Parallel Computing (COMITCon) 2019* (p. 3539). https://doi.org/10.1109/COMITCon.2019.8862451

19. Wyner, A. J., Matthew, O., Justin, B., & David, M. (2017). Explaining the success of AdaBoost and random forests as interpolating classifiers. *Journal of Machine Learning Research, 18*(48), 1–33.

20. Djemal, R., AlSharabi, K., Ibrahim, S., & Alsuwailem, A. (2017). EEG-based computer aided diagnosis of autism spectrum disorder using wavelet, entropy, and ANN. *BioMed Research International, 2017,* 1–9. https://doi.org/10.1155/2017/9816591.

21. Ali, N. A., Radzi, S. A., Jaafar, S., Shamsuddin, S., & Nor, N. K. (2021). LSTM-based electroencephalogram classification on autism spectrum disorder. *International Journal of Integrated Engineering, 13*(6), 321–329.

22. Anas, H., Mahmod, K., Mohammed, A., Hussein, M., Khalid, T., Foud, D., & Alsaggaf, E. A. (2012). EEG based autism diagnosis using regularized Fisher linear discriminant analysis. *International Journal of Image, Graphics and Signal Processing (IJIGSP), 4,* 35–41. https://doi.org/10.5815/ijigsp.2012.03.06.

23. Tawhid, Md. N. A., Siuly, S., Wang, H., Whittaker, F., Wang, K., & Zhang, Y. (2021). A spectrogram image based intelligent technique for automatic detection of autism spectrum disorder from EEG. *PLOS ONE, 16,* e0253094. https://doi.org/10.1371/journal.pone.0253094.

Fully Automatic Wheat Disease Detection System by Using Different CNN Models

Neha Kumari and B. S. Saini

Abstract In India, agriculture assumes a significant part in view of the rising number of individuals and extended interest in food. In the world, Wheat is the third most gathered and consumed grain. One critical effect on wheat crop yield is a disease because by parasites, contaminations, and microorganisms. So that is the justification behind a huge piece of the wheat crop becoming spoiled. Multiple dozen wheat contaminations are risky to the yields, so it is very important to detect these diseases at the exact time. The determination of the diseases is done with a visual investigation by specialists and an organic assessment is a subsequent option if fundamental. It is an extremely tedious and costly procedure. Deep learning is the only way by which one can solve such problems, this branch additionally considers the early recognition of wheat diseases by applying convolutional neural networks (CNNs) close to the well-known models. In the current work, we consider four different classes of wheat pictures which contain tan spot, fusarium head blight, stem rust, and healthy wheat. We execute the different CNN models to the gathered dataset. We use number of parameters to train these models are: loss function = "categorical cross-entropy", activation function = "softmax", optimizer = "adam", batch size = 64, epochs = 80. After applying these models to the dataset we examine that the ResNet model delivered the best calculable outcomes. The proposed approach has gotten the most noteworthy classification exactness of 98% utilizing the ResNet50 model when contrasted with MobileNet, and DenseNet models. This shows that deep learning has shown generally excellent execution for the order of various illnesses. We did this execution with the assistance of google colab.

Keywords Deep learning · Wheat disease · Convolutional neural network (CNN) · Pathogens

N. Kumari (✉) · B. S. Saini
NIT Jalandhar, Jalandhar, Punjab, India
e-mail: kumarineha57486@gmail.com

B. S. Saini
e-mail: sainibs@nitj.ac.in

© The Author(s), under exclusive license to Springer Nature Singapore Pte Ltd. 2023   351
S. Shakya et al. (eds.), *Sentiment Analysis and Deep Learning*,
Advances in Intelligent Systems and Computing 1432,
https://doi.org/10.1007/978-981-19-5443-6_26

1 Introduction

Wheat is perhaps the most internationally huge harvest species with a yearly overall grain creation of 800 million tons. The demand for wheat is now increasing world-wide. After increasing the demand for this crop, it started cultivating all over the world. Now, it is the second-most grown crop in the world. Infectious diseases and bugs are known to diminish grain yield potential and quality. Infectious diseases are the main restricting component in wheat creation. Leaf rust, fusarium head blight, and tan spot are the normal wheat diseases that influence the wheat crop gravely. Figures 1, 2, and 3 show the different wheat infections.

Fig. 1 A sample of leaf rust

Fig. 2 A sample of fusarium head blight

Fig. 3 A sample of tan spot

1.1 Leaf Rust

Leaf rust of wheat is also called brown rust and this disease is caused due to fungal infection which affects wheat, barley, and grains. Leaf rust is one of the major types of rust infection in wheat. Leaf rust causes the most damage when rust starts covering the upper leaves of the wheat plant before flowering. Early infection can result in weak plants and can prompt up to 20% yield misfortune, which affects crop productivity. The pathogen that causes the leaf rust of wheat is *Puccinia*. It is the most common wheat rust disease in most wheat-producing areas. It is an overwhelming occasional disease in India.

1.2 Fusarium Head Blight

Fusarium head scourge (FHB) is one of the really contagious infections of grain yields like wheat, grain, and maize. Fusarium head blight is caused by *Fusarium graminearum. F. graminearum* produces two unique spore types: one spread by wind and the other spread by water sprinkled from a downpour or water system. It is also known as a tombstone or scab. This disease is most prevalent in wheat, barley, and corn. FHB majorly affects kernel development thereby reducing the growth and in turn reducing yield. Mycotoxin, a fungal infection affects the harvested grain and causes many negative effects like reduction in livestock feed, quality of wheat for baking and milling, major reduction in ethanol production which is a biofuel. The basic spore production is affected which causes infection during ripening and produces mycotoxin which causes seedling blight.

1.3 Tan Spot

Tan spot is a wheat disease caused by *Pyrenophora tritici-rrepent* pseudothecia. Tan spot side effects are most common in early April. On leaves, the sickness initially shows up as little, tan to the brown point. The parasite survives and replicates on wheat straw. Prior to the developing season, spores called ascospores are let out of pseudothecia (little dull, raised fruiting plans formed on wheat straw) and spread by wind or blowing precipitation. They are the essential wellspring of disease of the lower, more developed leaves during tillering and early jointing. After the infection, the organism spreads inside a field or to adjoining fields by an alternate kind of spore called a conidium that is delivered in the tan spot lesions themselves.

These infections are extraordinary difficulties for the production of the wheat crop so recognition of the wheat disease is an important stage for accomplishing great harvest and expanding the amount and quality of yield. It's a big problem how we classify the different diseases of wheat because most of the variety of wheat diseases are the same in nature. Deep learning is the best way to deal with tackling such sorts of classification issues.

CNN is the best strategy for any forecast issue including input picture data and requires insignificant pre-processing. CNN can arrange enormous scope pictures. In this particular situation, a couple of investigation works have been done to upgrade the introduction of CNN on such classification tasks, because of the identification of different diseases. Progress in "CNN" can be improved in different ways, including activation, loss function, data enhancement, and optimization algorithm.

Jiang et al. [1] offer a deep learning strategy for real-time identification of apple leaf diseases founded on upgraded convolutional neural networks (CNNs). They proposed a deep-CNN-based apple leaf disease detection algorithm. The model is prepared to identify five common apple leaf sicknesses by presenting the GoogLeNet Inception structure utilizing rainbow concatenation. The INAR-SSD model accomplishes an identification exactness of 78.80% with a location rate speed of 23.13 FPS, as indicated by the outcomes. The outcomes demonstrate that the imaginative INAR-SSD model is an elite exhibition strategy for the early recognition of apple leaf diseases that can identify these diseases in real-time and with greater accuracy.

Hasan et al. [2] suggested a CNN-based model for maize disease categorization. They can integrate CNN with a bi-directional long short-term memory model to improve the model's efficiency and accuracy. They constructed a dataset of 29,065 maize disease pictures, using 80% of the samples for training and achieving a 99.02% accuracy. As a result, the model is particularly trustworthy for AI-based disease identification systems, and it may help boost crop output.

In 2017, Pavithra and Palanisamy [3] proposed a more precise and productive automated approach for identifying leaf rust illnesses. Image recognition techniques are utilized to improve image quality and diagnose the disease. Image acquisition, pre-processing, segmentation, inspection, and disease recognition are all part of the interaction. Oats, maize, barley, wheat, and paddy are among the five varieties of leaves that have been tested. The results of the experiments suggest that using a multiclass support vector machine to implement the proposed technique improves precision. Finally, the technique's performance is evaluated, and when compared to traditional tactics, it achieves a precision of 99.5%.

Shrestha et al. [4] developed a prevention and detection model. The model has been offered a total of 15 examples, 12 of which are infected plants leaves and three of which are healthier leaves. In this model, a CNN-based approach for detecting plant diseases was applied. The accuracy of the test was found to be 88.80%.

By applying a superior deep convolutional model, Goyal et al. [5] suggested a model for diagnosing and categorizing of wheat infection in 2021. This novel deep learning model was created to precisely identify wheat illnesses into ten categories. The suggested methodology has a good testing precision of 97.88%. It outperforms the other two well-known deep learning models, VGG16 and RESNET50, by 7.01% and 15.92%, respectively, in terms of precision. With a decent test exactness of 97.88%, the proposed strategy is used to categorize ten wheat infection classifications. When compared to other deep learning techniques, this shows a significant improvement. Exactness improves by 7.01% and 15.92%, respectively, as compared to VGG16 and RESNET50.

Kibriya et al. [6] utilized a convolutional neural network to develop a model for diagnosing tomato leaf disease [7–12]. They used two convolution neural network (CNN)-based models, GoogLeNet and VGG16, to classify tomato leaf disease in this article. Using a deep learning approach, the suggested work seeks to discover the optimum answer to the problem of tomato leaf disease detection [12–16]. On the plant village dataset with 10,735 leaf pictures, VGG16 achieved 98% accuracy while GoogLeNet achieved 99.23%.

The identification of wheat diseases is very important to save the yield and agriculture system [16–22]. For the further development of disease identification results, a few investigations have been led on various models. We plan a wheat infection identification technique that utilizes leaf pictures to distinguish between healthy and unhealthy wheat plants [22–28]. The pre-trained model assists us in the use of feature extraction [28–32]. Executing the models was done on Google collaborator. ResNet, DenseNet, and MobileNet are the different CNN models we use in this paper. Our point is to plan a framework through which farmers can detect the beginning phase contaminations of wheat with the combination of a digital camera.

2 Proposed Work

The entire procedure of developing a CNN model for the identification of various wheat diseases. This whole process is containing many steps beginning from gathering a dataset of sampled images to training the model, algorithm of the proposed strategy is displayed in Fig. 4.

Fig. 4 Flowchart of the
proposed algorithm

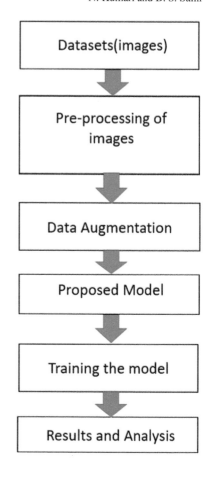

2.1 Dataset

We took the dataset online from the Kaggle site and information for the pictures come from a couple of sources. The majority of the data comes from public photos that are found on Images of google. The goal of this dataset is to make a model of deep learning out how to properly characterize the wheat in the photographs as healthy wheat, leaf rust, fusarium head blight, or tan spot. After downloading the dataset, firstly, we apply the CNNs-based model to the gathered dataset and observed that the model has given exceptionally awful outcomes after training and testing the dataset. We need a maximum amount of data to increase the exactness and efficiency of deep learning-based models. To create additional information from existing information the strategy is called Image augmentation. Image augmentation is a tool of deep learning, with the help of image augmentation we can create a huge amount of data by varying the different number of parameters as mentioned below:

2.2 Pre-processing of Images

Pre-processing is the first step performed on the dataset. The size of downloaded images is in different formats and resolutions. In order to get the exact size and resolution, pre-processing is important. This dataset contains 415 images of healthy wheat and 1700 images of unhealthy wheat which contain three different types of wheat diseases. The dataset is separated into two sections one is the training and another one is the validation set. The training and validation proportion considered for this study is taken 75:25 implies that 75% of information is available in the training set and 25% in the validation set according to the size of the dataset.

2.3 Data Augmentation

Data augmentation is a process to expand the dataset by producing new data from the current dataset. As we know deep learning needs a huge amount of trainable data to produce correct results. Image augmentation is a new technology that makes it easier to enlarge the dataset for deep learning research. The data augmentation utilized to train the model in this paper was not done separately, but rather while the model was being trained. Expanding the dataset for convolutional neural networks can reduce overfitting. Since the dataset for this study isn't wide enough, we perform data augmentation as mentioned below:

Image Data Generator: ImageDataGenerator() is a type of function used in Image augmentation to increase the amount of data. It has different parameters, for example, turn of the input picture at a specific point which will change the intensity value of the picture, height-wise, and width-wise shifting of the image. The parameters that we used during data augmentation are shown as:

Rotation_range = 30
Zoom_range = 0.15
Width_shift_range = 0.2
Height_shift_range = 0.2
Shear_range = 0.15
Horizontal_flip = True

2.4 Proposed Models

(a) **ResNet**: ResNet is called residual network. ResNet is a particular kind of neural network that was presented in 2015 by Kaiming He, Xiangyu Zhang, Shaoqing Ren, and Jian Sun. ResNet uses skip connection pattern and made some shortcuts to jump over the convolutional layers as shown in Fig. 5.

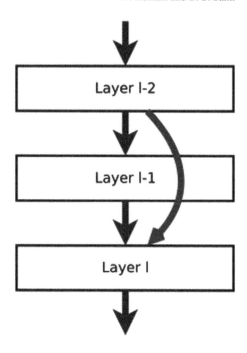

ResNet takes less time to train the model as compared to VGGNet. We used
the ResNet-50 model to train the dataset in this paper. ResNet50 contains fifty
convolutional layers. This model is able to train a million images and classify
the images into thousands of categories. The filter size of the ResNet model is
3 * 3 and takes input images in this format 224 * 224 * 3. The stride and padding
of one are often used in the CNN model. It's done through a sliding window,
with the neural network on which our model will be trained sliding every time.

(b) **MobileNet**: MobileNet is based on depth-wise separable convolution layers. We
use MobileNet and MobileNetV2 in this paper. A depth-wise convolution and
a point-wise convolution make up each depth-wise separable convolution layer.
A MobileNet contains 28 layers if depth-wise and point-wise convolutions are
considered differently. A basic MobileNet has 4.2 million parameters, which can
be further reduced by suitably tweaking the width multiplier hyperparameter.
The information image is 224 * 224 * 3 pixels in size.

(c) **DenseNet**: Dense convolutional network (DenseNet) is a plan that includes
deep learning networks that go a lot further while likewise making them more
proficient at training by utilizing less joints between the layers. DenseNet is
a convolutional brain network in which each layer is associated with the other
layers in the model; for example, the initial layer is linked to the model's second,
third, and fourth layers, while the subsequent layer is linked to the third, fourth,
fifth, and so on.. Maximum data flow between the layers through this process.
Each layer gets input from past layers and makes its own feature map for next
layers.

2.5 Experimental Setup

For training the model on the gathered dataset, the experiment has been carried out using Google Colab. Google Colab is an open-source platform for Python programming. In Google Colab, we can automatically download all libraries from the Internet; no external libraries must be installed. It is now widely used to conduct machine learning, deep learning, and computer vision studies. These tests need a lot of computing power, and training the deep learning model will take several days. If the dataset is large enough, deep learning produces superior outcomes. As a result, graphical processing unit (GPU) power is required to train the model in less time. Using their server, Google Colab supports GPU and TPU compatibility. Anyone who has a Google account can access the Colab for experiment purposes. Keras is a neural network toolbox with high-level APIs that is used to design and train deep learning models. Because its program is compiled in Python, it is more user-friendly than Tensorflow, which is an open-source library both for high-level and low-level APIs.

Operating System: Windows11 Graphic card: 4 GB of NVIDIA GEFORCE GTX, Intel(R) Core(TM) i7-10870H CPU @ 2.20 GHz 2.21 GHz, RAM: 16 GB RAM.

2.6 Results and Discussion

The study of healthy and diseased wheat plants was examined in this paper. In Google Colab, the experiment was conducted. This segment's outcome is related to the training of the entire dataset, which includes all original and enhanced pictures. Deep CNN models were used to create a model, as it was used to apply various transfer learning approaches to attain high accuracy. ResNet50 has a maximum accuracy of 98%. The outcomes are listed in Table 1.

With a batch size of 64, each CNN model was trained for 80 epochs. Figure 6a–d shows the accuracy and loss of each model against the number of epochs.

As demonstrated in Table 2, after training the various models, ResNet50 is more exact than the other deep learning models. In comparison to other models, ResNet50's confusion matrix produces better results. The classification model's performance is described using the confusion matrix. There are four potential combinations of expected and actual values in this matrix. It comes in handy for calculating recall,

Table 1 Accuracy of proposed models

Models	Accuracy (%)
ResNet50	**98**
DenseNet121	90
MobileNet	91
MobileNetV2	89

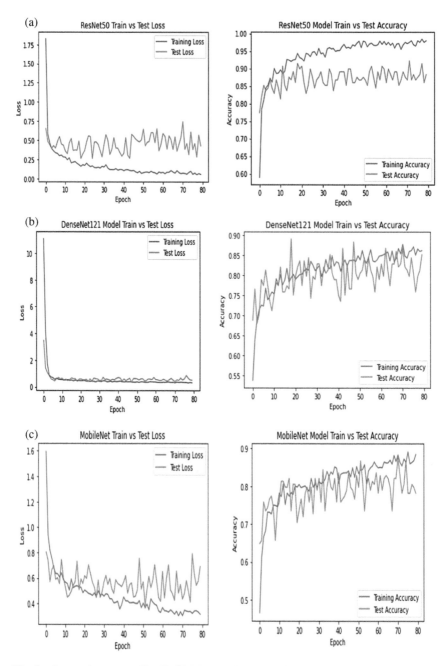

Fig. 6 **a** Loss and accuracy of ResNet50. **b** Loss and accuracy of DenseNet121. **c** Loss and accuracy of MobileNet. **d** Loss and accuracy of MobileNetV2

(d)

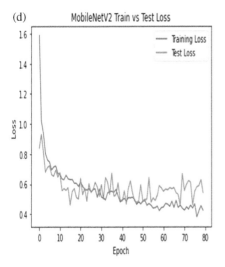

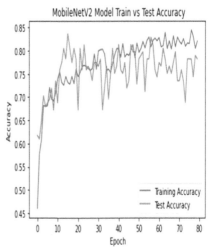

Fig. 6 (continued)

Table 2 Accuracy measures of proposed models

Models used	Precision	Recall	f_1-score	Accuracy
ResNet50	0.92	0.95	0.93	0.98
DenseNet121	0.90	0.89	0.8025	0.90
MobileNet	0.89	0.87	0.86	0.91
MobileNetV2	0.7875	0.78	0.785	0.89

precision, specificity, and accuracy. True Positive (TP), True Negative (TN), False Positive (FP), and False Negative (FN) are the terms used in the confusion matrix (FN). TP and TN are correct results predicted by the network, while FP and FN are errors predicted by the network. Type 1 error is FP, and type 2 error is FN.

- Precision $= \frac{TP}{TP+FN}$
- Recall $= \frac{TP}{TP+FP}$
- F_1-score $= \frac{2 \times Recall \times Precision}{Recall+Precision}$
- Accuracy $= \frac{TP+TN}{TP+TN+FP+FN}$

2.7 Prediction Results

Figure 7a–d show the predictions for tan spot, leaf rust, healthy wheat, and fusarium head blight, respectively. Some pictures have been selected for each prediction. The results show that using the majority polling strategy, the suggested method accurately places the correct labels on these frames.

(a)

(b)

Fig. 7 **a** Prediction of healthy wheat. **b** Prediction of fusarium head blight. **c** Prediction of leaf rust. **d** Prediction of tan spot

3 Conclusion

Using four different transfer learning strategies, this paper attempts to develop and analyze deep learning models for the diagnosis of wheat rust disease. Wheat disease, on the other hand, is a prevalent disease that can wipe out wheat crops. As a result, in order to cure diseased wheat plant, it is required to diagnose the problem early in order to obtain a good production. For this investigation, we used an open-source dataset from Kaggle, which is freely available online. Employing four distinct deep neural network algorithms ResNet50, MobileNet, MobileNetV2, and DenseNet121

(c)

(d)

Fig. 7 (continued)

architecture based on convolutional neural networks (CNNs), we further enhance this dataset by using the image data augmentation methodology to improve classification accuracy. During the experimentation, we discovered that the ResNet50 model outperforms the other model. With batch size $= 64$, optimizer $=$ "Adam," and epochs $= 80$, a 98% accuracy rate was achieved. Furthermore, this system will help farmers in recognizing infections and preventing disease in their crops. This research can further be improved by using different feature recognition techniques to remove noise from the input and to apply the identified feature to the model.

References

1. Jiang, P., Chen, Y., Liu, B., He, D., & Liang, C. (2019). Real-time detection of apple leaf diseases using deep learning approach based on improved convolutional neural networks. *IEEE Access, 7*, 59069–59080.
2. Hasan, M. J., Alom, M. S., Dina, U. F., & Moon, M. H. (2020, June). Maize diseases image identification and classification by combining CNN with bi-directional long short-term memory model. In *2020 IEEE Region 10 Symposium (TENSYMP)* (pp. 1804–1807). IEEE.
3. Pavithra, G., & Palanisamy, K. S. Computer based diagnosis of leaf rust disease using multiclass SVM. *International Journal of Engineering Research & Technology (IJERT)*, 2278–0181.
4. Shrestha, G., Das, M., & Dey, N. (2020, October). Plant disease detection using CNN. In *2020 IEEE Applied Signal Processing Conference (ASPCON)* (pp. 109–113). IEEE.
5. Goyal, L., Sharma, C. M., Singh, A., & Singh, P. K. (2021). Leaf and spike wheat disease detection & classification using an improved deep convolutional architecture. *Informatics in Medicine Unlocked, 25*, 100642.
6. Kibriya, H., Rafique, R., Ahmad, W., & Adnan, S. M. (2021, January). Tomato leaf disease detection using convolution neural network. In *2021 International Bhurban Conference on Applied Sciences and Technologies (IBCAST)* (pp. 346–351). IEEE.
7. Niu, X., Wang, M., Chen, X., Guo, S., Zhang, H., & He, D. (2014, August). Image segmentation algorithm for disease detection of wheat leaves. In *Proceedings of the 2014 International Conference on Advanced Mechatronic Systems* (pp. 270–273). IEEE.
8. Yuan, L., Huang, Y., Loraamm, R. W., Nie, C., Wang, J., & Zhang, J. (2014). Spectral analysis of winter wheat leaves for detection and differentiation of diseases and insects. *Field Crops Research, 156*, 199–207.
9. Dixit, A., & Nema, S. (2018). Wheat leaf disease detection using machine learning method—A review. *International Journal of Computer Science and Mobile Computing, 7*(5), 124–129.
10. Maid, M. K., & Deshmukh, R. R. (2018, May). Statistical analysis of WLR (wheat leaf rust) disease using ASD FieldSpec4 spectroradiometer. In *2018 3rd IEEE International Conference on Recent Trends in Electronics, Information & Communication Technology (RTEICT)* (pp. 1398–1402). IEEE.
11. Francis, M., & Deisy, C. (2019, March). Disease detection and classification in agricultural plants using convolutional neural networks—A visual understanding. In *2019 6th International Conference on Signal Processing and Integrated Networks (SPIN)* (pp. 1063–1068). IEEE.
12. Li, L., Zhang, S., & Wang, B. (2021). Plant disease detection and classification by deep learning—A review. *IEEE Access, 9*, 56683–56698.
13. Kumar, S. S., & Raghavendra, B. K. (2019, March). Diseases detection of various plant leaf using image processing techniques: A review. In *2019 5th International Conference on Advanced Computing & Communication Systems (ICACCS)* (pp. 313–316). IEEE.
14. Zhang, T., Yang, Z., Xu, Z., & Li, J. (2022). Wheat yellow rust severity detection by efficient DF-UNet and UAV multispectral imagery. *IEEE Sensors Journal*.

15. Huang, W., Guan, Q., Luo, J., Zhang, J., Zhao, J., Liang, D., Huang, L., & Zhang, D. (2014). New optimized spectral indices for identifying and monitoring winter wheat diseases. *IEEE Journal of Selected Topics in Applied Earth Observations and Remote Sensing, 7*(6), 2516–2524.
16. Hande, R., Ahuja, A., Watwani, A., Chichriya, P., & Shamnani, S. (2021, January). Krishi Manch: Disease detection in rice crops using CNN. In *2021 4th Biennial International Conference on Nascent Technologies in Engineering (ICNTE)* (pp. 1–6). IEEE.
17. Rashid, M., Ram, B., Batth, R. S., Ahmad, N., Dafallaa, H. M. E. I., & Rehman, M. B. (2019, December). Novel image processing technique for feature detection of wheat crops using python OpenCV. In *2019 International Conference on Computational Intelligence and Knowledge Economy (ICCIKE)* (pp. 559–563). IEEE.
18. Wang, L., Dong, F., Guo, Q., Nie, C., & Sun, S. (2014, October). Improved rotation kernel transformation directional feature for recognition of wheat stripe rust and powdery mildew. In *2014 7th International Congress on Image and Signal Processing* (pp. 286–291). IEEE.
19. Ashourloo, D., Aghighi, H., Matkan, A. A., Mobasheri, M. R., & Rad, A. M. (2016). An investigation into machine learning regression techniques for the leaf rust disease detection using hyperspectral measurement. *IEEE Journal of Selected Topics in Applied Earth Observations and Remote Sensing, 9*(9), 4344–4351.
20. Marzougui, F., Elleuch, M., & Kherallah, M. (2020, December). Evaluation of data augmentation for detection plant disease. In *International Conference on Hybrid Intelligent Systems* (pp. 464–472). Springer.
21. Guan, X. (2021, April). A novel method of plant leaf disease detection based on deep learning and convolutional neural network. In *2021 6th International Conference on Intelligent Computing and Signal Processing (ICSP)* (pp. 816–819). IEEE.
22. Menon, V., Ashwin, V., & Deepa, R. K. (2021, June). Plant disease detection using CNN and transfer learning. In *2021 International Conference on Communication, Control and Information Sciences (ICCISc)* (Vol. 1, pp. 1–6). IEEE.
23. Asif, M. K. R., Rahman, M. A., & Hena, M. H. (2020, December). CNN based disease detection approach on potato leaves. In *2020 3rd International Conference on Intelligent Sustainable Systems (ICISS)* (pp. 428–432). IEEE.
24. Saeed, Z., Raza, A., Qureshi, A. H., & Yousaf, M. H. (2021, October). A multi-crop disease detection and classification approach using CNN. In *2021 International Conference on Robotics and Automation in Industry (ICRAI)* (pp. 1–6). IEEE.
25. Zhou, G., Zhang, W., Chen, A., He, M., & Ma, X. (2019). Rapid detection of rice disease based on FCM-KM and faster R-CNN fusion. *IEEE Access, 7*, 143190–143206.
26. Shrestha, A., & Mahmood, A. (2019). Review of deep learning algorithms and architectures. *IEEE Access, 7*, 53040–53065.
27. Jia, X. (2017, May). Image recognition method based on deep learning. In *2017 29th Chinese Control and Decision Conference (CCDC)* (pp. 4730–4735). IEEE.
28. Kamilaris, A., & Prenafeta-Boldú, F. X. (2018). Deep learning in agriculture: A survey. *Computers and Electronics in Agriculture, 147*, 70–90.
29. Zhang, X., Qiao, Y., Meng, F., Fan, C., & Zhang, M. (2018). Identification of maize leaf diseases using improved deep convolutional neural networks. *IEEE Access, 6*, 30370–30377.
30. Aravind, K. R., Raja, P., Mukesh, K. V., Aniirudh, R., Ashiwin, R., & Szczepanski, C. (2018, January). Disease classification in maize crop using bag of features and multiclass support vector machine. In *2018 2nd International Conference on Inventive Systems and Control (ICISC)* (pp. 1191–1196). IEEE.
31. Pawlak, K. (2016). Food security situation of selected highly developed countries against developing countries. *Journal of Agribusiness and Rural Development, 40*(2), 385–398.
32. Trivelli, L., Apicella, A., Chiarello, F., Rana, R., Fantoni, G., & Tarabella, A. (2019). From precision agriculture to Industry 4.0: Unveiling technological connections in the agrifood sector. *British Food Journal*.

Twitter Sentiment Analysis Using Naive Bayes-Based Machine Learning Technique

Priya Gaur, Sudhanshu Vashistha, and Pradeep Jha

Abstract "Computational" sentiment analysis can determine whether a sentiment is favorable, negative, or neutral. Another term for this approach is "opinion mining," or obtaining a speaker's sentiments. Businesses use it to develop strategies, learn what customers think about products or brands, how people react to campaigns or new product releases, and why they do not buy certain products. It is used in politics to keep track of political ideas and to check for contradictions between government claims and actions. It can even be used to predict election results! It is also used to track and analyze social phenomena like recognizing dangerous circumstances and evaluating blogging mood. In this paper, we look tackle the problem of sentiment categorization using the Twitter dataset. To analyze sentiment, preprocessing and Naive Bayes classifier approaches are utilized. As a result, we applied a text preprocessing classification accuracy classifying strategy and improved our classification accuracy score on the Kaggle public leaderboard. The aim of this paper is to classify the twitter sentiments using machine learning algorithm based on Naïve Bayes Classifier. The proposed model indicated better accuracy and precision based on performance parameters such as precision, recall and accuracy.

Keywords Sentiment analysis · Twitter · Facebook · Instagram · Machine learning · AI · Naive Bayes

1 Introduction

Data forecasting is a hot topic in research, whether in the financial or medical sectors, for example [1–3]. The process of predicting future values based on data from the present and the past is known as forecasting. A forecast is only an estimate of some factor or incentive at some predetermined future date and age [4]. Data forecasting can be used in a variety of ways, from weather forecasting to predicting whether a patient is at risk of developing a specific disease in the medical field. Data forecasting

P. Gaur (✉) · S. Vashistha · P. Jha
Department of CSE, Global Institute of Technology, Jaipur, Rajasthan, India
e-mail: priyagaur5757@gmail.com

© The Author(s), under exclusive license to Springer Nature Singapore Pte Ltd. 2023
S. Shakya et al. (eds.), *Sentiment Analysis and Deep Learning*,
Advances in Intelligent Systems and Computing 1432,
https://doi.org/10.1007/978-981-19-5443-6_27

is essentially utilizing authentic information to foresee future patterns. Forecasting is used by business people and companies to plan their budgets and future projects in order to maximize their profits, as well as to analyze their performance and potential risks. If the company's sales increase or decrease, investors try to predict whether that will have an effect on the company's stock price. Stock market analysts use forecasting to predict GDP, unemployment, or how it changes over a year in the Stock Market [5]. For a statistician, forecasting is a tool that they use in every situation where they need to make predictions [6]. To accurately predict something, a number of steps must be taken in the forecasting process, which is more complicated than most people imagine. As follows are the various steps:

- As a first step, we need to analyze and identify the variables that are relevant to the situation we want to forecast.
- We must first analyze the situation before selecting a dataset and making changes to it in accordance with our requirements.
- After that, the data is used to forecast the future value.
- As a final step, the procedure is validated by comparing predicted values to the actual results.

In any event, selecting a high-precision model purely based on a single model resulted in inadequate accuracy. As it turns out, the organizers of these events have begun to look into these modest internet journals (blogs) in order to have a better understanding of their product. They monitor and respond to client comments on minor internet forums on a regular basis. Coming up with fresh concepts for identifying and condensing universal emotions is one of the obstacles. Several people have recently been drawn into the phases of interpersonal connection by sites like Facebook and Twitter [7]. Most of individuals utilize web-based entertainment to offer their viewpoints, feelings, or suspicions about objects, places, or people. Twitter is a vast storehouse of public opinion on a wide range of issues, including personalities, products, services, and business initiatives [8]. A public evaluation of the public analysis system is known as sentiment evaluation.

2 Literature Survey

As we all know that everything cannot be predicted, but if factors that are related to what we want predict, and sufficient amount of data is there, then we can easily predict the thing [9]. Data Forecasting has wide range of applications like as follow:

Egan Forecasting, i.e., process of controlling the heat of building by calculating the demand of heat that should be supplied to the building according to climate and building requirement.

- Forecasting is used in stock market and foreign exchange.
- In business, to predict the customer demand according to the current market analysis.

- For earthquake prediction, etc.
- In supply chain management, which project should be at which time and at which place, it will help to maximize the profit.
- In Banking or financing, to check the defaulter risk, i.e., either there is probability to have a risk like particular person who applied for a loan can be defaulter or not.
- In Sales Forecasting, predicting whether sales will increase or decreased in future.
- Political Forecasting, predicting the results of elections.
- Weather forecasting, flood forecasting

Some 25% of online adults use Twitter according to Pew Research Center findings. A global total of 313 million monthly active Twitter users is the result of this figure. Tweets are limited to 280 characters on the free, real-time messaging service (which was increased from 140-characters in November 2017). As a result, Twitter has seen substantial growth despite its 280-character constraint. The use of Twitter by Dell to notify customers of future product reductions, for example, has proven to be a great success. Twitter's business value is also appreciated by many marketers because it makes it simple for businesses to find out what customers are saying about their products [10].

Because of Twitter's great acceptability around the world, this project focuses on using its data. As mobile technology has proliferated, it has opened up a whole new world of views and perspectives that were previously unavailable. The ideas and feelings expressed on Twitter, when gathered and studied, can reveal a wealth of information about a people [11]. By gaining a deeper understanding of their target market, companies can benefit from this information [12]. A deeper understanding of a population can help governments make better decisions for that group. A topic modeling and lexicographical strategy will be used to assess sentiment and opinions in English text in order to determine a general sentiment, such as positive or negative, during the course of this research, in order to obtain tweets as the basis for data. Using social media, this application may be utilized to get a sense of what people are thinking and feeling [13].

Opinion mining can be used in a variety of contexts. Use it for advertising, hotspot identification in discussions, web crawlers, recommendation systems and email filtering, as well as for questions and notes. Using opinion mining to improve human–computer interactions, corporate intelligence, opinion surveys, government intelligence, citation examination, etc. is the most intriguing use of opinion mining in daily life. Opinion mining's potential uses can be better understood by looking at the example questions that go along with it.

- In general, what are people's opinions on government strategies?
- Which aspects of a product are most popular or least popular with the general population?
- Is there anyone in the general race body that is a good competitor?
- What may be the reason for a decrease in sales?

When it comes to opinion mining there are some issues that need to be overcome, such as detecting which sections of text are opinions and identifying the opinion-holders. Human surveys, feelings, and emotional exchange are all part of the scope of a sentiment analysis. All of us have our own thoughts and concerns regarding a given issue or topic. Inaccurate, irrelevant, or ambiguous information can all be found in opinionated text. It is much more difficult to depict opinions than facts.

3 Proposed Methodology

While the Bayes theorem determines the likelihood of one event occurring given the chance of another event occurring, Naive Bayes alters the procedure by "naively" expecting that every occasion is restrictively autonomous of the others. Instead, the following equation can be applied for independent events.

$$P(\text{spam} | \text{viagra, enlarge})$$
$$= \frac{P(\text{spam}) * P(\text{viagra}|\text{spam}) * P(\text{viagra}|\text{spam, enlarge})}{P(\text{viagra}) * P(\text{viagra}|\text{enlarge})}$$

The good news is that the equation grows linearly with each extra variable, allowing you to process massive amounts of data at once without any issues. Because of its simplicity, Naive Bayes is a quick and versatile technique that beats more complicated models as long as your informational index does not get excessively enormous.

In the event that you meet information with a variable with zero likelihood, Naive Bayes will make your condition breakdown when increased with different factors. This can be fixed, however, on the off chance that you smooth the information in advance, eliminating zero likelihood.

Let's imagine you have to decide whether or not to go to the beach. Our minds can examine all of the pros and disadvantages of traveling in a fraction of a second, however if a computer were asked to make the decision for us, it would need to follow a set of rules to arrive at the most beneficial answer. To make the problem easier to understand, the machine will calculate the outcome using three different variables (weather, air temperature, and where friends are going). The following principles are used to construct a relationship between the different variables so that the computer can make sense of it. The rales are read in order from top to bottom.

The variables of the row in Table 1 is compared to the criteria of each set of rules. When a rule's conditions are met, the final value is determined, and the remaining rules are not controlled.

The variables of the row in Table 1 is compared to the criteria of each set of rules. When a rule's conditions are met, the final value is determined, and the excess standards are not controlled.

Table 1 A couple of scenarios used to decide if you want to go to the beach or not

Scenario ID	Weather	Air temperature (°C)	Friends are going
1	Sunny	20	Yes
2	Cloudy	23	Yes
3	Rainy	17	No
…	…	…	…
n	Rainy	26	Yes

Let's imagine you have to decide whether or not to go to the beach. Our minds can examine all of the pros and disadvantages of traveling in a fraction of a second, however if a computer were asked to make the decision for us, it would need to follow a set of rules to arrive at the most beneficial answer. To make the problem easier to understand, the machine will calculate the outcome using three different variables (weather, air temperature, and where friends are going).

Classification is a technique used in data mining and machine learning to determine which categorical class a data instance belongs to. Figure 1 depicts a categorization example, in which data is split down into individual components until a precise classification is identified. It could be anything from a single data point to a 500-page book that needs to be categorized. This strategy is frequently combined with supervised learning. By assessing one or more sets of labeled training data and utilizing the data's unique properties, a classification model that appears to identify the correct class can be developed. This classification model can then predict which class they belong to using new data that is equivalent to the training data. Some of the most prevalent classification algorithms include Bayes/Naive Bayes, Decision trees, neural networks, Rule-based techniques, and Support Vector Machines.

A comma-separated values file with tweets and their corresponding sentiments is provided. As an example of the training dataset, the tweet id is a unique integer that identifies a specific tweet, and the tweet sensation is either 1 (positive) or 0 (negative). Our model will be cross-validated using just this one set of data.

4 Result Analysis

There the data has been gathered in the sentiment 140 dataset. It contains 1,600,000 tweets extracted from the Twitter API. The tweets have been labeled (0 = negative, 4 = positive) so that sentiment may be determined. Another essential phase in the sentiment analysis process is feature extraction, which entails processing tweets and creating a word cloud based on sentiment. Figures 2, 3, and 4 depict the various stages of the procedure.

For classification analysis and sentiment prediction from dataset analysis, a Naïve Bayes-based classifier is used. Figures of merit, such as the confusion matrix illustrated in Fig. 5, have been used to express the process. The procedure was carried

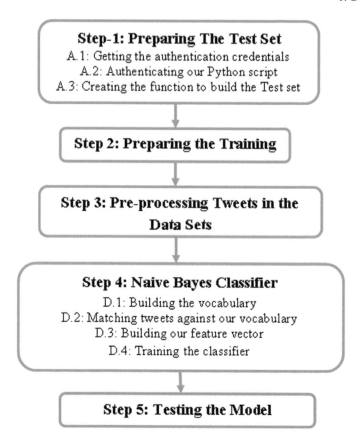

Fig. 1 Flow chart of proposed methodology

	polarity	text	processed_tweets
720002	0	@MOCAShop Wow, normally a Ferrell fan. This ma...	mocashop wow normally ferrell fan make sad wai...
262514	0	@netbender I got that enough from punk shows t...	netbender got enough punk show ringing never w...
812355	1	@ramblelite i got everyone noaw yo. simm's got...	ramblelite got everyone noaw yo simms got cat ...
385860	0	@Rayuen yeah a couple (like randall munroe of ...	rayuen yeah couple like randall munroe xkcd kn...
321526	0	I feel for the families of the plane crash dis...	feel family plane crash disaster one
438569	0	i dont feel well today	dont feel well today
29364	0	@LishaKatherine so true! Im miss talking to you	lishakatherine true im miss talking
577052	0	CANT READ	ant read
14386	0	UGH. don't wanna edit anymoreeee So lost.	gh dont wan na edit anymoreeee lost
254577	0	just got up and I have a toothache	ust got toothache

Fig. 2 Feature extraction

Fig. 3 For negative tweets word cloud formation

Fig. 4 For positive tweets word cloud formation

out in Python using the Jupyter notebook software. Ordering an enormous number of English tweets about unambiguous things into positive and negative feelings is now possible thanks to the research presented in this paper. Excellent accuracy is achieved when sentiment features are used instead of traditional text categorization from Tables 2 and 3. A business group can use this method to rank sentiment classifiers that are acceptable and to help them plan for the product's future business development.

5 Conclusion

Sentiment analysis can determine whether a piece of text has a good, negative, or neutral sentiment. Sentiment analysis is a subtype of natural language processing that includes information extraction. The choice of algorithms is an important component of a data scientist's job. The most effective technique is frequently to test a wide range of algorithms. Machine learning-based sentiment analysis algorithms are projected

Confusion Matrix

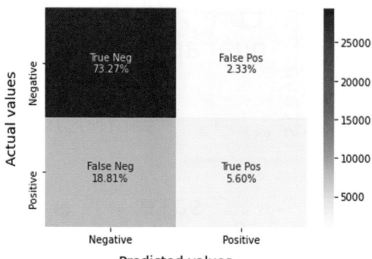

Fig. 5 Analysis of confusion matrix by Naïve Bayes classifier

Table 2 A couple of scenarios used to decide if you want to go to the beach or not

Parameters	Value
Accuracy of model on training data	84.44%
Accuracy of model on testing data	78.77%
Precision	0.80 (Negative) and 0.71 (Positive)
Recall	0.97 (Negative) and 0.23 (Positive)
Macro average accuracy	75%
Weighted average accuracy	77%

Table 3 Comparative analysis of accuracy

Parameter	Random forest method	Decision tree method	Proposed method (Naive Bayes classifier)
Accuracy of model on testing data	76%	75.89%	84.44% on training 78.86% on testing of unknown data

to be the most effective since they can be tuned to a specific type of data, such as tweets or reviews. Machine learning methods, on the other hand, require much larger datasets than the emotion lexicon algorithm or off-the-shelf algorithms. There should also be a set of tweets for practice. It should be observed that the three types of tweets in the training set were split unevenly: good, negative, and neutral. There were very few negative or indifferent comments. If the data had been dispersed more evenly, the machine learning systems would have found that the vast majority of tweets were positive and would have come to rely on that assumption for every tweet they received. There are various methods for classifying product analysis (which can appear as tweets) in view of the reactions communicated in Twitter to decide if the huge way of behaving is positive, negative, or unbiased, and afterward utilizing that data to assess the item market. Using data from Twitter, we were able to rate the "satisfied" sentiment classifier for online product evaluations. It is possible to compare numerous classifiers for classifying an enormous number of English tweets about unambiguous items into positive and negative contemplations because of this examination. At the point when feeling highlights are utilized rather than standard message arrangements, phenomenal precision is accomplished. Using this method, companies can rank the most acceptable sentiment classifiers and use the results to help them develop long-term product strategies.

References

1. Jha, P., Baranwal, R., Monika, & Tiwari, N. K. (2022). Protection of user's data in IOT. In *IEEE 2022 Second International Conference on Artificial Intelligence and Smart Energy (ICAIS)* (pp. 1292–1297)
2. Soni, G. K., Rawat, A., Jain, S., & Sharma, S. K. (2020). A pixel-based digital medical images protection using genetic algorithm with LSB watermark technique. In *Springer smart systems and IoT: Innovations in computing* (pp. 483–492)
3. Soni, G. K., Arora, H., & Jain, B. (2021). A novel image encryption technique using Arnold transform and asymmetric RSA algorithm. In *IEEE International Conference on Artificial Intelligence: Advances and Applications 2019. Algorithms for Intelligent Systems* (pp. 83–90). Springer.
4. Shahvaroughi Farahani, M., & Hajiagha, S. H. R. (2021). Forecasting stock price using integrated artificial neural network and metaheuristic algorithms compared to time series models. In *Soft computing* (pp. 1–31)
5. Kumar, G., Jain, S., & Singh, U. P. (2020). Stock market forecasting using computational intelligence: A survey. *Archives of Computational Methods in Engineering*, 1–33.
6. Jha, P., Biswas, T., Sagar, U., & Ahuja, K. (2021). Prediction with ML paradigm in healthcare system. In *IEEE 2021 Second International Conference on Electronics and Sustainable Communication Systems (ICESC)* (pp. 1334–1342).
7. Tripathi, M. (2021). Sentiment analysis of Nepali COVID19 tweets using NB, SVM AND LSTM. *Journal of Artificial Intelligence, 3*(03), 51–168.
8. Wongkar, M., & Angdresey, A. (2019). Sentiment analysis using Naive Bayes algorithm of the data crawler: Twitter. In *IEEE 2019 Fourth International Conference on Informatics and Computing (ICIC)* (pp. 1–5).
9. Roh, Y., Heo, G., & Whang, S. E. (2021). A survey on data collection for machine learning: A big data—AI integration perspective. *IEEE Transactions on Knowledge and Data Engineering (TKD), 33*(4), 1328–1347.

10. Goo, A. M. R. H. S. (2017). Twitter sentiment analysis using deep learning methods. In *2017 7th International Annual Engineering Seminar (InAES)*.
11. Prakruthi, V., Sindhu, D., & Anupama Kumar, D. S. (2018). Real time sentiment analysis of Twitter posts. In *IEEE 2018 3rd International Conference on Computational Systems and Information Technology for Sustainable Solutions (CSITSS)* (pp. 29–34).
12. Ahuja, K., Khushi, D., & Sharma, N. (2022). Cyber security threats and their connection with Twitter. In *2022 Second International Conference on Artificial Intelligence and Smart Energy (ICAIS)* (pp. 1458–1463).
13. Mandloi, L., & Patel, R. (2020). Twitter sentiments analysis using machine learning methods. In *IEEE 2020 International Conference for Emerging Technology (INCET)* (pp. 1–5).

A Survey on Cognitive Internet of Things Based Prediction of Covid-19 Patient

Lokesh B. Bhajantri⬤, Nikhil Kadadevar, Anup Jeeragal, Vinayak Jeeragal, and Iranna Jamdar

Abstract Nowadays, special care of patient systems are required for predicting or detecting Covid-19 patients ubiquitously. Also, there is a requirement for quarantine centers to set up for treating Covid-19 patients ubiquitously in a real-world environment due to the highly infectious virus. In a pandemic situation, it is difficult to keep track of the health condition of every individual patient. Also, doctors face problems to monitor and controlling controls patients' health conditions. In this regard, it is investigated the survey on cognitive Internet of Things based predicting Covid-19 patients using a machine learning algorithm. In this paper, it discusses a detailed survey on the proposed problem statement in terms of limitations, advantages and disadvantages, and performance parameters for various algorithms, and finally, it proposes system architecture for predicting and monitoring Covid-19 patients ubiquitously. Hence the proposed system is used to monitor the symptoms of patients like Temperature, SpO$_2$, and Cough rate of Covid-19 patients ubiquitously using intelligent sensors. The proposed system transmits data to the web server using Wi-Fi connectivity.

Keywords Internet of things · Cognitive internet of things · Cloud layer · Network layer · Sensing layer

1 Introduction

Internet of Things (IoT) is a technology that allows things or objects devices for sending and controlling the physical environment by making smarter things and connected through intelligent networks. In which IoT objects are connected through intelligent networks [1]. IoT devices can sense their surroundings and collect information. The architecture of IoT is shown in Fig. 1. This architecture comprises various layers such as the application layer, service layer, cloud layer, network layer,

L. B. Bhajantri (✉) · N. Kadadevar · A. Jeeragal · V. Jeeragal · I. Jamdar
Department of Information Science and Engineering, Basaveshwar Engineering College (Autonomous), Bagalkot, Karnataka, India
e-mail: lokeshcse79@gmail.com

© The Author(s), under exclusive license to Springer Nature Singapore Pte Ltd. 2023   377
S. Shakya et al. (eds.), *Sentiment Analysis and Deep Learning*,
Advances in Intelligent Systems and Computing 1432,
https://doi.org/10.1007/978-981-19-5443-6_28

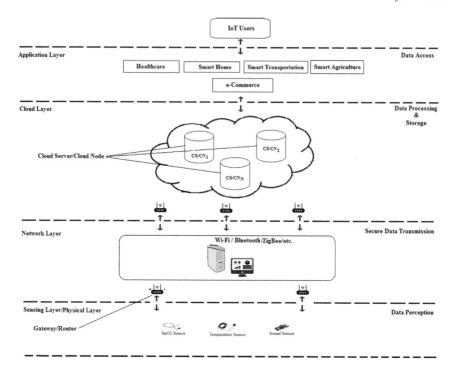

Fig. 1 Architecture of IoT

and sensing layer or physical layer for transmission of data over the Internet through objects or devices. Each layer of IoT is described below:

- *Application layer:* This layer is used to describe the connectivity between devices and their applications. It also supports the protocols for various applications like healthcare, smart home, smart transportation, smart agriculture, etc.
- *Cloud Layer:* This layer associates with cloud servers or cloud nodes required to process and store the data.
- *Network layer:* This layer is used for communication between things and endpoints. It is used to transmit secure data across the network. It includes wireless mesh IEEE 802.15.4 and wireless point to multipoint IEEE 801.1.11. It also includes wired connection IEEE 1901.
- *Cloud Layer:* This layer associates with cloud servers or cloud nodes required to process and store the data.
- *Sensing Layer/Physical Layer:* This layer comprises various types of sensors like temperature, sound, motion gesture, and SpO$_2$ sensors. These sensors are used to percept the data from the network and transmit the collected data to the network layer through IoT gateways.

In IoT architecture, smart objects or devices are used to communicate with each other. These objects are communicated through the field area networks (FAN) to

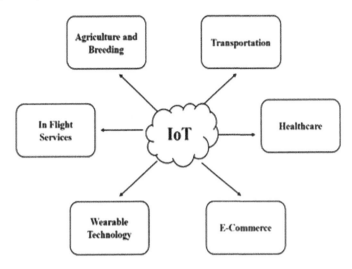

Fig. 2 Applications of IoT

use precise applications. It also uses gateway devices that provide communication between devices and network domain [2]. The IoT architecture comprises some of the challenges that are scales, security, privacy, data Analytics, etc. [3]. In the real-world environment, IoT is used for a variety of purposes ranging from improving quality of life to addressing urban challenges, food production, agriculture, the consumer, commercial, industrial, and infrastructure categories manufacturing, medicine, energy supply, water distribution, etc., [4]. The various applications are associated with IoT as shown in Fig. 2.

In an IoT network, the data collected from sensors/objects are not utilized efficiently in terms of data processing, data management, data aggregation, classification, security, and privacy constraints. Such collected data can be aggregated in real-time bases and meaningful information is extracted from IoT devices. Therefore, it is needed for developing intelligent or cognitive devices or objects or things to capture the various context data efficiently by the environment. Cognitive Internet of Things (CIoT) can perceive and analyze the data to make smart choices and increases network choices [5].

The second generation of IoT is known as CIoT. The CIoT comprises cognitive computing technologies and devices to make intelligent or cognitive decisions. The Cognitive Internet of Things (CIoT) has a set of intelligent things that are interconnected and will intercommunicate and collaborate collectively behave in an intelligent fashion over the Internet. This capacity of CIoT needs to be exploited for facilitating automated intelligent healthcare services effectively. Healthcare organizations are required to deploy cognitive technological solutions for monitoring, predicting diseases, patient behavior analysis, and healthcare services. CIoT is an active research domain whose applications can be seen in various fields. So CIoT has attracted more interest in healthcare systems as it provides efficient, reliable, cost-effective recommendations, and fault tolerance services to healthcare users ubiquitously. Some of the CIoT challenges are [6]:

- *Standardization:* It is one of the key steps for the development of CIoT networks. The required for increasing the standardization of CIoT network architecture to process the data efficiently. Also, it supports various applications.
- *Spectrum Efficiency:* This is required for the amount of data transmitting over the network bandwidth efficiently in the IoT environment.
- *Security and Privacy:* This is one of the most promising challenges in the IoT environment. Sometimes data is to be considered as a highly sensitive data by the network's things. It is required for processing of data securely over the CIoT networks. Hence, security and privacy mechanisms or algorithms are required at the IoT devices or things and network levels. Therefore, only trusted users can access the services over the CIoT networks. Also, a lot of research is going on the same challenges in the IoT networks to provide secure data.
- *Data Perception:* It is one of the significant issues of CIoT. The CIoT things or devices intelligently capture or percept the data from the environment. Therefore, cognitive techniques are required for perception of data by the devices efficiently.

The typical CIoT Architecture is as shown in Fig. 3. This architecture associates various elements like monitoring and control, context data consumption and integration, and data sense.

The despite the paper is as organized follows: the literature survey for the proposed work in illustrated in the Sect. 2. Section 3 explains the proposed work. Finally concludes the work in Sect. 4.

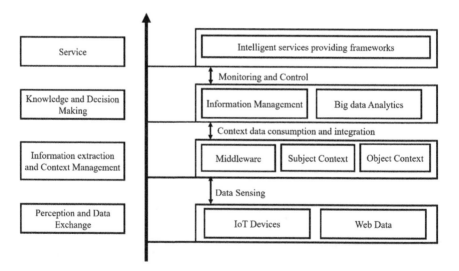

Fig. 3 Architecture of CIoT

2 Literature Survey

In this section, the literature survey for various works toward IoT and CIoT are described. Also, it is discussed some of the algorithms for the prediction of data in IoT/ CIoT environment. The following Tables 1, 2, and 3 show the approaches and limitations of the related works, advantages and disadvantages, and their performance parameters of various algorithms respectively, towards the proposed work.

As per the detailed survey of the literature, it is observed that inefficient monitoring and predicting the patient information. Therefore, it has been proposed to predict Covid-19 patients efficiently using cognitive approaches that provide better decisions in emergency situations. Also, it has been observed that no such kind of work exist in the present context. Hence, it has been planned to propose to Covid-19 patients

Table 1 The various approaches and limitations

References	Approach	Limitations
[7]	• Illustrate the differences between existing technologies and healthcare systems • CNN-based healthcare system is presented in terms of state of the heart conditions of the patient	• High power consumption—wireless sensors in CT-Scan device
[8]	• This approach describes the concept of detection system using IoT • The proposed system is used to protect individual users as well as providing medical assistance	• Regular connectivity between the clients and the server • Only user devices can detect, if the device are placed in patients' pocket
[9]	• Describes the healthcare assistance to the patients at critical conditions • Also, assist the advanced medical facilities	• Problem associated with having data online for data privacy/security
[10]	• This approach is presented on contactless medical services by the doctors due to spreading of infections	• Cough sensor is not accurate as expected
[11]	• The proposed approach is describing the tracking of Covid-19 areas or zones for monitoring and tracking of Covid-19 patients remotely	• Bluetooth communications technology
[12]	• Describes the health conditions of Covid-19 patients using wearable sensors • Also, to measure the health parameters of the patients	• High production cost • It is a cost-effective approach for the lifetime management of the patients
[13]	• To monitor the social distancing and mask detection	• Limitations in performances due to a number of processed frames per second

(continued)

Table 1 (continued)

References	Approach	Limitations
[14]	• This approach used to describe the design concept of medical devices using IoT • Also, illustrate the biomedical wearable APIs	• Design factors like economic cost and human precautions
[15]	• It describes about the technologies like IoT and CR and their applications • Discussed the importance of adding Cognitive capabilities to IoT devices	• They did not show how to add Cognitive capabilities to IoT devices and also did not discuss about requirements
[16]	• It includes cloud-based connected devices which will help doctors to monitor patient health remotely	• This device does not have multiple user access
[17]	• Patient's health is detected based on the lungs condition • Potential benefits are graphed using mechanical ventilator	• Distortion of collected data by healthy and unhealthy patients • Fault in the classification of data
[18]	• Spatial patterns are used for decision making process • It will analyze and collect the variable Bigdata	• There is an unlabeled bid data, difficult in data reliability
[19]	• Discussion on social distancing Covid-19 cases and death rates • Economic consequences of Covid-19	• It does not have much information about unemployment
[20]	• The proposed approach describes a deep learning-based prediction of patients with Corona virus automatically • Also, the proposed system examines the results based on the X-ray of the patient's chest	• Using CT-Scan, cannot able to access the individual degree of abstraction and CT-Scan is relatively expensive • A relatively high radiation
[21]	• It measures the SpO$_2$ and Heart rate through these data it predicts Covid-19	• Here there is no doctor's role to monitor a patient's health • Data need to be entered to system manually
[22]	• The approach is used to measure the body temperature of Covid-19 patients using thermal cameras	• Data storage is more expensive and there is a miss lead in the patient results
[23]	• The work discusses the cognitive radio networks of 5G and their impacts on the network • The main function of Cognitive radio goes beyond spectrum sharing, spectrum sensing, spectrum mobility, spectrum management	• The 5G network is still in design process in many countries, which will face lot of problems in designing
[24]	• Discussion on challenges of IoMT • The security solutions are used to examine the primary challenges with IoMTs security and privacy	• The IoMT devices have an issue that it sends false alerts on medical emergencies

Table 2 Data prediction algorithms

Algorithm	Advantages	Disadvantages
Linear regression	• It is simple to implement and easier to explain the output • This algorithm is best to use because of its less complexity	• Main disadvantage of linear regression is the assumption of linearity between the dependent and independent variables
K—Nearest Neighbors	• To classify the new data point KNN algorithm reads through the whole dataset to find out K- Nearest neighbors	• Algorithm is not well performed with a large dataset • Also, it will not perform with high dimensions • Feature scaling is required
Decision tree	• Normalization and scaling of data are not required • Data preparation during preprocessing with less effort	• Often, more time involved for training the model • It is poor to apply predicting continuous values and regression
Support vector machine (SVM)	• SVM is relatively memory efficient • SVM is effective where the number of dimensions is greater than the number of samples	• Long training time and difficult to interpret • Choosing and appropriate kernel function is difficult
Naive bayes	• Exceedingly scalable by the number of data points and predictors • It handles both continuous and discrete data	• In some cases, the estimations can be lead to wrong
Logistic regression	• It makes no assumption of distributions of classes • It not only provides coefficient size but also its direction of the association	• Non-linear problems cannot be solved • It is tough to obtain complex relationships using logistic regression

Table 3 Performance matrix for variable data prediction algorithms

References	Algorithm	Accuracy	Quantity
[25]	Linear regression	87.0.00	172,479
[26]	K-Nearest neighbors	87.60	1236
[27]	Decision tree	93.22	263,007
[27]	Naive bayes	94.30	263,007
[27]	Logistic regression	87.34	263,007
[7]	Support vector machine (SVM)	95.38	1282

for providing services ubiquitously. The next section describes the proposed work for the same.

3 Proposed Work

The proposed work objectives are as: (1) To analyze the patient data using CIoT devices, (2) to design a medical recommended system for Covid-19 patients in terms of Medicine doctors, etc., (3) to design and implement of predicting algorithm using machine learning approach, (4) to design a health monitoring device for predicting the health issues of human, (5) to provide high level applicability in hospitals, remote check-ups, workstations where many patients can be frequently monitored, and (6) to provide customer satisfaction services and implementation as low cost.

The proposed system architecture is as shown in Fig. 4. This architecture comprises intelligent sensors with high-speed network. These sensors are to sense and transmit the data over the CIoT networks Also, analyze the data by the cognitive sensors. In this work, it has been planned to propose appropriate sensors for detecting the symptoms of Covid-19 patients using algorithms. The sensors like temperature, SpO_2, and Sound. This system is to collect body temperature data, saturated oxygen level (SpO_2), and sound (cough) of the Humans. These collected data to predict Covid-19 patients using machine learning algorithm.

Nowadays, CIoT is a more promising technology to access data in anytime and anywhere in the real-world environment by intelligent sensors. Intelligent things or objects are for providing emergency services to users. The proposed system is to

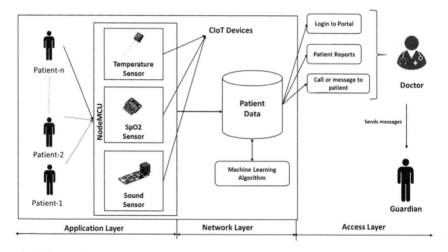

Fig. 4 Proposed system architecture

monitor the COVID -19 patients remotely. The device is able to monitor the temperature, SpO_2, and cough sounds of patients through various sensors remotely. The data will be classified according to our requirements using a machine learning algorithm. Doctors can view the patient data ubiquitously. Whenever the physiological matrices values go below or above the normal range and also the doctor will notify the condition of the patient to their guardian. The doctor can give emergency services to patients ubiquitously. According to the patient's report, the doctor can give a prescription to the patient.

Initially, deployment of CIoT network in a real-world environment. The device will read the health parameters of patients like temperature, Oxygen level, and cough sounds. These values of the sensor are fed as analog inputs to the NodeMCU. The analog signals are converted into digital signals. If the value goes above or below the normal ranges are as follows:

Case 1: Non-symptomatic parameters

- $SpO_2 \geq 95\%$
- Cough Rate: NIL
- Temperature ≤ 37.2 °C

Case 2: Mild symptoms parameters

- $92\% < SpO_2 < 95\%$
- Cough Rate ≤ 7/min
- 36 °C $\leq$ Temperature ≤ 38 °C

Case 3: Major symptoms parameters

- $SpO_2 \leq 92\%$
- Cough Rate ≥ 15/min
- Temperature > 38 °C.

4 Conclusion

In this paper, it has been presented the detailed survey of the proposed work. Also, it has presented the new architecture of CIoT-based prediction of the Covid-19 system in a real-world environment. Also, the proposed architecture illustrates the monitoring of a patient's health remotely to reduce contact between Covid-19 infected patients and doctors to avoid the spreading of the virus. It has been planning to propose a wearable device to monitor Covid-19 patients ubiquitously in future work. Therefore, the proposed system is proficient, trustworthy, and accurate in healthcare data.

References

1. Bhardwaj, V., Joshi, R., & Gaur, A. M. (2022). IoT-based smart health monitoring system for COVID-19. *Journal Computer Science, 3*(137), 1–11.
2. Goncalo, M., Nuno, G., & Pombo, N. (2019) A survey on IoT: Architectures, elements, applications, QoS, platforms and security concepts. *Journal Electronics, 8*(10), 1–27.
3. Choudary, A., & Godara, S. (2017). Internet of things: A survey paper on architecture and challenges. *Journal International Journal of Engineering Technology Science and Research IJETSR, 4*(6), 442–447.
4. Hussein, A. H. (2019). Internet of things (IOT): Research challenges and future applications. *Journal (IJACSA) International Journal of Advanced Computer Science and Applications, 10*(6), 77–82 (2019).
5. Bhajantri, L. B., & Balugari, P. (2019). A survey on data perception in cognitive internet of things. *Journal of Telecommunications and Information Technology, 3,* 75–86.
6. Park, J. -H., Salim, M. M., Jo, J. H., Sicato, J. C. S., Rathore, S., & Park, J. H. (2019). CIoT-Net: a scalable cognitive IoT based smartcity network architecture. *9*(29), 1–20.
7. Song, Y., Zheng, S., Li, L., Zhang, X., Zhang, X., Huang, Z., Chen, J., Wang, R., Zhao, H., Chong, Y., Shen, J., Zha, Y., & Yang, Y. (2020). Deep learning enables accurate diagnosis of novel-coronavirus with CT images. *IEEE/ACM Transactions on Computational Biology and Bioinformatic, 18*(6).
8. Rashidi, N. A. (2020). Covid-19 detection based on CT-scan. *Journal Computer Science, 9*(20)
9. Nooruddin. (2019–2020). Covid-19 prediction using real time data.
10. Palanisamy, R., Kartik, M., Rohit, H., Jay, S., Puranik, A., & Vaidya, A. (2019). IoT based patient monitoring system. *International Journal of Recent Technology and Engineering (IJRTE), 8*(2S11), 1–6.
11. Mukhtar, H., Rubaiee, S., Krichen, M., & Alroobaea, R. (2021). Screening of COVID-19 using real time data from wearable sensors. *International Journal Environment Research Public Health, 18*(8), 1–17.
12. Bassam, N. A., Hussain, S. A., Qaraghuli, A. A., Khan, J., Sumesh, E. P., & Lavanya, V. (2021). IoT based wearable device to monitor the signs of quarantined remote patients of Covid-19. *Journal Elsevier Public Health Emergency Collection, 9,* 1–16.
13. Petrovic, N., & Kocic, D. (2020). IoT based system for covid-19 indoor safety monitoring. In *IcETRAN* (pp. 1–7).
14. Fayez, Q., & Krishnan, S. (2018). Wearable hardware design for IoT medical things. *Journal Department of Electrical, Computer and Biomedical Engineering, 18*(11), 1–22.
15. Shah, M. A., Zhang, S., & Maple, C. (2013). Cognitive radio networks for internet of things; application, challenges and futures. In *19th International Conference on Automation and Computing* (pp. 1–6).
16. Shreerang, J., Pranav, M. S., Jitendra, P., More, M., Prayag, S., Satish, P., & Marathe, S. (2020). IoT based patient health care for covid-19 center. *International Journal Recent Technology and Engineering. 9*(3), 258–263.
17. Acho, L., Vargas, A. N., & Pujol-Vazquez, G. (2020). Low cost open source Mechanical ventilator with pulmonary monitoring for Covid-19 patients. *9*(3), 1–14.
18. Dagazany, A. R., Stegagno, P., & Mankodiya, K. (2018). Variable internet of things and deep learning for big data analytics. *Journal Mobile Information System.* 1–20.
19. Brodeur, A., Gray, D., Islam, A., & Bhuiyan, S. (2021). A literature review of the economics of covid-19. *Journal Economics Survey, 35*(2), 1007–1044.
20. Salehi, A. W., Baglet, P., & Gupta, G. (2020). Review on machine and deep learning models for detection and prediction of coronavirus. In *Proceedings Conference on Nanotechnology* (pp. 3896–3901).
21. Revar, D. S., Sevaniya, J. S., & Joshi, V. R. (2020). Pulse oximeter design to predict covid-19 possibilities on patients health using machine learning. *GRD Journal,5*(10), 9–14.

22. Abhadji, I. E., Awuzi, B. O., Ngowi, A. B., & Millham, R. C. (2020). Review of big data analytics, artificial intelligence and nature inspired computing models towards accurate detection of covid-19 pandemic cases and contract tracking. *International Journal Environment Research Public Health, 17*(15), 1–16.
23. Valanarasu, R., & Christy, A. (2019). Comprehensive survey of wireless cognitive and 5G networks. *Journal of Ubiquitous Computing and Communication Technologies (UCCT)*, 23–32.
24. Smys, S., & Raj, J. S. (2022). Future challenges of the internet of things in the health care domain-an overview. *Journal of Trends in Computer Science and Smart Technology, 3*(4), 274–286.
25. Cavovean, D., Ioana, I., & Nitulescu, G. (2020). IoT system in diagnosis of covid-19 patients. In *Informatic Economic* (Vol. 24(2), pp. 75–89). Bucharest University of Economic Studies.
26. Gothai, R., Thamilselvan, E., & Sakthivel, R. (2020). Prediction of COVID-19 growth and trend using machine learning approach. In *Proceeding of International Virtual Conference on Sustainable Materials* (pp. 1–5).
27. Theerthagiri, P., Jacob, I. J., Ruby, A. U., & Yendapalli, V. (2020). Prediction of COVID-19 possibilities using KNN classification algorithm.

An Expert System for the Detection and Mitigation of Social Engineering Attacks (Sea) Using Machine Learning Algorithm

Rathnakar Achary and Chetan J. Shelke

Abstract Social engineering is a mechanism of convincing an Internet user to disclose their private information. It is a technique of influencing people, so that they pretend to share their private information. Using social engineering attack (SEA) individual's or the systems are targeted by the criminals, trying to access the password and bank information's or secretly install malicious software, this also gain access to your computer. The attackers not only capture an individual's information, also access information in various business environments such as discussion with the corporate clients, interaction between the employees through a chat service. This can be identified, and its effect can be analyzed using advanced machine learning models. In the research work we have proposed and demonstrated the implementation of a model using machine learning, to analyze the malicious behavior of the attackers and for the detection of SEA. This method is applied for both off-line text and in real-time situations to identify whether a human Chatbot or off-line chat is performing a SEA or not. In the proposed method the text message is analyzed and verified for linguistic errors and the model used to detect and separate the chances of any attacks. The technique proposed is also analyzed using both factual and semi-synthetic dataset to obtain the result with better accuracy. The model developed will detect phishing links (uniform resource locator -URLs) from the webpage and classify the URL whether it is legitimate or illegitimate URL using machine learning algorithm.

Keywords Phishing URL · Legitimate URL · Machine learning · Logistic regression · Prediction · Social engineering attack

R. Achary (✉) · C. J. Shelke
Alliance College of Engineering and Design Alliance University, Anekal, India
e-mail: rathnakar.achary@allaince.edu.in

C. J. Shelke
e-mail: chetan.shelke@alliance.edu.in

© The Author(s), under exclusive license to Springer Nature Singapore Pte Ltd. 2023   389
S. Shakya et al. (eds.), *Sentiment Analysis and Deep Learning*,
Advances in Intelligent Systems and Computing 1432,
https://doi.org/10.1007/978-981-19-5443-6_29

1 Introduction

SEA is a method of manipulating Internet or mobile users by performing activities like getting in to their confidential information. The rapidly increasing social engineering attacks are the threat to a cyber system. Attackers perform this to manipulate individuals' and enterprises confidential information. The social engineering attacks are mainly depending on the facts that the people are unaware of their acts. The attacker uses the psychological tricks to gain trust of the victim, to gather sensitive information such as password and bank details. SEA poses a security challenge for all the Internet users regardless of the security mechanisms used such as confidentiality, intrusion detection and prevention mechanisms, firewalls and antivirus software [1]. The behavior of the Internet user is one of the main challenges for the security and privacy regarding to social engineering observations. To understand the implications of psychology and personal traits a five-factor model is used. These factors are organized among a series of five statistically independent factors such as openness, conscientiousness, extroversion, agreeableness and neuroticism as in Table 1.

Social engineers take the advantage of poor awareness of the valuable information and depend on the fact that people are unaware of their valuable information and not care about their personal information also careless about protecting that information. Among the several types of SEA it was identified that the attack may be both from internal and external sources, where insider intentionally violating rules of their organization, when they are connected to the organization network. This sort of attack generally done by a third party or disgruntle employee for the financial gain or revenge.

1.1 Related Work

Several research implementations targeted to identifying the repeated callers, considering their voice signature, the proof-of-concept model to detect the attack. The researcher in [2] analyzed the characteristics of social engineers, they use the

Table 1 Psychology and personal traits

Openness	For an individual it is to experience measures creativity and curiosity, as opposed to caution and consistency
Conscientiousness	Measures how organized or rigid one is as opposed carelessness
Extroversion	Refers to how open and interactive one is with other people
Agreeableness	Reflects friendliness, kindness and empathy as opposed to manipulation and lack of cooperation
Neuroticism	Is correlation with stress nervousness, langer and frustration as well as other kind of negative emotions

psychology of Internet users to exploit the people for their own use. In social engineering the criminals smear the trick for the benefit of gathering confidential data, fraud and identity theft, or maliciously access the user's computer system. In the enterprises the SEA is highly challenging because they depend upon the behavior of the employees. The attacker will gain access to the vulnerable employee accounts. In the same paper it is also presented about the social engineering attacking using e-mail and other attack tools. Authors in [3] identified the pattern of the requests for retrieving private details for gaining control over the user activities. This mechanism offers a manually obtained list of verb-noun pairs to blacklist the attacker. An enhancement of this work is carried out in paper [4]. The researchers have identified four main attack vectors including the type of the questions, urgency of the dialog, and urgency of the dialog and the method of greeting [5]. Naive Bayes classifier is used to blacklist a large corpus of fishing e-mails. In [6] a semantic-based approach used to identify the social engineering attack. Human factors and behaviors are considered as weakest link for most of the SEA cases [7]. Authors in [8] proposed an Island-based search technique to detect the social engineering attack. The authors proposed in their algorithm a highly disruptive polynomial mutation technique. The entire research work in this paper is organized as follows: Sect. 2—explained the types of social engineering attacks. Section 3—represents the different methods of social engineering attack detection. Section 4—use of artificial neural network for classification of attacks. Section 5—explained the experimental evaluation of the dataset followed with the conclusion.

2 Types of Social Engineering Attacks

Based on the types of attacks, the SEA is classified into three categories as: technical-based, social-based and physical-based attack. In technical-based attack, the attacker uses Internet or web as a medium for attack. In a physical-based attack, the attacker performs the attack by physical actions. A detailed classification of SEA is shown in the Fig. 1. This denotes that the attack surfaces are increasingly large and varied. To counter this attacker targeting to the highly vulnerable link in the cyber security chain, to exploit the infrastructure as an attack vector. The other two types of attacks are human-based attack and computer-based attacks categories. The computer-based attacks the user mobile phones are also targeted. They will target many victims simultaneously. A computer-based attack technique such as social engineering toolkit (SET) is used for spear phishing e-mails. In mobile-based attack, malicious third parties Apps are used to gain access to a mobile device. By adopting new technologies in enterprises, the security architecture and the attack perimeter have changed.

The four phases of SEA are information gathering, trust, elicitation and pretexting.

Information gathering: Information gathering is useful for establishing the possible attack vector to perform the attack. It is a process of gathering all information about the victims. Sources for gathering information are Facebook, Twitter, LinkedIn,

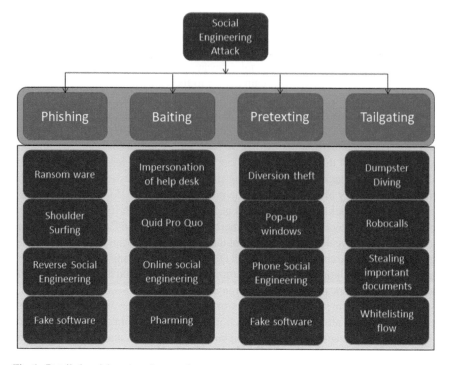

Fig. 1 Detailed social engineering attack

google and many more public websites. The attacker may use the other sources such as key-logger and shoulder surfing.

Trust: Only when the people develop trust with someone will disclose the information. In this phase the attacker seeks to establish trust relationship between the victims.

Elicitation: Elicitation is a process of getting the information without asking it. Some of the techniques generally used for gathering information are understand how to communicate with people, adapt communication to fit situation and building bond with target.

Phishing could also be a type of SEA targets at exploiting the vulnerabilities found within the system at the end nodes. Phishing is a criminal act of acquiring personal information by sending out forged emails with fake websites and fraudulent web links in web pages. In Phishing attack the attacker forwards an e-mail message or a URL link and pretending himself as a legitimate user for extracting the sensitive information's form the target user. The receivers on receiving this message click the URL or respond to the email with sensitive or personal information's like credit card data, username and password which they are likely to urge cheated. The purpose of this attack is to fraudulently acquire confidential information from intended target using different modes like e-mail or phone calls. Attackers mislead the victims by tricking them using fake website, awards, e-mails, free offers, adds, antivirus software and scareware. The different categories of phishing attacks are,

- *Spear phishing*, which target a specific individual or selected group using their name to extract the information.
- *Whaling phishing*, it is a type of spear phishing attack for victimizing superior officials in the enterprises, named big "fishes".
- *Vishing phishing*, its derived based on the voice mode and the attacks are performed voice over IP (VoIP).
- *Interactive voice response phishing* is a voice-based attack carried out by using an interactive voice-based system and pretending as it is from a legitimate source such as banks or business to make the victim to disclose the sensitive data.
- Business email compromising phishing is also like whale phishing attack to high profile individuals in corporate to fetch their corporate email, calendar, accounting related details and other private information's.

Pretexting: The process of pretexting takes advantage of identification of weakness of the victim using the voice transactions. The attacker uses different techniques to identify their clients. Using this method, they also start extracting information's such as date of birth, residence details and account number. In pretexting the attacker attract the victims by gaining their confidence and assuring a job offer, perform a service or winning lottery, and start extracting their personal details.

Baiting Attacks: In these types of attacks the attacker urge the users to click on a URL, to get an assured gift. These links act as a trojan horse, where the attacker targets to the unsecured IT infrastructure or devices connected to the computer, like USB devices. A baiting attack named controller area network (CANDY) in [9] the attack on an automotive system, launch a trojan horse, which manipulate the message exchange between the vehicle and the controller.

Tailgating Attacks: This types of attack the intruder is trying to access the restricted areas of the user, i.e., physical access to the site or the restricted areas.

3 Methods for Social Engineering Attack Detection

A message exchange in a social media platform or chat is analyzed to estimate the dialog is amounting to social engineering attack or not. For the analysis of the message exchanged, its content is preprocessed and converted into a dataset for classification by logistic regression method. The algorithm verifies the case-based on reasoning system for checking the URL and other details by using Sklearn machine learning library to import classification matric, classification report and accuracy, also Seaborn for visualization. Each score is classified as:

$$S_L = \text{Link Score}, \quad S_P = \text{Spelling score and } S_I = \text{Intent score}.$$

Table 2 Web site nature—categories and subcategories

1XX negative	2XX questionable	3XX neutral	5XX positive
101—Malware 103 - Phishing 104—Scam 105—Politically illegal	201—Misleading claims or unethical 205—Spam 207—ads/popups	301—On-line tracking 302—Controversial 303—political	501 a good site

Their normalized values range between 0 and 1. Once the target message is preprocessed the classification dataset has four labels as; Intent, Spelling, Link and social engineering attack or not.

A classifier uses these inputs for the classification mechanism.

1. Use a regular expression pattern finder to extract the URL of the message link.
2. If the message contains the URL it is sent to a case-based reasoning system to assess the link associated with the message is malicious or not.
3. The web of trust (WOT) API used verifies the link and return the reputation of the site, the confidence of the given reputation. Their value ranges between 0 and 100. Also identifying the categories and subcategories (1 out of 17) to determine the nature of the website. The subcategories are listed as in Table 2.

4. With categories IXX or 2XX, the $S_L = 1$.
5. Else divide the reputation by 100 take it away from 1 as given in Eq. (1).

$$S_L = 1 - \frac{\text{reputation value}}{100} \tag{1}$$

6. We can check the spelling by using SymSpelly library.
7. In this step we analyze the spelling quality, which is represented as an Eq. (2). Where x—*represents the misspelled words.* The numerical value to represent this is between 0 and 1 using the equation.

$$S_{SP} = 1 - e^{-\alpha x} \tag{2}$$

The exponential function is used to find the greater amount of spelling errors. To accommodate the score to equate to 1, the value α can be adjusted to impact how you want to penalize grammatical errors. With multiple iterations allowed to use 0.5 as a small number of mismatches without creating a score of higher value.

$$\text{If } \alpha = 0.5 \text{ and } x = 1, \text{ then } S_{SP} = 0.39$$
$$\text{If } x = 5 \text{ then } S_{SP} = 0.92$$
$$S_{SP} = 1 - e^{-\alpha x}$$

8. The spellings of the message received are then matched to a group of prohibited text, which are acquired from a library of 48 security policy style words. This

can be filled easily with environmental, or company related words such as passwords credentials database and others. The number of prohibited words is given by M_B.

9. Find the purpose of the verbs and adjective such as necessary, must, urgent. It is characterized by M_I.

10. To get an accurate value, the values M_B and M_I are multiplied by the weights W_B and W_I weighted at values of 2 and 1 altogether. This is represented in Eq. (3).

$$x = (M_B \times W_B) + (M_I \times W_I) \tag{3}$$

11. The normalized value of x is given in Eq. (4)

$$S_l = 1 - e^{-\alpha x} \tag{4}$$

A satisfactory output may be achieved when $\alpha = 0.4$. A larger value of S_l represents that there is a higher concentration of the prohibited words in the message.

12. In this step the new dialog dataset is examined attack and allocate the true (1) or false (0) value to the new dataset used for categorization. Using a valid dataset and Convolution Neural Network model the attack message received has been trained over a corpus of sentences that have been tagged [11].

4 Artificial Neural Network for Classification of Attacks

Once the dataset is inhabited the multi-layer perceptron classifier is applied, based on the output the status of the attack is decided [11]. It is considered as an attack if the output is HIGH otherwise not. An activation function $f(z)$ defined in the beginning, with the x input value and w weights as inputs. The activation function is,

$$z = w_1 x_1 + w_2 x_2 + \ldots \ldots w_n x_n$$

$f(z)$ is greater than a given threshold Ø the output is 1 or -1 otherwise

$$f(z) = \begin{cases} 1 & \text{if } z > \varphi \\ -1 & \text{else} \end{cases}$$

When $z = w_1 x_1 + w_2 x_2 + \ldots w_n x_n$

$$z = \left\{ \sum_{j=1}^{n} x_j w_j = w^T x \right.$$

Using the perceptron rules the weights are updated as follows.

i.   The weights are initialized with a random number
ii. In every integration, the weights are updated for each value of x as,

$$w^+ = w_j + \Delta w_j$$

where $\Delta w_j = e(t \text{ arg } et(i) - \text{output}(i))x_j^i$ and e is the learning rate

5   Experimental Evaluation

The evaluation of the social engineering attack is analyzed by using a data set with eleven thousand records and 32 features, classified as an attack or not. The dataset is preprocessed and derived for the four classical labels as intent, spelling, link and attack. Finally, the datasets are transformed into numerical values for classifications with four entries each. Three label $(-1, 0, 1)$ for the respective data and one with a (1 or 0) value to determine the SE conversation is an attack or not. The classification models are evaluated by accuracy metric as in Eq. (5) to calculate the small percentage of the estimate that each model got right.

$$\text{Accuracy} = \frac{T_P + T_N}{T_P + T_N + F_P + F_N} \tag{5}$$

where T_P—for true positive
T_N—True negative.
F_P—for false positive.
F_N—for false negative.
The accuracy matrices are used to evaluate the multi-layer perceptron classification model and compared to the logistic regressions, which is mainly depends on the quality of the dataset with an appropriate validation. For a final cross validation, the training and testing dataset set are selected at a rate of 80:20. The algorithms logistic regression and neural network multi-layer perceptron's with the Sklearn machine learning library for classification and seaborn for visualization are used. The complete analysis of the dataset use for comparison is given in the tables.

The dataset is analyzed the different parameters such as IP address as in Tables 3, 4, 5, 6, 7, 8, @symbol, sub-domain title, domain registration, linkage tags and mouse. The parameters in the Tables (9, 10, 11, 12, 13, 14, 15 and 16) below are analyzed based on the three parameters as: -1 (attack), 0 (suspicious) and 1 (non-malicious). The percentage of malicious traffic from different sources is shown in Figs. 2, 3.

The Classification value on mouse-over, request URL web traffic is shown in Fig. 4 and classification report showing the precision and accuracy is shown in Table 17.

Table 3 IP address

index having_IPhaving_IP_AddressURL URL_Length \			
Count	11,055.000000	11,055.000000	11,055.000000
Mean	5528.000000	0.313795	−0.633198
Std	3191.447947	0.949534	0.766095
Min	1.000000	−1.000000	−1.000000
25%	2764.500000	−1.000000	−1.000000
50%	5528.000000	1.000000	−1.000000
75%	8291.500000	1.000000	−1.000000
Max	11,055.000000	1.000000	1.000000

Table 4 @ symbol

Shortining_Service having_At_Symbol double_slash_redirecting \			
Count	11,055.000000	11,055.000000	11,055.000000
Mean	0.738761	0.700588	0.741474
Std	0.673998	0.713598	0.671011
Min	−1.000000	−1.000000	−1.000000
25%	1.000000	1.000000	1.000000
50%	1.000000	1.000000	1.000000
75%	1.000000	1.000000	1.000000
Max	1.000000	1.000000	1.000000

Table 5 Sub-domain details

Prefix_Suffix having_Sub_Domain SSLfinal_State \			
Count	11,055.000000	11,055.000000	11,055.000000
Mean	−0.734962	0.063953	0.250927
Std	0.678139	0.817518	0.911892
Min	−1.000000	−1.000000	−1.000000
25%	−1.000000	−1.000000	−1.000000
50%	−1.000000	0.000000	1.000000
75%	−1.000000	1.000000	1.000000
Max	1.000000	1.000000	1.000000

6 Conclusion

In this paper a detailed presentation of SEA detection method using logistic regression. The method gathers the data from different sources and used for the classification purpose. The model recommended in this paper has been verified by using a valid dataset with eleven thousand records and 32 features. On preprocessing the data are

Table 6 Domain registration

Domain_registration_length Favicon port HTTPS_token \			
Count	11,055.000000	11,055.000000	11,055.000000	11,055.000000
Mean	−0.336771	0.628584	0.728268	0.675079
Std	0.941629	0.777777	0.685324	0.737779
Min	−1.000000	−1.000000	−1.000000	−1.000000
25%	−1.000000	1.000000	1.000000	1.000000
50%	−1.000000	1.000000	1.000000	1.000000
75%	1.000000	1.000000	1.000000	1.000000
Max	1.000000	1.000000	1.000000	1.000000

Table 7 Linking tags

Request_URL URL_of_Anchor Links_in_tags SFH \			
Count	11,055.000000	11,055.000000	11,055.000000	11,055.000000
Mean	0.186793	−0.076526	−0.118137	−0.595749
Std	0.982444	0.715138	0.763973	0.759143
Min	−1.000000	−1.000000	−1.000000	−1.000000
25%	−1.000000	−1.000000	−1.000000	−1.000000
50%	1.000000	0.000000	0.000000	−1.000000
75%	1.000000	0.000000	0.000000	−1.000000
Max	1.000000	1.000000	1.000000	1.000000

Table 8 Mouse over

Submitting to email Abnormal_URL Redirect n_mouseover \			
Count	11,055.000000	11,055.000000	11,055.000000	11,055.000000
Mean	0.635640	0.705292	0.115694	0.762099
Std	0.772021	0.708949	0.319872	0.647490
Min	−1.000000	−1.000000	0.000000	−1.000000
25%	1.000000	1.000000	0.000000	1.000000
50%	1.000000	1.000000	0.000000	1.000000
75%	1.000000	1.000000	0.000000	1.000000
Max	1.000000	1.000000	1.000000	1.000000

classified into three categories as; attack, suspicious and non-malicious to detect SEA at a very high accuracy. The experimental result obtained out of this research specifies that the approaches achieved significant accuracy of social engineering detection compared to other related works. It also indicates that the social engineering attacks can still be effectively tackled with appropriate selection of feature vectors. We have

Table 9 IP address

index	having_IP	having_IP_Address URL	URL_Length	Shortining_Service \
11050 11051	1	−1	1	
11051 11052	−1	1	1	
11052 11053	1	−1	1	
11053 11054	−1	−1	1	
11054 11055	−1	−1	1	

Table 10 Redirecting prefix

	Having_At_Symbol	double_slash_redirecting	Prefix_Suffix \
11050	1	1	1
11051	−1	−1	−1
11052	1	1	−1
11053	1	1	−1
11054	1	1	−1

Table 11 Domain registration

	having_Sub_Domain	SSLfinal_State	Domain_registration_length \
11050	1	1	1
11051	−1	−1	−1
11052	1	1	−1
11053	1	1	−1
11054	1	1	−1

Table 12 HTTPs_token

	Favicon	port	HTTPS_token	Request_URL URL_ of_Anchor	Links_in_tags \
11050	−1	−1	1	1	1
11051	−1	−1	1	−1	−1
11052	1	1	1	0	−1
11053	−1	1	1	−1	1
11054	1	1	1	−1	0

attained 72% accuracy. Further there is a scope for future enhancement by collecting more real data with expanded attributes.

Table 13 SFH submitting_to_email

SFH Submitting_to_email Abnormal_URL Redirect on_mouseover \				
11050	−1	−1	1	1	1
11051	−1	−1	1	−1	−1
11052	1	1	1	0	−1
11053	−1	1	1	−1	1
11054	1	1	1	−1	0

Table 14 Rightclick popup window

RightClick popUpWindow Iframe age_of_domain DNSRecord web_traffic \				
11050	−1	−1	−1	−1	−1
11051	1	−1	1	1	1
11052	1	1	1	1	1
11053	1	−1	1	1	1
11054	1	1	1	1	1

Table 15 Page rank

Page_Rank Google_Index Links_pointing_to_page Statistical_report \				
11050	−1	−1	−1	−1	−1
11051	1	−1	1	1	1
11052	−1	1	1	1	1
11053	−1	−1	1	1	1
11054	−1	1	1	1	1

Table 16 Result

Result	
11050	1
11051	−1
11052	−1
11053	−1
11054	−1

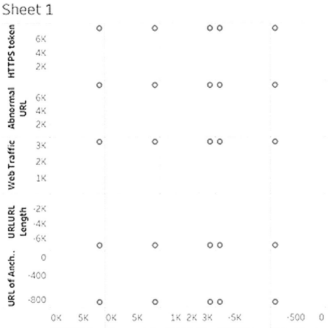

Fig. 2 Classification at URL, web traffic, http, etc.

Abnormal_URL, age_of_domain, DNSRecord, double_slash_redirecting, Domain_registeration_length, Google_Index, having_At_Symbol, HTTPS_token, on_mouseover and web_traffic

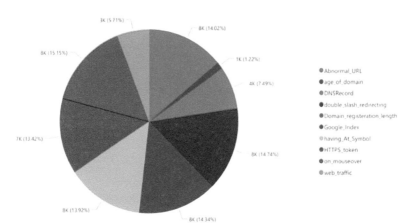

Fig. 3 Percentage of malicious traffic

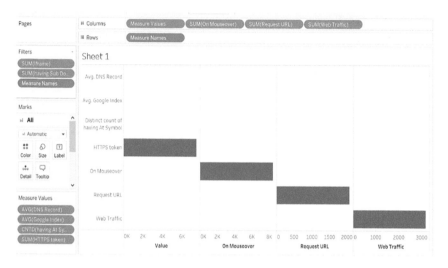

Fig. 4 Classification value on mouse-over, request URL web traffic

Table 17 Classification report

	Precision	Recall	F1-score	Support	Accuracy	0.72	3649		
−1	0.65	0.47	0.58	1632	Macro avg	0.66	0.65	0.65	3649
1	0.68	0.75	0.68	2017	Weighted avg	0.71	0.67	0.71	3649

The Confusion Matrix is: $\begin{bmatrix} 606 & 1026 \\ 503 & 1514 \end{bmatrix}$

The Accuracy is: 0.7209810907097835.

References

1. Mouton, F., Leenen, L., & Venter, H. S. (2015). Social engineering attack detection model: Seadmv2. In *2015 International Conference on Cyberworlds (CW)* (pp. 216–223).
2. Carlander-Reuterfelt Gallo, M. (2020). Estimating human resilience to social engineering attacks through computer configuration data. Examensarbete Inom Teknik, Grundnivå, 15 HP Stockholm, Sverige 2020.1
3. Sawa, Y., Bhakta, R., Harris, I. G., & Hadnagy, C. (2016). Detection of social engineering attacks through natural language processing of conversations. In *2016 IEEE Tenth International Conference on Semantic Computing (ICSC)* (pp. 262–265)2
4. Peng, T., Harris, I., & Sawa, Y. (2018). Detecting phishing attacks using natural language processing and machine learning. In *2018 IEEE 12th International Conference on Semantic Computing (ICSC)* (pp. 300–301).3
5. Orgill, G. L., Romney, G. W., Bailey, M. G., & Orgill, P. M. (2004). The urgency for effective user privacy-education to counter social engineering attacks on secure computer systems. In *Proceedings of the 5th Conference on Information Technology Education*, (pp. 177–181). ACM, 11.
6. Bhakta, R., & Harris, I. G. (2015). Semantic analysis of dialogs to detect social engineering attacks. In *Proceedings of the 2015 IEEE 9th International Conference on Semantic Computing*

(IEEE ICSC 2015) (pp. 424–427).4.

7. Heartfield, R., & Loukas, G. (2018). Detecting semantic social engineering attacks with the weakest link: Implementation and empirical evaluation of a human-as-a-security-sensor framework. *Computers and Security, 76*, 101–127. https://doi.org/10.1016/j.cose.2018.02.0205

8. Abed-alguni, B. H. (2019). Island-based cuckoo search with highly disruptive polynomial mutation. *International Journal of Artificial Intelligence, 17*(1), 57–82 6.

9. Lansley, M., Polatidis, N., & Kapetanakis, S. (2019). Seader: A social engineering attack detection method based on natural language processing and artificial neural networks. In N. T. Nguyen, R. Chbeir, E. Exposito, P. Aniorté, & B. Trawiński (Eds.), Computational collective intelligence (pp. 686–696). Springer International Publishing. 7

10. Tripathi, S., Acharya, S., Sharma, R. D., Mittal, S., & Bhattacharya, S. (2017). Using deep and convolutional neural networks for accurate emotion classification on DEAP dataset. In *Proceedings of the Twenty-Ninth IAAI Conference*, San Francisco, CA, USA, 6–9 February 2017. (pp. 15).

11. Atwell, C., Blasi, T., & Hayajneh, T. (2016). Reverse TCP and social engineering attacks in the era of big data. In *Proceedings of the IEEE International Conference of Intelligent Data and Security*, New York, NY, USA, 9–10 April 2016, (pp. 1–6). 13.

12. Jamil, A., Asif, K., Ghulam, Z., Nazir, M. K., Mudassar Alam, S., & Ashraf, R. (2018). Mpmpa: A mitigation and prevention model for social engineering-based phishing attacks on facebook. In *2018 IEEE International Conference on Big Data (Big Data)* (pp. 5040–5048). 8

13. Uma, M., & Padmavathi, G. (2013). A survey on various cyber-attacks and their classification. *IJ Network Security, 15*(5), 390–396, 10.

14. Granger, S. (2017). Social engineering fundamentals, part I: Hacker tactics. https://www.symantec.com/connect/articles/social-engineering-fundamentals-part-i-hacker-tactics. December, 2001. Accessed 7 March 2017. 12

15. Achary, R., Naik, M., & Pancholi, T. (2022). Prediction of congestive heart failure (CHF) ECG data using machine learning. Intelligent Data Communication Technologies and Internet of Things. https://doi.org/10.1007/978-981-15-9509-7_28

16. Xiangyu, L., Qiuyang, L., & Chandel, S. (2017). Social engineering and Insider threats. In *Proceedings of the International Conference on Cyber-Enabled Distributed Computing and Knowledge Discovery*, Nanjing, China, 12–14 October 2017, (pp. 25–34). 14.

17. Achary, R., & Shelke, C. (2021). An expert system for detection and prevention of social engineering attacks using machine learning algorithm. SPAST Abstracts.

Combining the Knowledge Graph and T5 in Question Answering in NLP

Sagnik Sarkar and Pardeep Singh

Abstract The NLP relies heavily on question answering. It provides an automated method for retrieving responses from a context. Finding context from several documents, studies, and testimonials is becoming increasingly important. Because NLP models for question answering are quite sophisticated and resource intensive, the purpose of this study is to provide a novel lightweight method to question answering in NLP by introducing the knowledge graph as input to the transfer learning model of T5. This study provides an overall holistic perspective of the suggested model's various parameters.

Keywords Knowledge graph · Transformers · Transfer learning · Text to text transfer transformer · BERT

1 Introduction

Answering questions (QA) is a challenge and a research base for indigenous language analysis. The main idea behind the question-and-answer task is to provide an appropriate response to the questions as per the passage [1]. In other words, this QA system enables asking a question and getting answers using the questions processed in natural language [2]. The following diagram tried to give a block diagram for question answering system (Fig. 1).

The latest technology in NLP and in-depth learning models has shown great potential in the QA area. Models such as BERT [3], knowledge graph [4] , XL-net [5], T5 [6], ALBERT [7], GPT-2, GPT-3, ELECTRA [8], RoBERTa [9] transformed the NLP environment, and the flexibility of these modifications changed the overall NLP community status:

S. Sarkar (✉) · P. Singh
NIT Hamirpur, Hamirpur, Himachal Pradesh, India
e-mail: 20mcs011@nith.ac.in

P. Singh
e-mail: pardeep@nith.ac.in

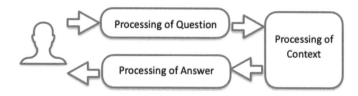

Fig. 1 Question answering method

- Structured data
- Unstructured data

2 Drawbacks in Traditional Model

Along with the successful feat, the main drawbacks faced by the state-of-the-art models are:

- The traditional state-of-the-art model fails to answer the questions where the order of the facts is not important in producing the correct answer
- The traditional state-of-the-art model produces huge attention overheads which increases the size of the model
- The traditional model works good on structured data but it does not perform well on unstructured data.

Also the traditional models are so huge that sometimes it is impossible to train the model itself. According to Hans Peter Luhn [9],

> The weight of a term that occurs in a document is simply proportional to the term frequency—Hans Peter Luhn [9]

As the resource is limited in practical scenario so it is necessary to use a model which is not that resource consuming and be able to perform good in such situation. So to overcome such situation, we have tried to come up with a solution for the problem in this paper.

3 Methodology

The method we are trying to propose consists of the concepts of the knowledge graph [4] and T5 [6] transfer model combined. The knowledge graph helps to reduce the context from which the model T5 tries to capture the essence of the answer. Figure 2 shows our proposed model: The observations obtained from our proposed model that are presented in this paper is mainly focused on the BioASQ and SQuad2.0 dataset.

Convert the context into knowledge graph triples → T5 Embeddings → T5 Model → Output

Fig. 2 Proposed model

Table 1 Analysis of various training models

Model used	BioASQ trainable?	SQuad2.0 trainable?	BioASQ F1	SQuAD2.0 F1
T5 base	Yes	Yes	76%	51%
BERT	Yes	Yes	48.6%	44%
Our model	Yes	Yes	59.97%	47.8%
T5 small	Yes	Yes	52.45%	45.67 %
T5-3B	No	No	–	–
T5-11B	No	No	–	–

4 Results

All the model is trained in Google Colab Pro offered by Google. Many models used here could not even handle 10th epoch with batch size 6. The observations from various models are given in Table 1.

4.1 GPU Power Utilization

Now, let us compare then T5 base and T5 small model with our model on the criteria of GPU utilization in Fig 3. Here, we can see that its accuracy is closer to the base T5 model.

4.2 CPU Power Utilization

The CPU utilization of these models is shown in Fig. 4.

Here, we can see that our model performs good in the CPU calculations as well. It tries to keep up with the T5 base model and in long run is able to hold less than 55% CPU utilization benchmarks. Thus helping in making the model resource friendly and performs good in the overall comparison of the models. So we can see that it satisfies our problem statement.

The overall performance of the models is compared here in Table 2. The following table shows the evaluation metrics of various models involved with a comparison to

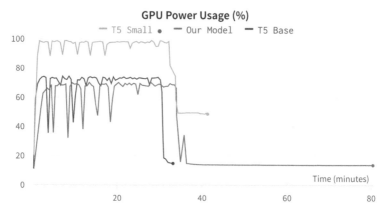

Fig. 3 GPU utilization of various models

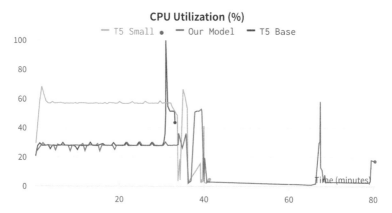

Fig. 4 CPU utilization of various models

Table 2 Various Utilization Models

Model used	GPU utilization	CPU utilization	Memory utilization
T5 base	75%	100%	17%
BERT	115–135%	135%	40%
Our model	60%	60%	19%
T5 small	100%	65%	23%
T5-3B	No	No	–
T5-11B	No	No	–

our model. Most of the models are trained in Google Colab Pro except for BERT as it is trained outside as it is so large that it cannot be run in Colab. The main dataset that we have focused this research is the BioASQ dataset and SQuad2.0.

5 Conclusion

The NLP-based QA system, like other forms of information retrieval systems, is a very complex process. A closed domain QA model can provide more accurate results than an open domain QA system. It is, however, limited to a single domain. Following the screening steps, it is clear that the attention-based model is preferred by the researchers.

To make their task easier, researchers have also turned to hybrid techniques like graph attention and overlaid several forms of mechanism over the underlying model. The knowledge graphs are in primitive state of research. With more and more extensive research on this knowledge graph, we can improve the accuracy of models and can even use the largest transformer model as a driving backbone for desired model. The option should be explored in the future.

References

1. Allam, A., & Haggag, M. (2012). The question answering systems: A survey. *International Journal of Research and Reviews in Information Sciences, 2,* 211–221.
2. Abdi, A., Idris, N., & Ahmad, Z. (2018). QAPD: An ontology-based question answering system in the physics domain. *Soft Computing, 22,* 213–230.
3. Devlin, J., Chang, M., Lee, K., & Toutanova, K. (2018). BERT: Pre-training of deep bidirectional transformers for language understanding. arXiv:1810.04805
4. Hogan, A., Blomqvist, E., Cochez, M., D'amato, C., Melo, G., Gutierrez, C., Kirrane, S., Gayo, J., Navigli, R., Neumaier, S., Ngomo, A., Polleres, A., Rashid, S., Rula, A., Schmelzeisen, L., Sequeda, J., Staab, S., & Zimmermann, A. (2022). Knowledge graphs. *ACM Computing Surveys, 54,* 1–37. https://doi.org/10.1145/3447772
5. Yang, Z., Dai, Z., Yang, Y., Carbonell, J., Salakhutdinov, R., & Le, Q. (2019). XLNet: Generalized autoregressive pretraining for language understanding. In *Advances in neural information processing systems* (Vol. 32).
6. Raffel, C., Shazeer, N., Roberts, A., Lee, K., Narang, S., Matena, M., Zhou, Y., Li, W., & Liu, P. (2019). Exploring the limits of transfer learning with a unified text-to-text transformer. ArXiv:1910.10683
7. Lan, Z., Chen, M., Goodman, S., Gimpel, K., Sharma, P., & Soricut, R. (2019). ALBERT: A lite BERT for self-supervised learning of language representations. arXiv:1909.11942
8. Clark, K., Luong, M., Le, Q., & Manning, C. (2020). ELECTRA: Pre-training text encoders as discriminators rather than generators. arXiv:2003.10555
9. Turing, A. (2004). Can digital computers think? (1951)

Classification of the Severity of Attacks on Internet of Things Networks

Bhukya Madhu and M. Venu Gopalachari

Abstract In the Internet of Things (IoT) technology, there are several issues to overcome, including network security. New attacks have been developed to target vulnerabilities in IoT devices, and the Internet of Things on a huge scale will exacerbate present network threats. Machine learning is becoming increasingly used in applications such as traffic classification and intrusion detection. This paper offers a technique for detecting network fault severity and identifying Internet of Things (IoT) devices based on several attributes. By pushing information to the network area, the proposed framework captures different features from each network flow in order to identify the source of traffic and the type of traffic generated is used to detect different network attacks. Different machine learning algorithms are tested to find the better suited algorithm that deliver best results. The following are some of the examples: Random Forest (RF) algorithm, decision tree, SVM, etc. After completing the experimentation, it was found that the random forest classifier and decision tree are the best performing ML algorithms to classify the network severity with an accuracy of about 96.98% and GNB is found as the least performing machine learning model with an accuracy of about 54.41%.

Keywords Network security · IoT · Attack severity · Classification

1 Introduction

In literature, the IoT [1] and its properties have been extensively studied, the adjustment in security constraints and standards is particularly important to us at this point

B. Madhu (✉)
Department of Computer Science & Engineering, Osmania University, Hyderabad, Telangana, India
e-mail: madhu0525@gmail.com

M. V. Gopalachari
Department of Information Technology, Chaitanya Bharathi Institute of Technology, Hyderabad, Telangana, India
e-mail: mvenugopalachari_it@cbit.ac.in

© The Author(s), under exclusive license to Springer Nature Singapore Pte Ltd. 2023   411
S. Shakya et al. (eds.), *Sentiment Analysis and Deep Learning*,
Advances in Intelligent Systems and Computing 1432,
https://doi.org/10.1007/978-981-19-5443-6_31

of time. Not only does the Internet of Things [IoT] present new vulnerabilities, it also provides an extremely rich platform from which cyber-attacks can be conducted.

IoT is prone to security threats because it has many properties that make it popular, including heterogeneity of linked devices, power, processing limitations and scalability. They all contribute to the distinct and multifaceted nature of IoT security, which makes it difficult to implement. It is vital to identify IoT devices in order to deploy security measures as well as to determine Quality of Service (QoS). It is difficult to operate in this heterogeneous network without a unique validated identification because most identifying tools ([1] Zigbee ID, IP addresses, MAC addresses, Bluetooth ID and so on) are susceptible to spoofing. Another technique is to attempt to identify Internet of Things devices by searching for characteristics that describe their behaviour.

When we look at some of the numbers to understand the network attacks [2] and the need to implement machine learning to stop these attacks, when we look at the data related to small businesses there is a steep growth in frequency of attacks i.e. more than 66% of the small businesses have experienced the cyber-attack in last one year and these attacks are mainly in a targeted method.

When we classify the attacks based on its type, there is more than 57% of increase in phishing type of attacks [2], more than 33% of increase in attacks due to the compromised security of the devices and 30% increase in attacks due to the theft of the credentials. There will be various consequences of these attacks, e.g. Loss of reputation, financial losses, loss of productivity, etc.

For the classification of traffic [1] and the detection of aberrant traffic patterns, machine learning-based algorithms have been developed. As a result, a new research direction has emerged. Devices and the types of traffic they generate can be detected but we will be trying to analyse different Machine Learning (M) algorithms to identify the severity of faults, we have tried to implement five different machine learning (ML) algorithms (SVM, decision tree, random forest, KNN and GNB).

After conducting detailed comparative analysis of different machine learning algorithms with each other, we found that random forest classifier is the best performing amongst all the ML algorithms for classifying the network severity with an accuracy of greater than 97% and GNB is the worst performing algorithm with an accuracy of less than 54%.

In reality, the characteristics of network traffic are determined by the device's operating system, apps, memory capacity [3], suitable network protocols and CPU power. As a result, traditional Internet of Things devices and its assaults can theoretically be distinguished. Several machine learning-based intrusion detection systems (IDSs) have been developed to simplify and automate the detection process. Although adding additional traffic control features to an existing network is conceivable, it is a time-consuming process, and given the scale of IoT networks, a scalable management solution is required.

1.1 How is Severity Different from Intrusion

There is a lot of work has been done on the network intrusions and there is abundance of literature is available but where as in network intrusion only the false data or anomaly is detected but in severity the effect of the intrusion is quantified and it is categorized into different levels of severity, e.g. Level_1, level_2, level_3, which will be helpful in identifying the losses occurred.

For identifying the severity, first the previously occurred intrusions need to be studied and then after a detailed analysis, the dataset is labelled in a categorical manner to remain helpful in identifying the severity.

2 Network Fault Severity Prediction Using Network Disruption

2.1 Overview

Since the invention of modern technology and the dramatic increase in the breadth of network communication [4], there have been an increasing number of network interruptions in implementing the traffic and private protocols. Identifying unexplained network problems can aid in providing a provision of support and even the maintenance of backup systems. Researchers are also investigating different ways of identifying and classifying unknown network disturbances [5], which can help us to overcome the challenge of detecting unlawful network monitoring and intrusion detection [6]. They are also investigating different ways of providing day-to-day network analysis, which can eventually help us to ensure network behaviour. Network disruptions can be detected by using a variety of approaches, including the classic method of using fixed port numbers, which can be easily misled by altering the port numbers in the system, which is quite easy to do. Although Deep Packet Inspection (DPI) is a widely used protocol identification approach, it does have some drawbacks, such as resource utilization when dealing with the feature database that is used to identify protocols. It is unable to appropriately identify the types of protocol being used. Using association rule mining to find unknown protocols is an alternate method that has its own set of disadvantages like computational complexity, where the analysis of large-scale real time networks can be rather significant. To detect network disturbances [6], Machine Learning (ML) can also be utilized as an alternative method. With its excellent adaptive and learning characteristics, researchers across the globe have begun to use Machine Learning (ML) to detect network disruption and perform protocol analysis as a result of its widespread adoption. Depending on the requirements, machine learning can be divided into many different branches, each of which can be concentrated separately [4].

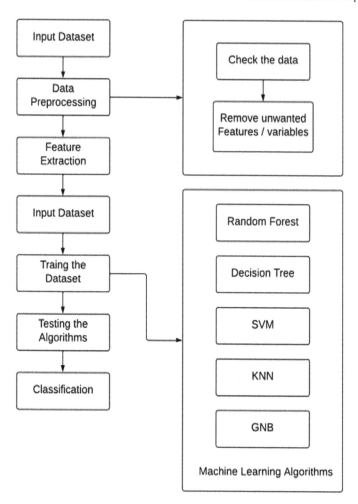

Fig. 1 Block schematic

2.2 Block Schematic

See Fig. 1.

2.3 Objective

This paper proposes to take the research into new direction, where the intrusion can not only be identified but also identify the severity of attack based on different parameters [3]. When trying to conduct a comparative analysis on the proposed work and existing research literature, the existing research works mainly focus on the anomaly detection and network intrusion identification by using the network traffic analysis method but the proposed method will be initially focussed on identifying the network attacks and thereby quantifying the losses occurred due to the attacks, which will help in labelling the data where the severity of attack can be classified into different levels and also in different categories like location based, type of resource, etc.

The new era of technology is rapidly emerging, and as the number of people using various networks increases, more network disruptions will occur [7], posing an immediate threat to enterprises. A network disruption classification and identification method based on artificial intelligence was used in this paper to investigate autonomous categorization and network protocol identification of unknown networks will be significant in reducing the time and labour costs associated with network disruption classification and identification. The dataset will be divided into rows, each row representing a location and time point, which we will utilise as a starting point [7]. Three machine learning algorithms are employed to pre-process the data and model it before it is presented to the user. As a result, the three algorithms that can deliver the highest level of accuracy can be determined.

The primary objective of the research is to predict the severity of a network fault at a specific location by analysing the log data. The proposed study has used information from smart home data repositories, which included a variety of features that allowed us to determine the severity of a network fault present in the network.

2.4 Proposed Work Flow

See Fig. 2.

2.5 Dataset Description:

The data is taken from Telstra corporation limited's website [8], Telstra is a telecom operating company, which deals with different services like mobile, internet, television, etc. This is a tabular data, which has lots of information on networks and with effects of disruption on the networks, we will be trying to explore data and predict the severity based on different features.

- train.csv—the fault severity training set
- test.csv—the fault severity test set
- sample_submission.csv—a sample of the right input format
- event_type.csv: event type related to the main dataset
- log_feature.csv—log files features extracted
- resource_type.csv: associated resource type to the primary dataset

Fig. 2 Proposed work flow

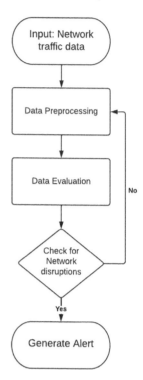

- severity_type.csv: The severity level of a warning message from the log

In this case, the datasets/log files used here are event type.csv, log feature.csv, resource type.csv, severity type.csv and the target class variable (event type.csv) as mentioned in Table 1. The severity type contains three classes: 0, 1 and 2, which describe the severity of network faults. "Fault severity" is measure of the number of actual problems reported by network users, and it is the goal variable [8]. All of the CSV files listed above, with the exception of train.csv, test.csv and sample submission.csv, have been merged into a single CSV file that is based on a specified primary key.

The proposed analysis is using the network disruption data and identify the disruptions in the network base on different parameters like location, fault severity, event type, etc. [8]. for preparing this data set we have merged various dataset on network disruption into one file, then we will check the data for missing values and finally clean the dataset after that we will be conducting the analysis, above in Table 2 we can see the first five rows of data by using the python function called *head* and size of the data set is (7381, 19).

After doing the exploratory data analysis we understand that (Synthetic Minority Over sampling Technique), by doing over sampling of minority classes we will make the balanced data, because of which we will not be facing the problem of over fitting or under fitting.

Table 1 Data Set Over view

id	Location	Fault_ severity	Severity_type	Event_type	Resource_type	Log_ feature	Volume
14,121	location 118	1	severity_type 2	event_type 34	resource_type 2	feature 312	19
9320	location 91	0	severity_type 2	event_type 34	resource_type 2	feature 315	200
14,394	location 152	1	severity_type 2	event_type 35	resource_type 2	feature 221	1
8218	location 931	1	severity_type 1	event_type 15	resource_type 8	feature 80	9
14,804	location 120	0	severity_type 1	event_type 34	resource_type 2	feature 134	1

Table 2 Futures extracted by the model

0	Location
1	Event_type
2	Log_feature
3	Volume
4	Severity_type_1
5	Severity_type_2
6	Severity_type_3
7	Severity_type_4
8	Severity_type_5
9	Resource_type_1
10	Resource_type_2
11	Resource_type_3
12	Resource_type_4
13	Resource_type_5
14	Resource_type_6
15	Resource_type_7
16	Resource_type_8
17	Resource_type_9
18	Resource_type_10

Below are the 19 features considered for our experimental analysis:

Table 3 Overall accuracy results of the various classifiers on the test set

	Accuracy (%)	Precision (%)	Recall (%)	F1 Score (%)
SVM	64.82	33.33	21.61	26.22
DT	96.98	93.82	97.48	95.54
RF	96.98	95.55	95.67	95.61
KNN	75.90	60.47	71.79	64.13
GNB	54.41	44.71	48.07	43.03

2.6 Experimental Results

After completing the analysis and when we compare all the 5 different models, it is observed that the GNB (GaussianNB) is the least accurate method, with an overall accuracy of 54.41% and the Random Forest classifier along with decision tree is the best performing algorithm with an accuracy of 96.98% with equivalent precision, accuracy and recall scores.

For conducting the experimental analysis a 3 step approach is taken in the first step the data collection is done, in second step we will train our ML algorithms [9] on training dataset further for evaluation the computations are applied on the test dataset and the results in the Table 3 are the outcome of evaluation conducted on test dataset.

2.7 Visual Analysis

The above figure shows the plotting of the count of fault severity, based on its different classes (0, 1, 2), by looking at the graph it can be inferred that the highest number of fault fall under 0th class, i.e. just less than 5000, second highest number of faults fall under 1st class, i.e. just less than 2000 and least number of faults fall under 2nd class, i.e. just less than 1000 (Figs. 3, 4 and 5).

The above figure shows the plotting of intensity of attack based on location and resource type, looking at the graph we can infer that the highest intensity of around 85 and majority of the intensity ranges between 0 and 30 (Fig. 6).

The above bar graph shows the plotting of top 5 intensity of attacks based on location and resource type, looking at the graph we can infer that the highest intensity of around 85 and the top 5 locations that have the most cases are: 821, 1107, 734, 126 and 1008, respectively (Fig. 7).

Network Disruption data provides very good insights the above plotted chart shows top 14 locations with type 1 fault the chart is plotted location against the count of type 1 faults here by looking at the visualization we can tell that location 1100 has the most number of type one faults which is just over count of 30.

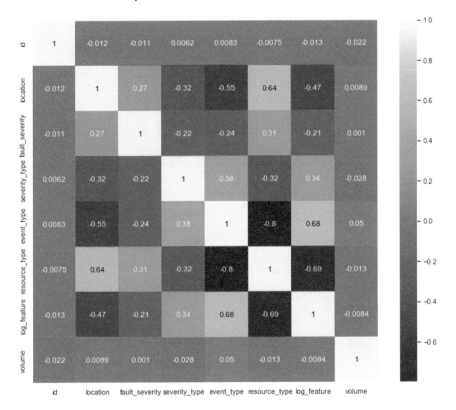

Fig. 3 Heat map

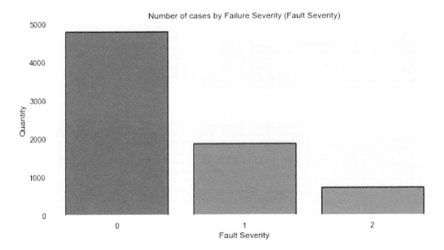

Fig. 4 Number of cases by type of severity (Severity Type)

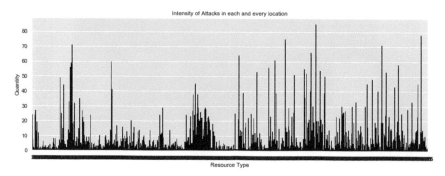

Fig. 5 Intensity of attack location wise

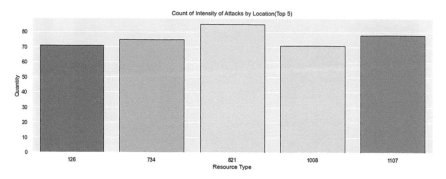

Fig. 6 Count of intensity of attack by location (Resource type)

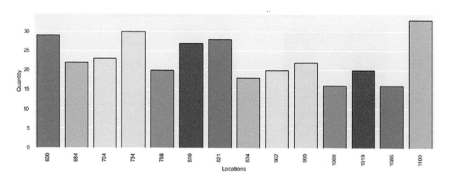

Fig. 7 Count of intensity of attack by location

2.8 Features of Dataset Based on Their Importance

Extraction of features is done by a pre-defined function in random forest called **feature_importances_** the features are extracted by applying mathematical method

[9], the probability of each node can be calculated by the number of samples that reach the node dividing it by total samples. The higher the value of the applied computation will directly increases the importance of the feature as in Table 4. The feature importance is calculated by using below equation.

$$RFfi_i = \frac{\sum_{j \in \text{all trees}} \text{norm fi}_{ij}}{T}$$

where

- RFfii(i) is the importance of feature i
- Normfi(ij) is the normalized feature importance for i in node j
- T is total number of nodes

There are total of 19 features extracted by our model for doing prediction as Random Forest and decision tree are performing the best amongst all the models with accuracy of greater than 96% we are tabulating the contributions of our important features and we can analyse that the top 4 features, i.e. location, volume, log_feature, evet_type makes up to almost 90% of importance and all the remaining will be contributing around 10% of the importance (Fig. 8).

Table 4 Feature importance for random forest model

Column name	Importance
Location	33.82
Volume	25.30
Log_feature	20.39
Event_type	10.02
Resource_type_2	2.39
Severity_type_1	2.14
Resource_type_8	2.00
Severity_type_2	1.51
Severity_type_4	0.92
Resource_type_7	0.38
Resource_type_6	0.36
Resource_type_4	0.28
Resource_type_9	0.18
Resource_type_1	0.10
Resource_type_3	0.09
Resource_type_10	0.07
Severity_type_5	0.05
Resource_type_5	0.00
Severity_type_3	0.00

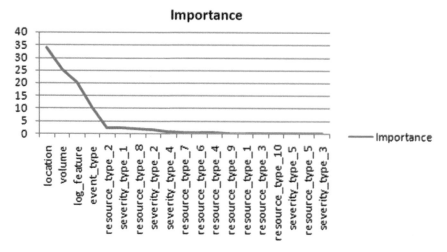

Fig. 8 Features of dataset based on their importance

3 Discussion

There is a very little amount of study is conducted in IoT Network Disruptions because of lack of publically available dataset [10]. As per the experimental results we understand that the analysis of network disruption data set can provide very good insights in protecting IoT networks the main features contributing to our models are location, volume, log_feature, evet_type which contribute almost 90% of importance and all the remaining will be contributing around 10% of the importance.

Identifying network disruptions based on locations and categorizing them based on fault severity will help in identifying irregularities in network and its causes thereby helping the users in increasing the safety measures, for doing this various comparisons were conducted amongst different ML algorithms and these comparisons provide very good insights for understanding performance of each model with the other model based on various parameters like accuracy, precision, recall, etc.

4 Conclusion

As per the main objective of the project, classifying and identifying the unknown network disruptions based on ML algorithms is being discussed throughout the paper. First, the network traffic's interrupted data information is extracted. Then, the dataset has been successfully sent for cleaning and data pre-processing to bring the data to the same scale, which should be understandable to the machine and in the process of that all the files are merged as one file to get a better understanding of the data to further assist in classifying and identifying the fault severity. Finally, feature engineering is used to intelligently select the feature vectors in order to classify and identify unknown network disruptions quickly and effectively. Machine Learning methods were fully utilized in this method. It avoided the complex stages of manually extracting features and lowered the training time of the intelligent algorithm as well as the amount of labelled data necessary required by ensuring the classification and identification accuracy [9].

The applied machine learning algorithms has efficiently classified the attacks based on its severity [5], after performing the analysis, it is finally concluded that the random forest classifier and decision tree algorithms are the best performing ML algorithms for classifying the network severity with accuracy greater than 96%, KNN is the second best performing algorithm with accuracy greater than 75% and GNB is least performing model with accuracy greater than 54%.

References

1. Sterbenz, J. P. G. (2017). Smart city and IoT resilience, survivability, and disruption tolerance: Challenges, modelling, and a survey of research opportunities. jpgs@{ittc.ku.edulcomp.{lancs.ac.uklpolyu.edu.hk}}
2. https://www.fundera.com/resources/small-business-cyber-security-statistics
3. Matisziw, T. C., Murray, A.T. & Grubesic, T. H. (2008). Exploring the vulnerability of network infrastructure to disruption. Received: 6 October 2006/Accepted 6 April 2008/Published online: 24 April 2008 © Springer.
4. Sterbenz, J. P., Hutchison, D., Çetinkaya, E. K., Jabbar, A., Rohrer, J. P., Scholler, M., & Smith, P. (2010). Resilience and survivability in communication networks: Strategies, principles, and survey of disciplines. *Computer Networks, 54*(8), 1245–1265.
5. Kaluza, B., Mirchevska, V., Dovgan, E., Gams, L. M. (2010). An agent-based approach to care in independent living. In *Ambient intelligence* (pp. 177–186). Springer.
6. Aazam, M., Zeadally, S., & Harras, K. A. (2018). Deploying fog computing in industrial internet of things and industry 4.0. *IEEE Transactions on Industrial Informatics 14(10),* 4674–4682.
7. Kilkki, K., Mantylä, M., Karhu, K., Hämm äinen, H., & Ailisto, H. (2018). A disruption framework. *Technological Forecasting and Social Change, 129* 275–284.
8. Data set https://www.telstra.com.au/data collected from the website
9. Brika, B., Bettayebb, B., Sahnouna, M., & Duvala, F. (2019). In *The 2nd International Conference on Emerging Data and Industry 4.0 (EDI40)*, April 29—May 2, 2019, Leuven, Belgium Towards Predicting System Disruption in Industry 4.0: Machine Learning-Based Approach.
10. Atzori, L., Iera, A., & Morabito, G. (2010). The internet of things: A survey. *Computer Networks, 54,* 2787–2805.
11. Lade, P., Ghosh, R., & Srinivasan, S. (2017). Manufacturing analytics and industrial internet of things. *IEEE Intelligent Systems, 32*(3), 74–79.
12. Chen, B., Wan, J., Shu, L., Li, P., Mukherjee, M., & Yin, B. (2018). Smart factory of industry 4.0: Key technologies, application case, and challenges. *IEEE Access, 6* 6505–6519.
13. What is industry 4.0. https://www.forbes.com/sites/bernardmarr/2018/09/02/what-is-industry-4-0-heres-a-super-easy. Accessed 07 Nov 2018.
14. Lin, C., & Yang, J. (2018). Cost-efficient deployment of fog computing systems at logistics centers in industry 4.0. *IEEE Transactions on Industrial Informatics, 14*(10), 4603–4611.
15. Peralta, G., Iglesias-Urkia, M., Barcelo, M., Gomez, R., Moran, A., & Bilbao, J. (2017). Fog computing based efficient Iot scheme for the industry 4.0. In *2017 IEEE International*

Workshop of Electronics, Control, Measurement, Signals and their Application to Mechatronics (ECMSM), (pp. 1–6).
16. Kilkki, K., M˙antyl˙a, M., Karhu, K., H˙amm˙ainen, H., Ailisto, H. (2018). A disruption framework. *Technological Forecasting and Social Change, 129* 275–284.
17. Sejdovic, S., & Kleiner, N. (2016). Proactive and dynamic event-driven disruption management in the manufacturing domain. In *2016 IEEE 14th Inter-national Conference on Industrial Informatics (INDIN)*, (pp. 1320–1325).
18. Modarresi, A., & Sterbenz, J. P. (2017). Toward resilient networking with fog computing. In *Reliable Networks Design and Modeling (RNDM), 2017 9th International Workshop on, IEEE*.
19. Sterbenz, J. P., & Hutchison, D. (2016). ResiliNets: Multilevel resilientand Survivable networking initiative. http://wiki.ittc.ku.edu/resilinets
20. Zanella, A., Bui, N., Castellani, A., Vangelista, L., & Zorzi, M. (2014). Internet of things for smart cities. *IEEE Internet of Things Journal, 1*, 22–32.
21. Clark, D., Sollins, K., Wroclawski, J., Katabi, D., Kulik, J., Yang, X., Braden, R., Faber, T., Falk, A., Pingali, V., Handley, M., & Chiappa, N. (2003). New arch: Future generation Internet architecture," technical report, DARPA, MIT, ISI, February 2003.
22. Bhattacharjee, B., Calvert, K., Griffioen, J., Spring, N., & Sterbenz, J. P. G. (2006). Postmodern internetwork architecture. Technical Report ITTC-FY2006-TR-45030-01, The University of Kansas, Lawrence, KS, February 2006.
23. Al-Fuqaha, A., Guizani, M., Mohammadi, M., Aledhari, M., & Ayyash, M. (2015). "Internet of things: A survey on enabling technologies, protocols, and applications. In *IEEE Communications Surveys Tutorials*, (Vol. 17, pp. 2347–2376). Fourthquarter.
24. R. T. M. (chair). Critical Foundations: Protecting America's Critical Infrastructure. tech. rep., President's Commission on Critical Infrastructure Protection.
25. Feldman, Z., Fournier, F., Franklin, R., & Metzger, A. (2013). Proactive event processing in action: A case study on the proactive management of transport processes (industry article). In *Proceedings of the 7th ACM International Conference on Distributed Event-based Systems*, DEBS '13, ACM, New York, NY, USA, (pp. 97–106).
26. Metzger, A., Franklin, R., & Engel, Y. (2012). Predictive monitoring of heterogeneous service-oriented business networks: The transport and logisticscase. In *2012 Annual SRII Global Conference*, (pp. 313–322).
27. Engel, Y., & Etzion, O. (2011). Towards proactive event-driven computing. In *Proceedings of the 5th ACM International Conference on Distributed Event-based System, DEBS '11*, ACM, New York, NY, USA (pp. 125–136).
28. Brzezinski, J. R., & Knafl, G. J. (1999) Logistic regression modeling for context-based classification. In *Proceedings. Tenth International Workshop on Database and Expert Systems Applications*. DEXA 99, (pp. 755–759).
29. Martinez-Arroyo, M., & Sucar, L. E. (2006). Learning an optimal naive bayes classifier. In *18th International Conference on Pattern Recognition (ICPR'06)* (Vol. 4, pp. 958–958).

Rainfall Forecasting System Using Machine Learning Technique and IoT Technology for a Localized Region

P. Sathya and P. Gnanasekaran

Abstract The logic of Weather Prediction and the associated climate monitoring is the most important concern to every country to handle and avoid disasters from occurring. Also, weather prediction scheme is helpful for farmers to forecast the rainfall details and the associated climatic conditions in time. Many machine learning schemes with Artificial Intelligence (AI) enabled logic are widely used in weather prediction. In previous works, public datasets used for rainfall prediction does not give correct predictions. And some existing work utilizes sensor signal data for localized regions implemented in different types of machine learning algorithms. In this paper, a new logic in machine learning technique namely, Enhanced Learning Scheme for Weather Prediction (ELSWP) is introduced, which is based on the conventional machine learning Logistic Regression (LR) model. The proposed approach is designed in association with the help of advanced Internet of Things (IoT) technology. The main purpose of this paper is to initiate a novel approach for predicting the weather and climate conditions by using learning schemes and advanced technologies. This kind of weather prediction scheme is helpful in local crop monitoring and improves the production level by allowing farmers to take necessary precautions and prevent unwanted agricultural losses. In this method, a specific sensor unit which consists of DHT11 and the Rainfall identification sensor provides continuous weather details to the NodeMCU module, which forwards it to the server end for data manipulation. The server end machine learning model ELSWP acquires the details and processes it for any dense case occurrence and alerts farmers to take appropriate actions and avoid disaster scenarios. The result section of this paper provides adequate proof to show that the performance of the proposed approach yields more accurate results.

P. Sathya (✉) · P. Gnanasekaran
Department of Computer Applications, B.S. Abdur Rahman Crescent Institute of Science and Technology, Chennai, India
e-mail: Sathya_ca@cresccent.education

P. Gnanasekaran
e-mail: gnanasekaran@crescent.education

© The Author(s), under exclusive license to Springer Nature Singapore Pte Ltd. 2023   425
S. Shakya et al. (eds.), *Sentiment Analysis and Deep Learning*,
Advances in Intelligent Systems and Computing 1432,
https://doi.org/10.1007/978-981-19-5443-6_32

Keywords Agriculture · Internet of Things · Enhanced Learning Scheme ·
Weather prediction · Logistic Regression

1 Introduction

The weather prediction schemes are important to monitor the climate conditions and
avoid unwanted disasters and associated losses. In India, major disasters usually occur
due to heavy rainfall and the associated climatic conditions. To avoid these hurdles
a proper climate or weather prediction scheme is compulsory to protect people and
their belongings. Weather prediction scheme enables farmer to protect their agricultural field by taking necessary preventive actions like manage local irrigation and
prevent crop flooding and so on. Rainfall forecasting is critical for Indian farmers
as rainfall is heavily dependent on monsoon seasons and they are quite erratic. Our
current weather prediction model relies majorly on global weather modeling data
and mostly they fail when it comes to a localized prediction. Experts from across the
globe have created a variety of theories to forecast monsoon rainfall, which are mostly
based on pseudo randomized principles and are comparable to weather information.
Multiple regression frameworks is used to build the proposed approach and the
suggested technique forecasts rainfall using Indian weather information. Typically,
machine learning techniques are broadly classified into two classes, such as: unsupervised machine learning and supervised machine learning. All segmentation methods
are considered to as the supervised learning algorithms. The following graphic,
Fig. 1, illustrates how machine learning techniques are classified under the proposed
learning scheme. The next image, Fig. 2, illustrates the weather forecasting investigation for an Indian setting using a computational model [1]. While several algorithms
have been established, it is important to conduct analysis using machine learning
techniques in order to obtain reliable predictions. Accurate prediction enables more
effective management in agricultural and industrial activities.

2 System Study

Shivang et al. used machine learning to build a modeled system to predict the meteorological conditions in the Indian subcontinent. Information for learning was obtained
from "http://data.gov.in", "http://ncdc.noaa.gov" and the University-of-California,
Irvine's machine learning information library. The technology trains on information using a Linear-Regression-Model (LRM) [2]. To calculate weather predictions,
Zaheer-Ullah-Khan et al. used a range of knowledge mining techniques, including K-
nearest neighbor and decision trees, which incorporate diverse behavioral elements.
When compared to other computations, the decision tree method has shown to be
one of the most effective ordering techniques. Researchers obtained an accuracy
of 82% in this paper [3]. Siddharth S. et al. used a knowledge extraction method

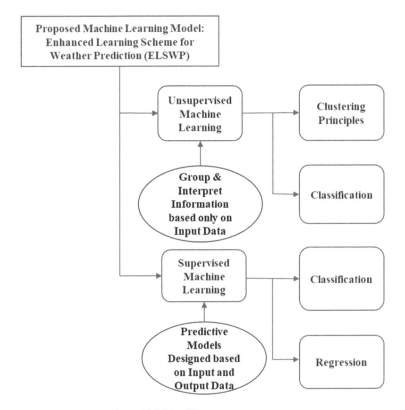

Fig. 1 Artifacts empowered by artificial intelligence

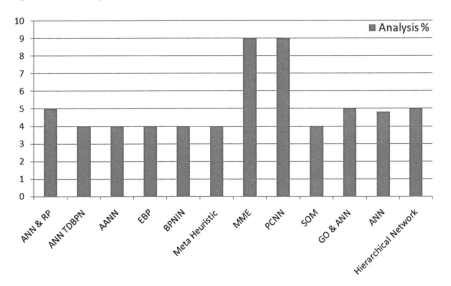

Fig. 2 Weather prediction computational analysis [1]

and decision tree analysis to classify climatic characteristics such as the lowest and highest temperatures by particular day, particular month and particular year exactly [4]. Radhika and her colleagues presented a work on using support vector machines to forecast climate change. The information request was created by dissecting the time arrangement information for each day's maximum temperature in a specific region in order to predict the most extreme temperature for the following day in that area based on the daily most extreme temperatures for the previous n days. To prepare the SVM for this application, a non-straight relapse strategy was utilized. Information mining is a process that forecasts future behaviors and trends, allowing businesses to make proactive decisions [5]. The goal of this article is to provide an overview of data mining approaches for weather prediction, as well as the benefits of adopting them. This study could be used to forecast meteorological data linked to climate predictions. The study reviews publicly available papers on a few computations that have been used by various professionals to develop various information mining techniques for weather prediction [6]. The machine learning model was determined to be taken from an external source after reviewing all of the papers. However, in this study, all data is obtained using a variety of sensors in a variety of situations. Two sensors, the rain identification sensor and the DHT11, were utilized in this research to monitor temperature, humidity and rainfall density.

3   Related Study

Wei-Chu et al. [7] conducted a study on the use of Convolutional Neural Networks to anticipate rainfall and climate conditions., in which the authors specified in this paper such as: rainfall forecasting, specifically brief weather forecasting, is a critical issue in the area of weather research. In practice, the majority of current research focuses on forecasting using spatial information or satellite photos. Conversely, there's also another case in which a collection of weather characteristics is made by many sensors located at multiple monitoring stations. Although data at a location are frequently inadequate, they give critical indications for weather forecasting at neighboring sites that have not been completely explored in previous studies. To address this issue, we present a multi-task learning algorithm capable of autonomously extracting features from time series collected at sampling points and exploiting the correlation across many sites for weather forecasting via multi-tasking. To the understanding, it is the first endeavor to forecast short-term rainfall amounts using multi-task training and learning techniques approaches based on multi-site characteristics. To be more precise, we define the learning problem as a final multiple site assisted neural-network approach that enables the transfer of knowledge acquired from one site to certain other associated sites as well as the modeling of correlations between various sites. Extensive experimentation demonstrates that the learnt site correlations are informative, and the suggested model beats a diverse collection of baseline models, such as the European Union agency for Moderate Meteorological Predictions approach.

　　Geetha et al. [8], proposed a paper on utilizing deep learning networks to forecast rainfall in which the authors specified in this paper such as: monsoon is a significant supply of rainwater for all organisms on the planet. Rainfall forecasting models offer information on the effect of numerous global weather factors on average rainfall. The methodology of deep learning has capacity factor self-learning information categories, allowing for the creation of an information framework for a response variable dataset. This system [8] enables the identification of anomalies/changes in time series analysis and moreover forecasts effectively for future information that relates to past occurrences. This article discusses developing models of rainfall precipitation utilizing deep learning Configurations such as LSTM, Convolutional Neural Network and deciding which architectural style is superior, with a root mean square error of 2.55 for LSTM as well as 2.44 for Convolution operation, asserting that deep learning models will be effective and efficient for modelers for every kind of data series raw data.

　　Channabasava et al. [9] submitted a paper on the subject of weather predictions using a climate information dataset for rainfall, in which the authors specified in this paper such as: weather is a random phenomenon, and forecasting is always been a difficult task for meteorologists worldwide. There are a variety of methods for forecasting this weather using air data obtained in a variety of ways. Our research focuses on the data mining techniques applied to forecast rainfall in a region based on some dependent variables such as rainfall and weather recurrence. Rather of using current data, we will use historical data gathered by the weather agency. We are examining a data mining approach that employs a linear regression analysis on data gathered for the frequency of wet days, temperature and rainfall in Bangalore, India from 1901 to 2002. The prediction model built was trained and verified against with the region's real precipitation and was then often used forecast future amount of rain. In this study, Divide the original dataset into two sections are training and testing set. Once all of the iterations for each sub-sample have been finished. To produce a single output, the results from each iteration must be averaged and aggregated. All the data from both sections were executed and processed. Using this way helps to increase the computation efficiency was further improved by employing the ensemble approach with a K-fold multiplier.

　　Shakila et al. [10] wrote an research related to the introduction of novel algorithm to predict weather in real-time manner, in which the authors specified in this paper such as the creation of information has accelerated in recent years and is anticipated to continue to do so in the future, making it a laborious process to evaluate large chunks of meteorological conditions while doing advanced analytics using conventional methods. The project's objective is to estimate the likelihood of rainfall using Hadoop. This frame work estimate the rainfall prediction accuracy using Naïve Bayes classification algorithm. The probability function used to predict the future chances of the rain. The suggested system acts as a tool that accepts rainfall data from a huge quantity of data as input and efficiently forecasts future rainfall with minimum, maximum and average values. Predictive analytics models using Pearson correlation coefficient technique, to identify correlations between several variables in a data collection in order to estimate risk associated with a certain set of conditions and assign a score or weight. The total points correlations discovered in historic information can be utilized to anticipate the future.

Verma et al. [11] did a research related to the pollution monitoring scheme using intelligent MQ2 series sensor with Internet of Things interfacing, in which the authors specified in this paper such as: in the car industry, technological developments assist suppliers in producing smart automobiles. These intelligent cars are intended to give society with a variety of benefits. However, certain enhancements are necessary to make these cars more intelligent in terms of pollution control. The other major causes of air pollution are harmful gases generated by industry. Despite the government's numerous attempts, environmental contamination continues to grow steadily. Numerous techniques for reducing and monitoring environmental contamination are accessible in the literature. The analysis reveals the need for a sensor-based embedded system capable of screening and controlling air pollution with the production of challans from any location in the globe via IOT. An embedded prototype of the system was created around the notion of an IoT scenario, utilizing sensors and actuators on the Raspberry Pi board. The system prototype is written in Python and makes use of several open source frameworks and libraries on Adafruit and github. Additionally, a web page is created to monitor the amount of gases remotely from any location on Earth. The results indicate that the entire system has been tested and implemented effectively. Bhalaji et al. [12] proposed the first stage is to build a dropout prediction model using a deep learning algorithm and calculate the expected individual student dropout probability. Additional enhancements are made to increase the efficiency of the dropout prediction model and provide relevant interventions to course providers based on a temporal prediction mechanism. Smys et al. [13] In the current system, the cost-effective network based on long-range technology is used to automatically analyze the degree of fire risky and forest fire rural areas and broadcast to the website for public viewing via the things network server. The analysis found that providing better coverage, battery life, latency, cost and efficiency improved service quality.

4 Methodology

In this study, the Enhanced Learning Scheme for Weather Prediction (ELSWP) is introduced, which adapts the learning scheme to create a novel prediction scheme to help agricultural fields and avoid calamities, thereby saving people's lives. This paper utilizes data from dynamic dataset acquired by using the sensor unit presented into the real-time crop/agricultural field monitoring region. The objective of this study is to develop a web-based weather forecasting application that forecasts 10 days weather for Tanjore in Python programming language. In this study, logistic regression is being for the categorizing the data and to predict the probability of rainfall. The associated sensors are used to monitor the rainfall status, temperature and humidity. Sensors such as the DHT11 and the rainfall monitoring sensor are used to intelligently monitor the agro fields. In this proposed technique, the specialized microcontroller

NodeMCU with embedded Wi-Fi enabled options is used to monitor agricultural field data and determine weather conditions in a timely manner, as well as report it to the appropriate individual to minimize unnecessary obstacles. NodeMCU is an open source, which can connect objects and let data transfer using the Wi-Fi protocol. Very low cost and high features which makes it an ideal module for Internet of Things (IoT). It can be used in any application that require it to connect a device to local network or internet. The logic of Internet of Things (IoT) is utilized in this proposed approach to accumulate the weather data and send it to the remote server unit for processing. In the server end the logical learning-based scripts are available to process the data accumulated from the sensor unit. In this study proposed an enhanced logistic regression machine learning technique to predict the rainfall prediction and compared with existing system. The data obtained in real-time is referred to as testing data, and the data that is tested is stored on the server for further processing. The data maintained into the server end are considered to be training data, in which it will be loaded into the server scripting end to cross-validate the testing input. If the testing input contains any values above or below from the mentioned threshold ratios, will be notified immediately to the respective farmer or individual to take an appropriate actions to avoid the disasters or crop field losses. Using heat sinks in the control to reduce the circuit heating time in this study. NodeMCU and DHT11 need 3.3v–5v for the process. When power is supplied to the sensor, do not send any instruction to the sensor in within one second in order to pass the unstable status. Stop the power supply to the circuit, after that disconnecting the control. Figure 3 illustrates the perception of proposed block diagram and Fig. 4 illustrates the perception of proposed sensor unit with the associated sensors with controller specifications.

Pseudocode:

Data: Threshold Temperature $OT_{a1}^{(\text{base value})}., OT_{a2}^{(\text{base value})}$

Result: Operated Signals

itialization:

forall premises within the premises list do

forall devices in the premises do

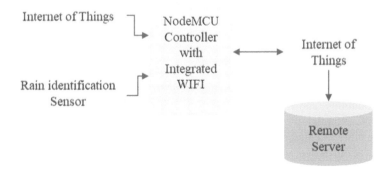

Fig. 3 Block diagram

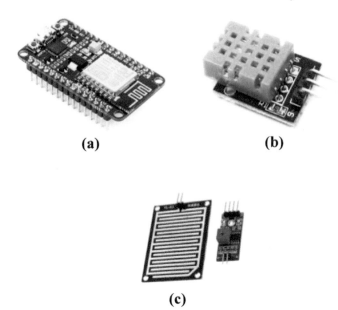

Fig. 4 Sensor unit: **a** NodeMCU, **b** DHT11 sensor and **c** rain identification sensor

if $OT_{a1} > OT_{a2}^{(\text{base value})}$ $\&\&OT_{a2} > OT_{a2}^{(\text{base value})}$
| Block the device (switch off);
end
if $OT_{a1} < OT_{a1}^{(basevalue)} \&\&OT_{a2} > OT_{a2}^{(basevalue)}$ then
| reduce heating time;
end
if $OT_{a1} < OT_{a1}^{(base)}$ $\&\& OT_{a2} < OT_{a2}^{(basevalue)}$ then
| disconnecting control
end
end
end

5 Results and Discussion

Two separate sensors are attached to the NodeMCU, one of which measures temperature, humidity, and rainfall ratio, and the other of which measures temperature, humidity, and rainfall ratio. One sensor is linked to an analog Input/output (I/O) pin and the other to a digital I/O pin. Figure 3 depicts the system's entire block diagram. As seen in Fig. 3, data is transmitted to the remote server through NodeMCU. Furthermore, the remote server transmits data to the NodeMCU in the form of a JSON file, which can now be shown on an HTML Web Page, as seen in Table 1, which displays

Table 1 Remote server record maintenance perception

Days	Temperature	Humidity (%)	Rainfall
1	153° F(67.2 °C)	50	0 (Clear)
2	155° F(68.9 °C)	35	0 (Clear)
3	154°F(68.9 °C)	55	0 (Clear)
4	154°F(68.7 °C)	36	0 (Clear)
5	155°F(69.2 °C)	60	1 (Raining)
6	152°F(67.1 °C)	29	0 (Clear)
7	152°F(68.0 °C)	31	0 (Clear)
8	155°F(69.7 °C)	42	1 (Raining)
9	155°F(69.6 °C)	43	1 (Raining)
10	154°F(68.1 °C)	39	0 (Clear)

the measured values of numerous sensors. Using a Google spreadsheet, the varied temperature, humidity, and rainfall intensity information is kept in a CSV file format. This data is used to train the machine learning algorithm, and a model based on the specified learning scheme is used and trained in Jupyter Notebook utilizing the pre-stored values in Google Drive (Python IDE). Additionally, NodeMCU collects real-time temperature, humidity, and rainfall ratio information for a specific area or location, which are utilized to validate the model and make decisions. Serial connection between the NodeMCU and the Jupyter Notebook is used to send these values to the Jupyter Notebook. The model is used to attempt to forecast the viability of a match. Following that, the Jupyter Notebook sends either logic '0' or '1' to the NodeMCU as an anticipated outcome. A LED is used to forecast the outcome of the match. This model has an accuracy of around 96% when monitored continuously for ten consecutive days across a real-time agricultural field.

Figure 5 illustrates the perception of rainfall sensor output checking scenario over the Arduino IDE using NodeMCU controller in direct connection, in which this type of testing is used to identify the working ability of the respective sensor in clear manner.

Figure 6 illustrates the perception of proposed learning strategy ELSWP's prediction accuracy over training phase and the figure, Fig. 7 illustrates the perception of training loss ratio in clear manner with proper specification.

Figure 8 illustrates the perception of proposed machine learning approach called ELSWP's validation accuracy over testing phase, in which it is cross-validated with the conventional learning scheme called Logistic Regression (LR) and prove the proposed method efficiency ifar beer than the conventional scheme. Evaluation metrics accuracy and root mean squared error used to perform the comparison between Logistic regression and ELSWP as referred Table 2.

$$R^2 = 1 - \frac{\sum(y_i - \hat{y})^2}{\sum(y_i - \overline{y})^2} \tag{1}$$

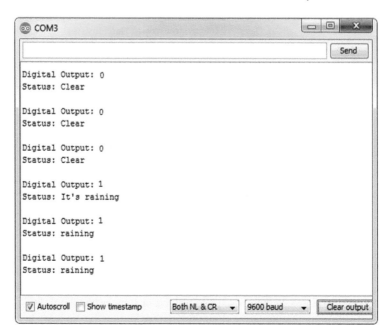

Fig. 5 Rain identification sensor working status test

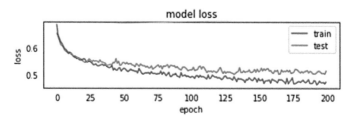

Fig. 6 Model training accuracy

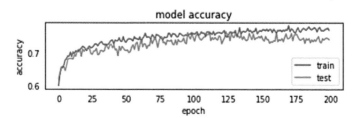

Fig. 7 Model training loss ratio

$$RMSE = \sqrt{MSE} = \sqrt{\frac{1}{N}\sum_{i=1}^{N}(y_i - \hat{y})^2} \qquad (2)$$

y_i = Actual value
$\hat{y}$ = Predicted value of $y_{i\wedge}$
$\overline{y}$ = Mean value of y
N = Number of data.

In this analysis, the existing and proposed logistic regression performance is compared. In existing evaluation it gives 93% accuracy. After parameter tuning in

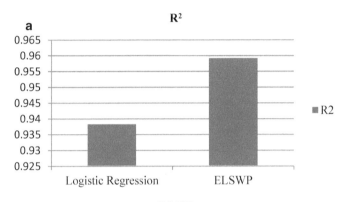

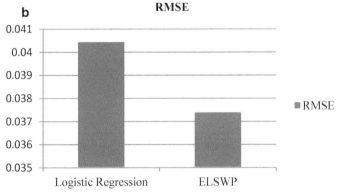

Fig. 8 **a** Accuracy comparison, **b** error rate comparison

Table 2 Comparative analysis of logistic regression and ELSWP	Evaluation metrics	Logistic regression	ELSWP
	R^2	0.9382	0.9591
	RMSE	0.04043	0.03738

this proposed work it gives 96% accuracy. Proposed study work more efficient than existing method.

6 Conclusion and Future Work

The actual weather forecasting technology described in this paper, called ELSWP, was built around with a limited expense Internet of Things module called NodeMCU and the associated sensors. The three critical characteristics based on the sensor units are observed and transmitted to a remote server, such as: temperature, rainfall status and humidity. The device was installed into the agricultural field for continuous monitoring and the measured values were recorded into the server end. The suggested model called ELSWP is implemented using a specialized machine learning processing OpenSource tool called Jupyter notebook context, where it was trained using pre-recorded input variables and then utilized to forecast meteorological data in real-time. The model's output is also compared to a traditional categorization model found in the literature and the suggested method is somewhat more accurate. Additionally, the system may be adapted for commercial usage and has a wide range of applications in home automation, buildings, sports and healthcare, among others. The application can predict temperature, humidity and the probability of rainfall in Tanjore for coming 10 days. Rainfall probability is calculated on a scale of 0–100 percent, resulting in 90% accurate forecasts.

In future, the work can further be enhanced by means of adding some crypto enabled security features over the transmission end for avoiding the injection attacks while communication. As well as this kind of security features enhance the performance of the system and creates trustability to the customers to utilize the model in real-time purpose.

References

1. Kingsy Grace, R., & Suganya, B. (2020). Machine learning based rainfall prediction. In *6th international conference on advanced computing and communication systems* (pp. 227–229). IEEE Xplore.
2. Shivang, J., & Sridhar, S. S. (2018). Weather prediction for Indian location using machine learning. *International Journal of Pure and Applied Mathematics, 118*, 1945–1949.
3. Khan, Z. U., & Hayat, M. (2014). Hourly based climate prediction using data mining techniques by comprising entity demean algorithm. *Middle-East Journal of Scientific Research, 8*, 1295–1300.
4. Bhatkande, S., & Hubballi, R. G. (2016). Weather prediction based on decision tree algorithm using data mining techniques. *International Journal of Advanced Research in Computer and Communication Engineering, 5*, 483–488.
5. Radhika, Y., & Shashi, M. (2009). Atmospheric temperature prediction using support vector machines. *International Journal of Computer Theory and Engineering, 1*, 1793–8201.

6. Chauhan, D., & Thakur, J. (2018). Data mining techniques for weather prediction. *International Journal of Computer Science Trends and Technology (IJCST), 6*(3), 249–254.
7. Qiu, M., & Zhao, P. (2017). A short-term rainfall prediction model using multi-task convolutional neural networks. In *IEEE international conference on data mining* (pp. 395–400).
8. Aswin, S., & Geetha, P. (2018). Deep learning models for the prediction of rainfall. In *International conference on communication and signal processing* (pp. 0657–0661).
9. Mohapatra, S. K., & Upadhyay, A. (2017). Rainfall prediction based on 100 years of meterological data. In *International conference on computing and communication technologies for smart nation* (pp 162–166).
10. Navadia, S., & Yadav, P. (2017). Weather prediction: A novel approach for measuring and analyzing weather data. In *International conference on I-SMAC (IoT in Social, Mobile, Analytics and Cloud)* (pp. 414–417).
11. Gautam, A., & Verma, G. (2019). Vehicle pollution monitoring, control and challan system using MQ2 sensor based on Internet of Things. Wireless personal communications. *International Journal of Springer.*
12. Muthukumar, & Bhalaji, N. (2020). MOOCVERSITY-deep learning based dropout prediction in MOOCs over weeks. *Journal of Soft Computing Paradigm*, 140–152.
13. Smys, S., & Jennifer, S. (2020). Assessment of fire risk and forest fires in rural areas using long range technology. *Journal of Electronics*, 38–48.

Infrastructure as Code (IaC): Insights on Various Platforms

Manish Kumar, Shilpi Mishra, Niraj Kumar Lathar, and Pooran Singh

Abstract In the present-day tech-stack, cloud computing is evolving as a successful and one of the popular fields of technology where the new businesses are achieving success by deploying their functionalities, products, data, and services on cloud instead of on-premises system and that also without depending on any physical component. Infrastructure as code (IaC) is a set of methodologies which uses code to set up the install packages, virtual machines and networks, and configure environments. A successful IaC implementation and adoption by developers requires a broad set of skills and knowledge. It is DevelopmentOperations' tactic of provisioning an application's infrastructure and managing it through binary readable configuration files, instead of any hardware configuration.

Keywords DevOps · Infrastructure · Server · Cloud · Ansible · Iac · Scalability · Terraform

1 Introduction

Cloud computing is a methodology through which programs and data can be stored and accessed even without storing or accessing them on any of the physical media. It is a recent means to save an organization's internal IT resources, as the data are not stored in premises house, instead it is stored in a cloud from where it can be accessed anytime and from anywhere by using the credentials[1, 2]. The digital data are mainly audio, video, text, and image [3, 4]. In the current scenario, using the cloud computing and its services helps large organizations to make huge savings of cost and time. It is because they do not need to think about the cost of needed

M. Kumar
Department of CSE, Arya Institute of Engineering and Technology, Jaipur, Rajasthan, India

S. Mishra · N. K. Lathar (✉) · P. Singh
Department of CSE, Arya College of Engineering and Research Centre, Jaipur, Rajasthan, India
e-mail: nirajlathar28@gmail.com

© The Author(s), under exclusive license to Springer Nature Singapore Pte Ltd. 2023 439
S. Shakya et al. (eds.), *Sentiment Analysis and Deep Learning*,
Advances in Intelligent Systems and Computing 1432,
https://doi.org/10.1007/978-981-19-5443-6_33

software, [5] rather, they just select a cloud service and gets the required services within a few clicks.

As a component of the DevOps group of practices, IaC advances dealing with the information and experience inside reusable contents of foundation code, rather than generally holding it for the manual-escalated work of framework heads which is commonly sluggish, tedious, exertion inclined, and frequently even blunder inclined. While IaC addresses, an always expanding broadly took on training these days, still a new technology and increasing in various aspects, expediently advance, and consistently work on the code behind the IaC system, but it is getting increasingly more foothold in the overwhelming majority of areas of society and industries, [5] and that is just the beginning. Well-known IaC advancements, like Chef and Puppet, give systems to consequently design and arrangement programming organization framework utilizing cloud occasions. Data innovation (IT) associations, such as the Ambit Energy, Mozilla Firefox, Netflix, and GitHub, utilize these components to make the arrangement for the cloud operations-based occasions, for example, Amazon Web Services (AWS), overseeing datasets, and overseeing client accounts both on neighborhood and remote processing occurrences.

2 Literature Review

The DevOps strategy is profoundly impacting how programming is planned and overseen these days. DevOps involves the reception of a bunch of authoritative and specialized rehearses, like constant integration/continuous deployment (CI/CD), mixing improvement and activity groups. The objective is, basically, to have the option to make due as an association in the cutting edge computerized biological system and advanced market, which requests for quick and early deliveries, ceaseless programming refreshes, steady development of market needs, and reception of adaptable innovations, for example, cloud figuring [6]. In this unique circumstance, infrastructure as code (IaC) is the development operations practice of portraying complex and cloud operations and services-based organizations through machine-level understanding code. The main driving force for the IaC practices has been the appearance and implementation of cloud processing, which, on add on of virtualization advancements, interestingly permitted the provisioning, design, and the board of computational assets to be performed automatically [7]. Along these lines, a wide range of dialects and comparing stages have been grown, every one of which manages a particular part of foundation the executives. From devices ready to arrangement and coordinate virtual machines (Cloudify, Terraform, and so forth), to those making a comparable showing as for holder advances (Docker Swarm, Kubernetes), to machine picture the executives' instruments (Packer), to design the board apparatuses (Chef, Ansible, Puppet, and so on.). Presently, the scene of IaC dialects and apparatuses is endangered by the innovation heterogeneity and by the gigantic number of accessible arrangements. From one angle of view, it is the after-effect of the extraordinary

Fig. 1 Architecture of Terraform. *Source* https://learn.hashicorp.com/tutorials/terraform/infrastructure-as-code

interest that infrastructure as a code has raised. Then again, it muddles the comprehension and reception of this new innovation [8]. Revealing insight into the IaC and the platforms it is utilized on, also, having issues and difficulties is accordingly essential toward carrying IaC to development and deployment.

3 Available Methodologies

3.1 Using Terraform

HashiCorp's framework Terraform is an as code instrument which allows a developer to characterize the assets and framework in knowledgeable, and definitive arrangement documents, and also deals with its foundation's lifecycle. Terraform helps to monitor an organization's genuine framework in a state document that goes about as a wellspring of truth for infrastructure's current circumstance [6]. It then utilizes the state record in order to decide the progressions to build up to the foundation with the goal that it will match the design. To send framework with Terraform, it has various step like scope, plan, author, apply, and initialize. Here, Fig. 1 depicts the working architecture of Terraform, which consists of a practitioner which designs and writes the code for the infrastructure, then that is planned and applied on a cloud platform such Amazon Web Services (AWS) or Google Cloud Platform (GCP) or Microsoft Azure and can be other as well.

3.2 Using Ansible

Ansible is an IT-computerized motor which is capable of application organization, mechanizing the cloud provisions, coordinating intra-services, setup executives, and

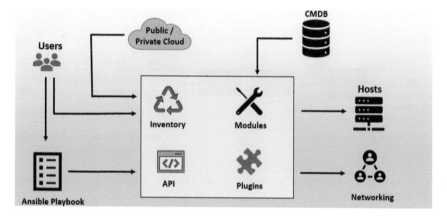

Fig. 2 Architecture of Ansible. *Source* https://www.educba.com/ansible-architecture/

various different functionalities. Since the starting when it came up, it is intended for multi-level arrangements, and it models the infrastructure of an organization by portraying how every one of the frameworks of an organization relate. It does not require any such specialization and no extra custom security framework, so it is not much difficult to convey, and above all, it utilizes an exceptionally basic language which is YAML as a Ansible Playbooks which permits one to depict their automation application such that it approaches the plain English. It works pushing out the project into little-little sized names as Ansible modules after Ansible interfaces with the hubs [9, 10]. These little projects are built to become the asset models of ideal state of framework. After this, Ansible executes those modules over SSH and terminates them when the task is done. Here, Fig. 2 depicts the architecture of Ansible in which we have a server which consists of inventory, modules, plugin, and APIs, and using these functionalities, the other services work in hand like the cloud or the host servers. The user can either contact directly to it or can also use Ansible Playbook for it.

3.3 Chef

It is a strong computerization platform which is capable of changing infrastructure to code. Regardless of the size and regardless of the thing you are working like be it on-premises or the cloud, or in any hybrid environment, Chef infrastructure monitors and automates the way infrastructure is designed, sent, and oversaw across any network [11]. Figure 3 shows the architecture of the Chef platform which consists of 3 main components that are as follows:

– Chef Workstation—it is where the consumers and clients connect with Chef Infra. With it, the developer can author and can perform cookbooks tests utilizing tools

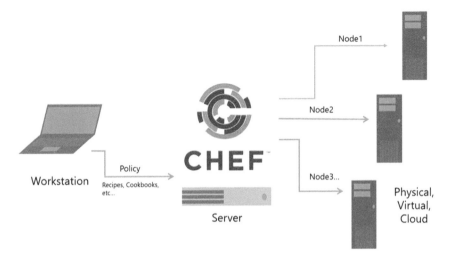

Fig. 3 Architecture of Chef. *Source* https://www.devopsschool.com/blog/what-is-chef-tools-ben efits-of-using-chef-tools/

like test kitchen and cooperate with the Chef Infra Server utilizing the knife and the chef command line tools.

- Chef Infra Client—it runs on frameworks that are overseen by Chef Infra. The Chef Infra Client executes on a timetable to arrange a framework to the ideal state.
- Chef Infra Server—it goes about as a center point for setup information. It is used to store cookbooks, metadata, and the policies which are applied to nodes. Nodes utilize the Chef Infra Client to ask the server for setup subtleties, like file distributions, recipes, and templates.

3.4 Puppet

It is a technological platform which helps an organization or a developer oversee and automate the server's configuration. While using Puppet, one needs to characterize the desired state of the system in the infrastructure of an organization that they want to manage. It can be done by composing infrastructure code in the Puppet Code which is Puppet's domain-specific language (DSL) that can be utilize with a wide cluster of OS and devices [12]. The Puppet code is declarative which means that the developers need to define the desired state of their systems instead of the steps to reach there, after which puppet automates the whole process and keeps the system in that desired state. It can be done by using a Puppet agent and its primary server. The code that is used to define the state is stored in Puppet primary server, while the Puppet agent is used to make an interpretation of the code into respected commands and afterward executes in the specified systems which is known as Puppet run. Figure 4 shows the architecture of Puppet platform which follows agent/server architecture and consists

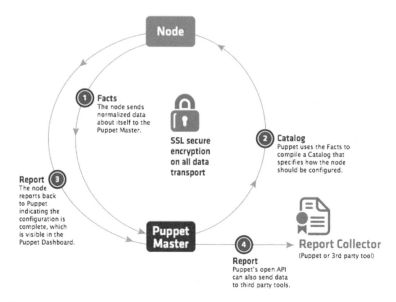

Fig. 4 Architecture of Puppet. *Source* https://www.trainingbangalore.in/blog/what-is-puppet/pup
pet-architecture

of a node and a Puppet master which handles and manages the whole flow. Both of
them use SSL to connect with each other.

3.5 CloudFormation

Amazon Web Service (AWS) CloudFormation is an assistance which helps to set
and manage the resources of AWS so efficiently that the developers can invest their
most of the focus on the application which runs on AWS cloud. In CloudFormation,
the organization first creates a template describing all the resources and services
required for the application, and it manages the configuration of those services.
These templates can be created either in JSON or YAML format. Because of AWS
CloudFormation, the developer doesn't need to define each of the services and their
properties individually every time [5]. We can see the architecture of CloudFormation
from Fig. 5, which shows that at first you need to either create or use an existing
template which can be then saved in a S3 bucket. And then, by creating stack based
on the templated we created, we can build the infrastructure using CloudFormation.

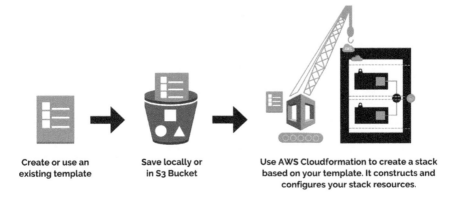

Create or use an   **Save locally or**   **Use AWS Cloudformation to create a stack**
existing template   **in S3 Bucket**   **based on your template. It constructs and**
                                                                   configures your stack resources.

Fig. 5 Architecture of CloudFormation. *Source* https://dustinward.cloud/what-is-aws-cloudform ation/

4 Analysis

In todays' market, there are of lot of platforms and tools which provides the facilities of making the infrastructure as code work such as Ansible, Chef, Puppet, Terraform, SaltStack, CloudFormation, and many more. The Puppet is the ones which we can say that started the journey of the IaC in the year 2005, and later on, Chef added in this league in the year 2009 and after those continuously new technologies are coming in which CloudFormation from AWS is the latest which is gaining popularity because of its integrity and work with AWS services although it is not compatible with any other cloud, but if we measure about the AWS community which is already very huge, we can rely on CloudFormation [13]. Given below, we have analyzed the different platforms on different-different perspective which gives an intel on which platform has the upper hand in the current market of infrastructure as code.

From Table 1, we can see how the various platforms are based on their type, which cloud technologies they support, type of their source and management, the languages they support and what type of architecture they holds [14].

From Table 2, we can make our views on what are the growth each of the platform is having in the recent years. If we observe in the table, we can see that Ansible and Terraform have a huge amount of increase in the contributors, and also, the number of libraries increased in Terraform is beyond any compare. As CloudFormation is limited to AWS and is not an open-source platform, it has no contributors, commits, or issues mentioned, but if we check the amount of increase in the library and the queries on stack overflow and the amount of job, they uphold is enormous regardless it is the latest among all in the market [14].

According to statista report of quarter 4 in the year 2021, from Fig. 6, we can see according to the market scenario in that part of time Amazon (AWS—Amazon Web Services) was leading the market of cloud infrastructure with a 33%. The total cloud market at that time was worth of $178-billion which is even 37% more than

Table 1 Various platform dependencies

	Chef	Cloud formation	Puppet	Terraform	Ansible	Salt stack
Code	All	AWS	All	All	All	All
Cloud	Open-source type	Close-source type	Open-source type	Open-source type	Open-source type	Open-source type
Architecture	Client–server architecture	Client–server architecture	Client–server architecture	Client-only architecture	Client-only architecture	Client–server architecture
Type	Configuration management	Configuration management	Configuration management	Orchestration	Configuration management	Configuration management
Infrastructure	Mutable	Mutable	Mutable	Immutable	Mutable	Mutable
Language Type	Procedural	Procedural	Declarative	Declarative	Procedural	Declarative

Source https://blog.gruntwork.io/why-we-use-terraform-and-not-chef-puppet-ansible-saltstack-or-cloudformation-7989dad2865c

Table 2 Platform growth on various aspects in the year 2019

	Source	Cloud	Jobs	Stars	Commits (1 month)	Contributors	Issues (1 month)	Stack overflow	Libraries
Chef	Open source	All	−22%	+31%	+139%	+18%	+48%	+43%	+26%
CloudFormation	Closed source	AWS	+249%	NA	NA	NA	NA	+441%	+57%
Puppet	Open source	All	−19%	+27%	+19%	+19%	+42%	+36%	+38%
Ansible	Open source	All	+125%	+97%	+49%	+195%	+66%	+223%	+157%
Terraform	Open source	All	+8288%	+194%	−61%	+93%	−58%	+1984%	+3555%
SaltStack	Open source	All	+257%	+44%	+79%	+40%	+27%	+73%	+33%
Heat	Open source	All	+2957%	+23%	−85%	+28%	+1566%	+69%	0%

Source https://blog.gruntwork.io/why-we-use-terraform-and-not-chef-puppet-ansible-saltstack-or-cloudformation-7989dad2865c

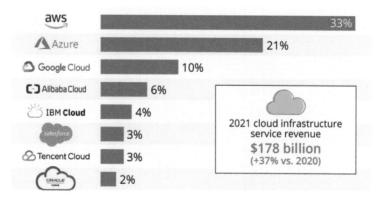

Fig. 6 Worldwide market share of leading cloud infrastructure service providers in Q4 2021. *Source* https://www.statista.com/chart/18819/worldwide-market-share-of-leading-cloud-infrastructure-service-providers/

the previous year 2020, and the leading platforms that are dominant in the market are AWS and Azure which is currently governing the marketing [15].

5   Conclusion

In the current scenario of technology, a lot of changes are going on, and in the field of cloud computing platforms, there are various big names in the market although AWS had made its impact in a huge amount and currently leading the cloud market. DevOps is a group of strategies that target speeding up organization and conveyance of huge scope applications. Basically, all the development operation automations are driven more by infrastructure as a code, which we can mention as the series of outlines that are outspreading the application infrastructure and all the involved middleware across a DevOps pipeline [16]. From one viewpoint, a few prescribed procedures exist yet they are for the most part focused on restricting the intricacies intrinsic inside IaC. Then again, many difficulties exist, from clashing accepted procedures, to absence of testability, lucidness issues, and there are some more as well. Thus, contingent upon the environment and elements the software engineers can pick any of the stages as indicated by their reasonableness. In spite of the fact that we can survey from the investigation that for the ongoing business sector, Terraform has acquired a great deal of prevalence in the market in light of its adaptability, cross-stage work, and versatility; then again, in the event that we are working with the most well-known cloud AWS, utilizing CloudFormation will be the most ideal choice [17].

References

1. How Cloud Computing Works. http://computer.howstuffworks.com/cloudcomputing/cloud-computing.htm
2. Arora, H., Mehra, M., Sharma, P., Kumawat, J., & Jangid, J. (2021). Security issues on cloud computing. *Design Engineering,* 2254–2261.
3. Soni, G. K., Arora, H., & Jain, B. (2020). A novel image encryption technique using arnold transform and asymmetric RSA algorithm. In *International conference on artificial intelligence: Advances and applications 2019.* Algorithms for Intelligent Systems. Springer.
4. Soni, G. K., Rawat, A., Jain, S., & Sharma, S. K. (2020). A pixel-based digital medical images protection using genetic algorithm with LSB watermark technique. In *Smart systems and IoT: Innovations in computing. smart innovation, systems and technologies* (Vol. 141). Springer.
5. Alam, A. B., Haque, A., & Zulkernine, M. (2018). Crem: A cloud reliability evaluation model. In *2018 IEEE Global Communications Conference (GLOBECOM)* (pp. 1–6). IEEE.
6. Artac, M., Borovssak, T., Di Nitto, E., Guerriero, M., & Tamburri, D. A. (2017). Devops: Introducing infrastructure-as-code. In *2017 IEEE/ACM 39th International Conference on Software Engineering Companion (ICSE-C)* (pp. 497–498). IEEE.
7. Rahman, A., Parnin, C., & Williams, L. (2019). The seven sins: Security smells in infrastructure as code scripts. In *Proceedings of the 41st international conference on software engineering,* in Press.
8. Bhalaji, N. (2021). Cloud load estimation with deep logarithmic network for workload and time series optimization. *Journal of Soft Computing Paradigm, 3*(3), 234–248.
9. Hasicorp Learn. *What is infrastructure as code with Terraform* (Online). Available: https://learn.hashicorp.com/tutorials/terraform/infrastructure-as-code
10. Kumar, R., Gupta, N., Charu, S., Jain, K., & Jangir, S. K. (2014). Open source solution for cloud computing platform using OpenStack. *International Journal of Computer Science and Mobile Computing, 3*(5).
11. Cowie, J. (2014). Customizing chef getting the most out of your infrastructure automation. *O'Reilly Media, 9.*
12. Fowler, M., Beck, K., Brant, J., Opdyke, W., & Roberts, D. (1999). *Refactoring: Improving the design of existing code.* Addison-Wesley Longman Publishing Co., Inc.
13. Brikman, Y. *Why we use Terraform and not Chef, Puppet and Ansible* (Online). Available: https://blog.gruntwork.io/why-we-use-terraform-and-not-chef-puppet-ansible-saltstack-or-cloudformation-7989dad2865c
14. Baldini, I., Castro, P., Chang, K., Cheng, P., Fink, S., Ishakian, V., Mitchell, N., Muthusamy, V., Rabbah, R., Slominski, A., et al. (2017). Serverless computing: Current trends and open problems. In *Research advances in cloud computing* (pp. 1–20). Springer.
15. Mishra, S., Kumar, M., Singh, N., & Dwivedi, S. (2022). A survey on AWS cloud computing security challenges & solutions. In *IEEE international conference on applied artificial intelligence and computing.*
16. Buyya, R., Srirama, S. N., Casale, G., Calheiros, R., Simmhan, Y., Varghese, B., Gelenbe, E., Javadi, B., Vaquero, L. M., Netto, M. A., et al. (2018). A manifesto for future generation cloud computing: Research directions for the next decade. *ACM Computing Surveys (CSUR), 51*(5), 1–38.
17. Andi, H. K. (2021). Analysis of serverless computing techniques in cloud software framework. *Journal of IoT in Social, Mobile, Analytics, and Cloud, 3*(3), 221–234.

K-Means Clustering Decision Approach in Data Hiding Sample Selection

Virendra P. Nikam and Sheetal S. Dhande

Abstract Nowadays, data science plays an important role in decision-making process. Data science application varies from organization to organization, where it plays a very important role in decision-making process in multidimensional data. Author suggests an application of k-means clustering data science algorithm in the sample selection process of carrier during data hiding behind carrier. Suggested approach helps to minimize the time complexity for sample selection and also cluster samples more accurately which helps to focus on specific cluster rather than entire samples in carrier. The experimental results are calculated far better than any other clustering technique discussed in the result section of this paper. The primary objective of author is to make data hiding process so simple and accurate to minimize quantization error. K-means clustering is an effective technique to group samples into different classes called as cluster based on Euclidean distance and MOI. An accuracy of cluster depends on how tightly items in clusters are bounded. Results are compared with existing approach which are better and hence proved that k-means clustering data science approach is the best match for sample selection process during data hiding.

Keywords K-means clustering · Centroid · Moment of inertia (MOI) · Euclidean distance (ED) · Root mean square (RMS) · Region of interest (ROI)

1 Introduction

Data security is a challenging task nowadays when data is stored on a local system or on a server system. Many organizations lose lots of money on data privacy, authentication and security. Existing techniques are good enough to protect data but are not sufficient. 100% security for sensitive data is not quite possible in this modern era with modern technology. Most techniques exist, like data hiding, watermarking, compression, cryptography and others, to protect the data from all illegal access. The

V. P. Nikam (✉) · S. S. Dhande
Department of Computer Science and Engineering, Sipna College of Engineering
and Technology, Amravati, Maharashtra, India
e-mail: virendranikamphd@gmail.com

© The Author(s), under exclusive license to Springer Nature Singapore Pte Ltd. 2023   451
S. Shakya et al. (eds.), *Sentiment Analysis and Deep Learning*,
Advances in Intelligent Systems and Computing 1432,
https://doi.org/10.1007/978-981-19-5443-6_34

major obstacle to data security is its multidimensional nature. Most of the time, data is collected from real-time scenarios like marketing data from people's feedback, which is in multidimensional form. There is a need to preserve this data so as to make good decisions in business processes and to enhance them more accurately by processing and storing it. Propose an approach that has the main objective of protecting sensitive data by hiding it behind another object, which may be either an image, audio, video or text. Authors implement the project strategy with images as the main object for data hiding.

An image is an integration of three colour components called red, green and blue that makes sense as a multidimensional data set. Each of these colour components consists of 8 bits, which make a pixel, and pixels make an image. During the process of data hiding, there is a need to extract all these three channel components, and a decision needs to be taken on where to hide secret data. As mentioned by the author, because of the three colour components of an image, it is difficult to take a decision or to reach a decision at a specific point which sample to choose for data hiding. If the data gets hidden behind the wrong carrier channel, it will create a huge difference called a quantization error and that will create an audio-visual perceptual quality difference between the original carrier (image) and the resultant carrier. The primary objective of the author is to minimize the quantization error by selecting the best region carrier samples from an image with a data science k-means clustering algorithm.

There are lots of data science algorithms that cluster as well as classify multi-dimensional data in decision-making, like k-nearest neighbour, k-means clustering, genetic algorithm and others. It is suggested to use k-means clustering to group carrier samples and help developers focus on specific regional carrier samples for the data hiding process. Before moving on to the k-means clustering process and its application in data hiding behind carrier objects, consider Fig. 1 to see where k-means clustering is applicable in the data hiding process.

A sender is any legitimate user or software program that selects carrier objects and secret data to hide behind a carrier object. A selected carrier object is a colour image made up of RGB components. Let us consider an image $m \star n$, where m is width and n is height. So the total number of pixels in an image is $m \star n$, which means the image has three times the data carrying capacity of its total number of pixels. The image as a carrier object is a choice by the user. Secret information is sensitive in nature and should not be visible to unauthorized persons. Once the secret data is chosen, it is converted into its equivalent binary representation, for example

$$(21)_{10} = (00010101)_2 \tag{1}$$

It is mandatory to convert secret data into its equivalent binary format because the data hiding process works at the bit level. Samples from carrier objects are extracted and given as an input to the k-means clustering algorithm for grouping of samples. All the samples are clustered well with a sufficient number of iterations and a strong MOI. Then they are given to the data hiding algorithm, which focuses on specific clusters that generate the minimum quantization error after hiding the secret bit.

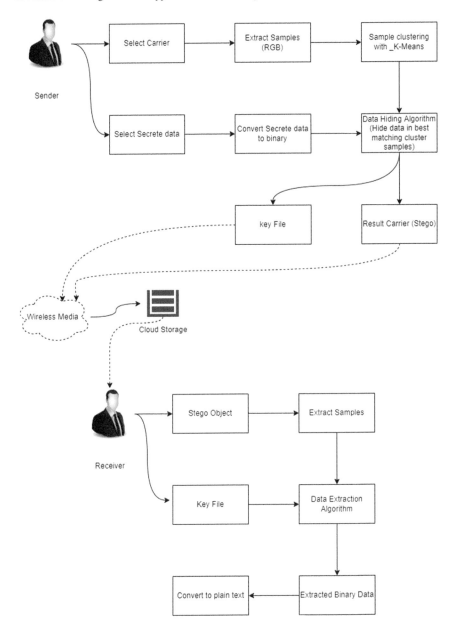

Fig. 1 Data hiding and extraction overflow

The data hiding algorithm inputs binary secret data and k-means cluster samples and hides data with the best samples by focusing on a specific effective cluster. The hiding process generates two product files, i.e. a stego object and a key file. A key file holds the positions of samples where the secret bits are hidden. As the k-means cluster algorithm outputs clusters or groups the samples from different positions of an image, the hiding process is also random, which means it needs to remember the positions where the secret bit is hidden. These two output files are sent via wireless network to public cloud storage, where data from the authenticated receiver is extracted. Initially, the proposed approach is intended to allow only legitimate users with the proper authentication code or permission to access public data, but later, all permissions or restrictions are removed and the data on public cloud is made available to everyone accessing cloud.

On the receiver side, it extracts or inputs stego objects and key files stored in the public cloud. Samples are extracted from stego carrier objects and given as an input to the data extraction algorithm, which produces secret data in binary format or extracts it with the help of a key file. The extracted binary data is then converted into its equivalent character representation. Clustering algorithms do not have any role in the data extraction process, but they affect exit data extraction by the means of generating key file on the sender side. The accuracy of extracted data is highly dependent on how accurately the data is recorded into the key file. One of the biggest advantages of k-means clustering data science algorithms is the unsupervised learning strategy where there is no requirement for any training data. Centroids are generated randomly based on which input data samples cluster. The accuracy of cluster data samples is measured by the means of the MOI.

K-means clustering is an unsupervised algorithm that is designed to partition unlabeled data into a certain number of groups called "clusters". One of the characteristics of the k-means clustering algorithm is that it finds and observes the input data that shares the important characteristics of input data and classifies them into different groups. One sign of a good clustering solution is the ability to find clusters in such a way that the observations within each cluster are more similar than the cluster themselves. The k-means clustering algorithm follows two steps.

1. **Assignment**: In assignment step, each data point is assigned to its nearest centroid, which is randomly inserted.

$$\min(c) = [\text{dist}(c_i, x)^2, \ldots, x)^2, \text{dist}(c_n, x)^2] \qquad (2)$$

An Euclidean distance is calculated for each centroid and data simple point. The process has been repeated for the data samples and randomly inserted centroid. A sample that has a minimum Euclidean distance with respect to the centroid will be assigned to the respective cluster of that centroid.

2. **Update step**: In the update step, the centroids are updated to their new positions as per iterative observation.

$$c_i = \frac{1}{|s_i|} \sum x_i \varepsilon c_i^{x^i} \qquad (3)$$

An algorithm continues its execution from step 1 to step 2 until the stopping criteria are not met. That means no data points change and the sum of distance, called the MOI, is minimized. The results obtained from the k-means cluster algorithm are the local optimum, meaning that the run of the algorithm with a randomized starting centroid gives a better outcome.

About two steps are repeated over and over until there is no further change in the cluster data samples. When all the samples in each cluster settle down and do not change even after the repetition of updations of the centroid, it means that samples are very well clustered and do not require any further iterations.

2 Literature Review

Abdul Nazeer et al. [1] found out the limitations of clustering that require you to specify the number of clusters regardless of the distribution of data points. It means the k-means clustering algorithm itself cannot find a good number of clusters until it is specified by the user.

Agrawal and Phatak [2] apply the k-means clustering technique to historic documents. Find out a few limitations of k-means clustering.

1 K-means clustering requires to specify the number of clusters.
2 Besides its activation, its convergence is in two different local minima and maxima.
3 It is comparatively slow and not fit for a large number of data set points.
4 It is not possible to handle cluster with no data set point with k-means clustering.

Mato et al. [3] applied the k-means clustering algorithm to solve geophysical problems that determine the geodynamic behaviour of the rapture zone. K-means clustering is an unsupervised technique that is mandatory for the identification of a geophysical problem. However, they have reported some problems with the algorithm.

1 Identification of cluster count and their special localization from geophysical and geodynamical point of view.
2 The author applies k-means algorithm incorporating to construct information within clustering data set points.

Their novel technique allows previously identified earthquake events to be associated with energy release points, which helps to perform geophysical clusters and calculate their centroid using k-means clustering.

Yang et al. [4] found out that k-means clustering is vulnerable to the impact of clustering count k. Their main objective is to improve speed and precision using k-means clustering. To achieve this, the author suggested a method that is accomplished by a subtractive clustering technique to find the optimal initial cluster count k. They perform their experiments on public databases like UCI and UT, which show that k-means clustering algorithm limitations can eliminate the sensitivity of the initial randomly generated cluster count k.

Kapil et al. [5] They suggested a k-means clustering technique that has been optimized using a genetic algorithm so that the problems that occur due to k-means should be minimized. The outcomes of these k-means and genetic algorithms are calculated and compared individually to come up with a conclusion about the effectiveness of the suggested approach. The author also reported some drawbacks which make use of k-means clustering along with genetic algorithms infeasible. Then, in terms of sum squared error, genetic algorithms outperform. The suggested approach requires comparing the performance of different clustering algorithms along with different machine learning techniques with different practices on a high-dimensional data set.

Shah and Singh [6] suggested a new algorithm that modifies the k-means clustering technique. In this technique, the author executes an algorithm like k-means and k-medoids and also tests several methods with different initial clusters with different data set points. The suggested modified k-means technique performs better in terms of the number of clusters and performance time. The suggested approach has been executed with real-time data, and the results of these are compared with k-means and k-medoids.

Gour et al. [7] suggested an approach that compares its result with the self-organizing map (som) and the k-means clustering technique. Their results show that the suggested clustering technique performs better than k-means and self-organizing map (som) clustering methods in terms of the MOI [8] and random centroid initialization. The author's main objective is to improve speed and precision in clustering, which is validated by comparing their results with other clustering techniques. The authors found that art1 is the best grouping fingerprint clustering technique compared to k-means and self-organizing map (som).

Li et al. [9] It turns out that even though the priority and randomicity of the cluster initialization centroid with k-means have solved many traditional problems, it is very difficult to apply widely and popularly due to computational complexity. The authors propose a new approach based on hierarchical k-means clustering that has good computational complexity. The efficiency of the proposed approach is validated using the iris data set. With the suggested approach, the distribution of different data set points is done through hierarchy clustering (hc) and then centroids are decided.

Sapkota et al. [10] suggested a method to analyse existing techniques used for data mining and find a way to maximize the accuracy of clusters using k-means. The author introduces a novel algorithm based on the combination of spectral clustering with k-means. The proposed system replaces the random initialization technique for cluster centroid, which solves some of the limitations of the k-means algorithm. Their algorithm selects an appropriate first centroid rather than selecting it randomly. The proposed system reduces errors but has the limitations of increasing processing time by up to 4 s.

Alvarez et al. [11] employed an adaptive clustering process to identify clients with the highest power consumption during peak times. The analysis indicates that appropriate control on search groups of users should be performed to reduce peak demand and meet requirements. The k-means clustering technique has been used for the clustering of those clients who are consuming more power during peak time. The author uses the silhouette method to estimate an appropriate number of clusters.

Yang and Sinaga [12] Most of the multi-view k-means clustering techniques do not provide feature reduction during the process of clustering, and irrelevant features may exist in multi-view data sets that may cause bad performance measures for the clustering algorithm. There are also multidimensional data sets, so it is necessary to consider reducing their dimension for accurate and precise clustering. The authors proposed a learning mechanism for the multi-view k-means technique to compute the feature centroid for a data set automatically. It also reduces irrelevant feature components in each cluster view. A suggested new type of multi-view k-means technique is also called a future reduction multi-view k-means.

Wang et al. [13] It suggested a fast adaptive k-means (FAKM) clustering technique where an adaptive loss function is specially designed to provide a flexible cluster centroid calculation mechanism, which is why the technique is most suitable for data sets under different distributions. The FAKM technique performs feature selection and clustering simultaneously without finding an eigenvalue decomposition. Extensive experiments on several data sets demonstrate that the fast adaptive k-means (FAKM) technique gets good results over other existing state-of-the-art clustering algorithms. The author exploits an efficient alternative optimization technique with FAKM.

Cheng et al. [14] This paper suggests a network security computational model based on a k-means clustering radial basis function neural network. The author uses k-means clustering to cluster input samples and get the nodes of the hidden layer of the radial basis function neural network to select centre points more quickly. The training data is set for neural network performance with neural network structure. The results obtained from the authors' approach have a great improvement in training speed as well as good accuracy compared to other neural networks.

Wu et al. [15] In this paper, the author proposed implementing a privacy-preserving learning as a service model and focusing on k-means clustering over encrypted cloud databases. The previously existing algorithms partially use homeomorphic encryption, which requires large numbers of protocols and high computational and communication costs, making it not practical in the real world. The author helps to solve this problem and proposes a new secure and efficient outsource k-means clustering that uses this fully homeomorphic encryption with ciphertext packaging technique. Schemes have preserved privacy in three aspects: database security, privacy of cluster results and data access pattern. The experimental results of the proposed approach require very little computational cost and are also suited for many large databases.

Liu et al. [16] The author suggested a fault diagnosis method for the control moment gyroscope based on k-means clustering that improves the fault diagnosis accuracy by considering the hidden correlation parameters between multidimensional data. Principal component analysis and random neighbouring embedding techniques are used to extract the features from control moment gyroscope and digital data. The fault diagnosis process is entirely based on k-means clustering, which is found to be very effective in actual orbit control moment gyroscope data.

3   Proposed Methodology

The entire process is divided into the below steps

1. Carrier selection
2. Extracting carrier samples
3. Secret data selection
4. K-means clustering
5. Data hiding
6. Data extraction.

3.1   Carrier Selection

The author implements the above-mentioned steps on the image as a carrier object, which is represented as

$$i = \int_{i=0}^{m} \int_{j=0}^{n} (r, g, b) \qquad (4)$$

An image is an integration of RGB colour components, which are the basic pillars of data hiding. A 24 bit RGB colour model image always has maximum data carrying capacity as compared to other image formats. The maximum data carrying capacity of a carrier object is represented with

$$C = m \star n \qquad (5)$$

which means C numbers of characters are possible to hide behind carrier object as per Eq. 4.

3.2   Extracting Carrier Samples

Table 1 shows the extracted RGB carrier sample with its average value. Only the average value is used to find or observe quantization error. Due to the length of the paper, only a few samples are given in Table 1. However, the total number of extracted samples highly depends on height and weight of carrier image as per Eq. 4.

Table 1 Carrier samples

Red	Green	Blue	Average
214	112	203	176.3333333
253	190	206	216.3333333
198	202	49	149.6666667
224	117	61	134
206	171	170	182.3333333
185	155	117	152.3333333
144	242	175	187
182	229	117	176
5	82	127	71.33333333
205	101	67	124.3333333
37	171	24	77.33333333
229	77	46	117.3333333
104	13	232	116.3333333
71	67	246	128
157	134	205	165.3333333
72	7	5	28
207	60	148	138.3333333
178	170	24	124
18	72	130	73.33333333
26	32	114	57.33333333
60	207	140	135.6666667
218	176	164	186
241	146	212	199.6666667
9	140	130	93
228	78	211	172.3333333
6	70	157	77.66666667
53	171	8	77.33333333
208	104	32	114.6666667
86	153	112	117
65	232	42	113
115	213	224	184
41	65	83	63
240	104	212	185.3333333
103	124	67	98
5	97	124	75.33333333
181	74	4	86.33333333
77	86	242	135
3	158	180	113.6666667

Algorithm 1

Result: Clusters of data set points.

1 Start
2 Input data set X
3 Initiate centroid randomly
  $C_{i,j} = Rand(X_{RoiMin}, X_{RoiMax})$
4 Calculate distance of data set points with respect to centroid
  $D_i = |C_i - X_i|$.
  or
  $D_i = \sqrt{\frac{(C_x - X_x)^2 + (C_y - X_y)^2}{2}}$
5 Calculate MOI
  $MOI = \sum_{i=1}^{n} |C_i - x_i|$.
6 **if** $MOI \mathrel{!=} Saturation$ **then**
  | Find new centroid
  | Repeat from step 4
  end
7 Stop

3.3 Secret Data Selection

The secret data may be an image, audio, video or text. Without converting secret data into its equivalent binary format, it is not possible to hide it behind a carrier object. And that is why the format of secret data is irrelevant. The authors here have proven results based on text as secret data. Let us consider carrier sample value 21 in decimal format, which can be represented into its equivalent binary format as per question 1. The binary conversion process is simple enough to remember with just a reminder and is easily available on the Internet.

3.4 K-means Clustering

The k-means clustering technique is further divided into the below steps.

1. Data collection
2. Data preprocessing
3. Random initialization of centroid
4. Calculating distance of data points from centroid
5. Calculating new centroid for each cluster
6. Analysing cluster strength based on its MOI
7. Finalizing number of clusters based on MOI.

Fig. 2 Input data

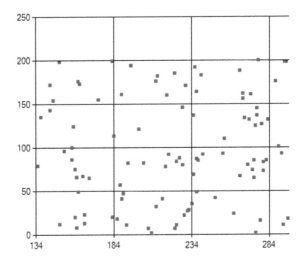

Data Collection The outcome of the proposed approach is highly dependent on the data collected from authentic sources. In the proposed approach, the authenticity of the carrier object does not matter because the author can choose any image as a carrier object. Table 1 represents the raw data which is input to the k-means clustering algorithm. If audio is considered as an input channel which consists of only left and right stereo channels, the total data hiding capacity is limited and directly proportional to the number of channels. Video has maximum data hiding capacity because of its integration of audio and image frames. Both audio and images are considered separate data types and hence almost have double data carrying capacity. Let input data set

$$X = x_1, x_2, x_3, \ldots, x_n \tag{6}$$

consist of n-th data set points. Data set points X contains all parameters required to process experiments with a proposed approach. All these parameters are required for preprocessing and filtration. A graphic representation of the data from Table 1 is shown in Fig. 2. The k-means clustering technique explains the limitations of representing only two-dimensional data in graphical format, which means that any two parameters, such as RG, GB or RB, are considered (Fig. 3).

Data preprocessing Data preprocessing is also called data filtering where only those parameters are considered which are highly precise and required for an implementation of the proposed concept. The data set points X is divided into ROI and non-ROI. The output of data preprocessing is the ROI points, whereas non-ROI point is kept aside from further processing.

$$X_{ROI} = x_{r1}, x_{r2}, x_{r3}, \ldots, x_{rn} \tag{7}$$

Fig. 3 Data preprocessing

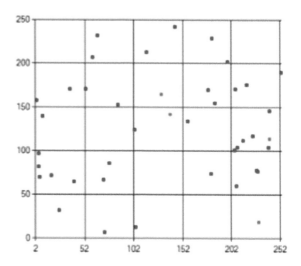

The selection of ROI points totally depends on programmer choice by considering gravity of parameter in the evolution of entire data set.

$$X_{ROI} = [R, G, B] \tag{8}$$

Factors considered in data preprocessing are highly important and best suited to the real-time scenario (Fig. 4).

Random initialization of centroid K-means clustering algorithm requires to initialize centroid randomly to start process. Random initialization of centroid is done with

$$C_{i,j} = \text{Rand}(X_{\text{Roi Min}}, X_{\text{Roi Max}}) \tag{9}$$

Calculating Distance of Data Points from Centroid To cluster data, distance of all data points is calculated from centroid. At initial stage, centroids are randomly inserted based on number of cluster count specified by the user. Distance measure performed is called as an ED represented with

$$D_i = |C_i - X_i| \tag{10}$$

A distance of every data point is calculated for every centroid C_i. A decision is taken based upon minimum value of $\text{Min}(C_i)$. A minimum value of $\text{Min}(C_i)$ divides all data set points into number of clusters specified by the user. If C_i and x_i are multidimensional, distance can be calculated by getting its RMS value.

$$D_i = \sqrt{\frac{(C_x - X_x)^2 + (C_y - X_y)^2}{2}} \tag{11}$$

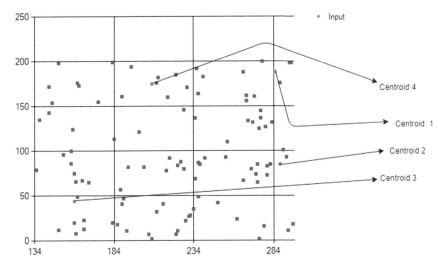

Fig. 4 Random insertion of centroid

Calculating New Centroid for Each Cluster Once clusters are created by randomly initializing centroid, new centroids are calculated by taking mean of each centroid. Ex. Suppose consider clustered 1 have n-th data points, then mean of all these data points is

$$C_{\text{new}} = \frac{\sum_{i=1}^{n} x_i}{n} \tag{12}$$

This C_{new} is considered as new centroid for cluster C_i. Likewise, new centroid for other clusters is also calculated and data points are again clustered into different groups.

Analysing Cluster Strength Based on its MOI Strength of cluster is measured with calculation of bonding between data set points and centroid. Minimum bonding distance implies that clusters are highly precise and accurate. This bonding is called MOI and calculated with

$$\text{MOI} = \sum_{i=1}^{n} |C_i - x_i| \tag{13}$$

Finalizing Number of Clusters Based on MOI The number of clusters and data points in clustering is always debated. Many times, it happens that a cluster with fewer than 4 points is not good for machine learning. Such a cluster is called an "overfit cluster". A mechanism should be required to decide the number of data points in a cluster. MOI is a good choice for deciding how much of the cluster should be for a given data set of points. MOI gets stabilized to the point where there is no further need for clustering data. This stability of the MOI is called a saturation level.

The MOI increases or decreases exponentially before becoming stable throughout the iteration.

Algorithm 2

Result: Decision of setting numbers of clusters
while *!MOI==Saturation* **do**
> Continue the process of finding new centroid
> Construct new cluster
> Calculate the MOI

end

3.5 Data Hiding

The data hiding process is not selective with any steganography method. As shown in Fig. 5, all data samples are clustered and now developers have to choose where to hide data. Which cluster to choose? The current approach is tested with least significant bit (LSB), spread spectrum, wavelet analysis, sample flipping and most significant bit (MSB). By spending lots of years in the development process of research analysis, it is observed that the samples in the dark regions of

$$0 \leq c_s \geq 20 \tag{14}$$

$$240 \leq c_s \geq 255 \tag{15}$$

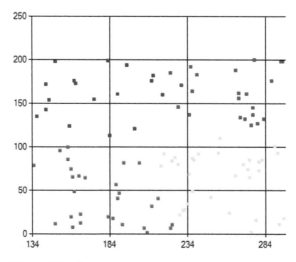

Fig. 5 Clustered data set $(C = 4)$

are good for data hiding. Even after data hiding, the visual perception quality of the carrier object remains the same because it represents the extreme dark or extreme white colour of an image. Another region is called the non-dark region, where samples are mostly representative of the entropy of an image where the hiding process generates an error and hence also disturbs the originality of the carrier object. It is suggested to not hide data in non-dark regions because of its sensitivity towards error. K-means clustering techniques easily find the samples from dark regions throughout the career area and help the developer easily select samples in the process of data hiding.

3.6 Data Extraction

To extract data from a stego object, the availability of a key file that holds the position of the sample where the data is hidden is mandatory. The extraction process simply finds out the position of the hidden sample as per the value in the key file and collects the data in binary format, which is later converted into character format. The process of data extraction is quite simple and not dependent on k-means clustering. This is considered one of the drawbacks because the clustering process cannot be sped up and you have to scan for the entire career area. The accuracy of data extraction depends on the measurement of comparison between original secret data and extracted data. Author implemented a comparative result analysis model that compares the input and extracted data and proves that there is no single loss of characters after data hiding and extraction.

4 Result Analysis and Discussion

The purpose of result analysis is to measure the performance of the proposed system with other similar algorithms with different parameters. In order to guarantee that the suggested approach is better and more advanced, it is mandatory to compare the outcomes of the algorithm with different parameters. The proposed concept suggests clustering of RGB data sets using k-means clustering. Its comparative parameters may include the MOI, the number of clusters, RMS value, cross-validation of clusters with the current scenario and others.

From Table 2, the value of MOI gets stable from iteration 8 onward. It still conducted iterations to find out exact number of clusters C_i for every data set. In other cases, if data set points are in thousands, MOI will be settled for 10 to 15 iterations. With the proposed k-means clustering, almost 20 iterations have been done to settle down exact and accurate data set points. From Table 3, it is found that the number of clusters gets perfectly set after 3 to 4 iterations. The ED is a measure of the distance of data points from its centroid. A data set point belongs to a cluster with a minimum distance from its centroid. Ed is calculated with an equation of RMS error D_i. A MOI

Table 2 MOI of cluster for iteration

Iteration	Cluster 1	Cluster 2	Cluster 3	Cluster 4
1	78.40	198.99	259.76	113.97
2	78.40	152.64	225.17	157.40
3	102.51	129.02	190.31	179.02
4	113.83	106.38	129.81	240.01
5	125.55	109.64	86.33	240.48
6	145.36	109.64	82.65	215.06
7	161.94	109.64	82.65	195.78
8	161.97	109.64	82.65	194.43
9	161.97	109.64	82.65	194.43
10	161.97	109.64	82.65	194.43
11	161.97	109.64	82.65	194.43
12	161.97	109.64	82.65	194.43
13	161.97	109.64	82.65	194.43
14	161.97	109.64	82.65	194.43
15	161.97	109.64	82.65	194.43
16	161.97	109.64	82.65	194.43
17	161.97	109.64	82.65	194.43
18	161.97	109.64	82.65	194.43
19	161.97	109.64	82.65	194.43
20	161.97	109.64	82.65	194.43

Table 3 Numbers of data set points in each cluster

Count_0	Count_1	Count_2	Count_3
29	24	26	20
27	25	27	20
25	24	30	20
28	25	27	19
29	25	27	18
29	25	27	18
29	25	27	18
29	25	27	18
29	25	27	18
29	25	27	18
29	25	27	18
29	25	27	18
29	25	27	18
29	25	27	18
29	25	27	18
29	25	27	18
29	25	27	18
29	25	27	18
29	25	27	18

is the sum of the distances between each data set point and the centroid. It is calculated to find out the bonding of cluster data set points. The minimum MOI means strong cluster binding. MOI is set to a minimum value through multiple iterations. From Table 5, it is found that after iterations 3 to 4, MOI settles to its minimum value. The proposed concept needs to make efforts to minimize iteration and to set the numbers of clusters adaptively based on cluster data set points. The proposed approach is tested not only in RGB scenarios but also in carrier formats. It is found that the proposed technique is very well suited to every kind of business sector.

5 Conclusion

From the result analysis section, it is true that k-means clustering is well suited for grouping up carrier samples, and based on that, it is very easy to make the decision where to hide data. The k-means clustering technique plays an important role only in the data hiding process, and hence, the complexity is automatically reduced on the receiver side because there is no requirement to share additional details with the receiver for data extraction. It aids in reducing errors that occurred after data hiding by selecting samples from the correct cluster. The precision and accuracy of

Table 4 ED iteration

Euclidean_0	Euclidean_1	Euclidean_2	Euclidean_3
32.42	56.72	45.7	53.98
26.72	57.15	45.82	59.56
20.95	51.23	54.61	66.33
18.64	56.53	49.55	67.47
16.79	56.24	69.67	89
22.84	61.94	42.17	64.43
20.12	55.34	50.09	66.05
21.05	57.72	74.58	94.05
16.13	65.48	69.61	96.85
26.58	60.31	42.68	60.43
5.09	67.9	58.19	90.38
30.1	55.16	83.51	98.27
12.28	59.19	65.45	88.24
29.79	66.2	36.41	60.16
16.17	68.03	40.43	73.42
11.64	63.73	65.38	92.25
18.15	49.31	61.61	73.86
19.08	58.66	72.67	93.41

Table 5 MOI for each iteration

Weight_0	Weight_1	Weight_2	Weight_3
402.70	864.34	573.94	677.28
451.68	828.12	465.08	712.78
482.04	759.27	418.74	783.75
485.56	722.70	416.55	824.84
485.56	723.28	416.55	827.68
485.56	723.28	416.55	827.68
485.56	723.28	416.55	827.68
485.56	723.28	416.55	827.68
485.56	723.28	416.55	827.68
485.56	723.28	416.55	827.68
485.56	723.28	416.55	827.68
485.56	723.28	416.55	827.68
485.56	723.28	416.55	827.68
485.56	723.28	416.55	827.68
485.56	723.28	416.55	827.68
485.56	723.28	416.55	827.68
485.56	723.28	416.55	827.68
485.56	723.28	416.55	827.68
485.56	723.28	416.55	827.68
485.56	723.28	416.55	827.68

k-means clustering are given in Tables 3, 4 and 5 in terms of iteration, Euclidean distance calculation and MOI, respectively. Clusters are settled with a minimum of iteration, even for highly multidimensional data.

6 Future Scope

The primary objective of future scope is to point out the drawbacks of the proposed approach. The author proposed k-means clustering techniques for carrier samples grouping and use of these groups for data hiding to reduce difference errors between the original and resultant carrier object. However, there are some limitations to k-means clustering in its processing only of two-dimensional data. Even though the input data is multidimensional, the k-means clustering technique accepts only two highly important parameters based on which the classification process takes place. An intention is required to modify existing k-means clustering to include multiple parameters in centroid calculation as well as with MOI. Future work is required to remove loosely coupled samples with clusters without significant loss in the process of data hiding.

References

1. Abdul Nazeer, K. A., & Sebastian, M. P. (2009). Improving the accuracy and efficiency of the K-means clustering algorithm. In *Proceedings of the World Congress on Engineering 2009* (Vol. I) WCE 2009, July 1–3, 2009, London, UK.

2. Agrawal, R., & Phatak, M. (2012. Document clustering algorithm using modified K-Means. In *Fourth international conference on advances in recent technologies in communication and computing (ARTCom2012)*.

3. Mato, F., & Toulkeridis,T . (2017). An unsupervised K-Means based clustering method for geophysical post-earthquake diagnosis. In *2017 IEEE Symposium Series on Computational Intelligence (SSCI)*.

4. Yang, Q., Liu, Y., Zhang, D., & Liu, C. (2011). Improved K-Means algorithm to quickly locate optimum initial clustering number K. In *Proceedings of the 30th Chinese control conference*.

5. Kapil, S., Chawla, M., & Ansari, M. D. (2016). On K-Means data clustering algorithm with genetic algorithm. In *2016 fourth international conference on Parallel, Distributed and Grid Computing (PDGC)*.

6. Shah, S., & Singh, M. (2012). Comparison of a time efficient modified K-mean algorithm with K-Mean and K-Medoid algorithm. In *2012 IEEE international conference on communication systems and network technologies*.

7. Gour, B., Bandopadhyaya, T. K., & Sharma, S. (2008). ART neural network based clustering method produces best quality clusters of fingerprints in comparison to self organizing map and K-Means clustering algorithms. In *2008 IEEE international conference on innovations in information technology*.

8. https://en.wikipedia.org/wiki/Moment-of-inertia

9. Li, W., Zhou, Y., & Xia, S. (2007). A novel clustering algorithm based on hierarchical and K-Means clustering. In *2007 IEEE Chinese control conference*.

10. Sapkota, N., Alsadoon, A., Prasad, P. W. C., & Elchouemi, A. (2019). Data summarization using clustering and classification: Spectral clustering combined with K-Means using NFPH. In *IEEE 2019 international conference on Machine Learning, Big Data, Cloud and Parallel Computing (COMITCon)*.

11. Alvarez, M. A. Z., Agbossou, K., Cardenas, A., Kelouwani, S., & Boulon, L. (2020). Demand response strategy applied to residential electric water heaters using dynamic programming and K-Means clustering. In *2020 IEEE transactions on sustainable energy*.

12. Yang, M.-S., & Sinaga, K. P. (2019). *A feature-reduction multi-view K-Means clustering algorithm*. IEEE Access.

13. Wang, X.-D., Chen, R.-C., Yan, F., Zeng, Z.Q., & Hong, C.-Q. (2019). *Fast Adaptive K-Means (FAKM) subspace clustering for high-dimensional data*. 2019 IEEE Access.

14. Cheng, P., Wang, Y., Yao, B., Huang, Y., Lu, J., & Peng, Q. (2020). Cyber security situational awareness jointly utilizing ball K-Means and RBF neural networks. In *2020 17th International Computer Conference on Wavelet Active Media Technology and Information Processing (ICCWAMTIP)*.

15. Wu, W., Liu, J., Wang, H., Hao, J., & Xian, M. (2021). Secure and efficient outsourced k-means clustering using fully homomorphic encryption with ciphertext packing technique. *IEEE Transactions on Knowledge and Data Engineering, 33*(10).

16. Liu, W., Yuan, L., Wang, S., & Liu, C. (2020). Research on fault diagnosis method of control moment gyroscope based on K-means algorithm. In *2020 39th Chinese Control Conference (CCC)*.

Noisy QR Code Smart Identification System

Ahmad Bilal Wardak, Jawad Rasheed, Amani Yahyaoui, and Mirsat Yesiltepe

Abstract The resurrection of the quick-response (QR) code has been made possible by the expansion of mobile network coverage combined with a rise in smartphone online content over the years. They have become much more accessible by integrating a code reader in smart devices, thus removing several unpleasant procedures and providing faster access to crucial information. However, noise in the printed images is unavoidable owing to printer processes and restricted printing technology, thus may decrease the quality of a QR code image during digital image collection and transmission which may eventually cause failure while scanning and extracting actual information. As a result, this study proposes an intelligent image classification strategy to correctly identify noisy and original QR code types. For this, a new dataset is built, containing 20,000 images pertaining to the original QR code and noisy QR codes. Later, the study exploited three well-known machine learning algorithms (logistic regression (LG), support vector machine (SVM), and convolutional neural network (CNN)) to segregate noisy images among original QR code images. The experimental results show that SVM outperformed others by attaining an overall performance accuracy of 97.5%, precision of 97.50%, recall of 97.5%, and F1-score of 97.5%, while LG almost competes by achieving 97.25% accuracy, 97.31% precision, 97.22% recall, and 97.25% F1-score.

A. B. Wardak
Department of Software Engineering, Istanbul Aydin University, Istanbul, Turkey
e-mail: ahmadwardak@stu.aydin.edu.tr

J. Rasheed (✉)
Department of Computer Engineering, Istanbul Aydin University, Istanbul, Turkey
e-mail: jawadrasheed@aydin.edu.tr

A. Yahyaoui
Department of Software Engineering, Istanbul Sabahattin Zaim University, Istanbul, Turkey
e-mail: amani.yahyaoui@izu.edu.tr

M. Yesiltepe
Department of Mathematical Engineering, Yildiz Technical University, Istanbul, Turkey
e-mail: mirsaty@yildiz.edu.tr

© The Author(s), under exclusive license to Springer Nature Singapore Pte Ltd. 2023
S. Shakya et al. (eds.), *Sentiment Analysis and Deep Learning*,
Advances in Intelligent Systems and Computing 1432,
https://doi.org/10.1007/978-981-19-5443-6_35

Keywords Noisy images · Quick-response code · Noise identification · CNN ·
SVM · Logistic regression

1 Introduction

A quick-response (QR) code is a machine-readable optical label that carries information about the associated item or product [1]. Information in barcodes is coded in just one direction or dimension. In contrast, information in a two-dimensional code, such as the QR code, is coded in two directions, horizontally and vertically. It is simple to read and can store a large amount of information. Denso Wave Incorporated in Japan invented the QR code in 1994 [2]. From then on, it was widely used as an identifying mark for a wide range of commercial items, marketing, and public notices. However, noise in the printed images is unavoidable owing to printer processes and restricted printing technology.

Various noises, such as Gaussian, salt and pepper, Poisson, speckle, and many more, may decrease the quality of a QR code image during digital image collection and transmission [3]. These noises are caused by poor environmental circumstances such as compression, a short focal length, incorrect memory allocation, and other issues. Noisy images and QR codes are detrimental elements in the training of models, lowering the networks' classification performance that's why some effective methods for distinguishing between original and noisy QR codes are required.

Machine learning, a popular topic in the technology business nowadays [4], and deep learning, a subfield of machine learning [5] and one of the biggest accomplishments and research hotspots lately, approaches and algorithms are very accurate for computer vision applications. As a deep neural network, convolutional neural network (CNN) learns feature detection by alternating between tens or hundreds of hidden layers, allowing us to stack deep layers and handle a wide range of image input properties [6]. With each layer, the complexity of the acquired properties rises, and it is a potent image categorization strategy [7, 8].

Support vector machine (SVM) is well-known pattern recognition and image classification technique [9, 10]. Based on a kernel function, SVM generates the best separating hyperplanes. Each data item in the SVM technique is plotted as a point in n-dimensional space, where n is the number of features, and the value of each feature is the value of a certain coordinate. In terms of binary classification and prediction, logistic regression has been extensively employed as a comprehensive data processing approach [11]. Logistic regression is mostly used to classify data [12]. Its data points are not organized in rows. It might be a lot, which is a pile, with each pile representing a category and each kind of data point having the same category name.

Various studies have attempted to categorize noise in images to the best of our knowledge. For instance, paper [13] used two different convolutional neural networks, VGG-16 and Inception-v3, to automatically identify noise distributions, whereas the paper [14] demonstrated how to recognize image noise of various types and intensities using a CNN method, as well as a backpropagation algorithm and stochastic gradient descent optimization approaches. Tripathi [15] created, implemented, and evaluated a CNN-based classifier to detect noisy pictures with high validation and training accuracy, as well as a UNET-based model to denoise images with ideal peak-signal-to-noise ratio and structural similarity index measure values.

Geng et al. evaluate image quality evaluation methodologies, fitting curves, mean opinion score, and creating two neural networks to provide an image noise level classification strategy for diverse application settings [16]. Momeny et al. [17] presents a noise-robust CNN for classifying noisy images without any pre-processing for noise reduction and to improve CNN classification performance for noisy images. Furthermore, the proposed CNN does not need any pre-processing for noise reduction, which speeds up the classification of noisy images.

When the user scans the printed visual QR code, it is accompanied by a noise phenomenon that interferes with recognition and causes failure. As a result, we are looking into constructing an intelligent image classification strategy for noisy and original QR code types. Although the capacity to minimize noise is crucial, it is equally important to determine if the QR code image contains noise or is an original one. We suggested CNN, logistic regression (LG), and SVM-based models, to successfully identify the original and noisy QR code images to overcome this problem. Following are the contributions of this study:

- Provides detailed literature work to show that no noisy QR dataset exists.
- Generated new dataset noisy QR code image dataset comprised of 20,000 images.
- Exploited four different noises while generating image.
- Analyzed three machine learning classifiers to distinguish noisy image and actual image.

The paper is further divided into sections as follows: The suggested technique is discussed in Sect. 2. Section 3 discusses the experimental findings as well as the comparison/discussion. Section 4 describes the conclusion.

2 Proposed Methods

The purpose of this research is to create a classification model that takes QR code images as input and classifies them as original or noisy QR codes. This paper develops its own dataset of QR code images and added several noises (Gaussian, Poisson, speckle, and salt and paper) to the generated images. The study's goal is to analyze generated QR code images, resize images, map noise to the original QR code images, encode the labels, train the proposed CNN, LG, and SVM models, detect the type of QR code image, and output the classified class/label whether it is the original QR code image or has noise.

Fig. 1 Sample of dataset images; the original image with four types of mapped noise as noisy images

2.1 Data Analyzation

The dataset we are using for this work includes 20,000 images of the original QR code and noisy ones. The dataset is highly varied; as our models take a fixed-size input that's why we pre-processed the images to resize them into a fixed size of (160 × 160) ratio for the CNN model and (50 × 50) for LG and SVM models. Figure 1 shows the original image and different noise types to the original image. Our dataset is divided into two distinct classes, the original one and the noisy one which we generated by adding four different types of noises such as Gaussian, salt and pepper, Poisson, and speckle to the original type.

2.2 CNN Model Architecture

We created a CNN model to classify the original and noisy QR code images. The developed CNN network architecture is depicted in Table 1. It is made up of four convolutional layers, four pooling layers, one dropout layer, one flatten layer, and a fully connected layer.

Table 1 Proposed CNN model network topology

Layers	Output shape	Parameters
Conv2D	(None, 158, 158, 16)	448
MaxPooling2D	(None, 79, 79, 16)	0
Conv2D	(None, 77, 77, 32)	4640
MaxPooling2D	(None, 38, 38, 32)	0
Conv2D	(None, 36, 36, 64)	18,496
MaxPooling2D	(None, 18, 18, 64)	0
Conv2D	(None, 16, 16, 128)	73,856
MaxPooling2D	(None, 8, 8, 128)	0
Flatten	(None, 8192)	0
Dropout	(None, 8192)	0
Dense	(None, 2)	16,386

Total parameters: 113,826
Optimizer: Adam
Epochs: 300
Batch size: 90
Loss: categorical_crossentropy
Metrics: accuracy

Table 2 Hyperparameteric values of the proposed logistic regression model

CV	Number of jobs	Random state
10	-1	1234

Max. iteration: 1000
Solver: liblinear
Class weight: balanced

2.3 LG Model Architecture

In the LG before feeding data to the model, we first resized and reshaped our data, then used the standard scaler to resize the distribution of values so that the mean of the observed values is 0 and the standard deviation is 1. That means it removes the mean and scales each feature/variable to unit variance. The developed logistic regression network architecture and hyperparameters tunning fitting tenfold cross-validation for each candidate, equating ten fits with parallel (-1) number of jobs using backend LokyBackend with eight concurrent workers are depicted in Table 2.

2.4 SVM Model Architecture

In the SVM model, we apply some pre-processing of data like resizing it to (50×50) ratio and shuffling the data to reorganize the order of the items. The developed SVM

Table 3 Proposed SVM model parametric values

C	Gamma	Kernel
1	Auto	poly

network architecture is depicted in Table 3. In the experimental setup for SVM, C (the penalty parameter of the error term) is set to 1 in order to control the error while gamma is auto-tuned to provide the decision boundary curvature weight. Moreover, the poly kernel is used to express the similarity of training samples in a feature set over polynomials of the original variables, enabling nonlinear models to be learned.

3 Result and Discussion

To evaluate the model accuracy, we used the dataset which contains 20,000 images of QR codes with four types of noises. The dataset was divided into the train and test sets with ratios of 70/30 correspondently. More details about the dataset split can be found in Table 4.

After successful training, the accuracy was computed using all images from the test dataset in each iteration. Figures 2 and 3 show the training accuracy and loss of the proposed CNN model. Our models' usefulness and performance were evaluated using four metrics: accuracy, precision, recall, and F1-score. Accuracy, precision, recall, and F1-score for two-class classification problem are measured by (1)–(4).

$$\text{precision} = \frac{tp}{fp + tp} \tag{1}$$

$$\text{recall} = \frac{tp}{fn + tp} \tag{2}$$

$$\text{F1-score} = \frac{2 * \text{recall} * \text{precision}}{\text{recall} + \text{precision}} \tag{3}$$

Table 4 QR code images dataset information

Label/class	Training set	Testing set	Total
Original type			
Normal/original QR	7000	3000	10,000
Noisy type			
Salt and pepper	1750	750	2500
Speckle	1750	750	2500
Poisson	1750	750	2500
Gaussian	1750	750	2500
Total	14,000	6000	20,000

$$accuracy = \frac{tp + tn}{fp + fn + tp + tn} \qquad (4)$$

Table 5 lists the performance of proposed models on test data in terms of various metrics. The confusion matrix in test data for all QR code images CNN, LG, and SVM models is depicted correspondingly in Figs. 4, 5 and 6.

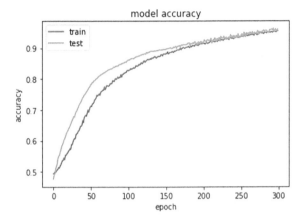

Fig. 2 Accuracy increment of training and validation data

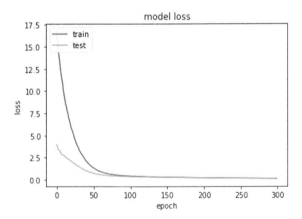

Fig. 3 Loss decrement of training and validation data

Table 5 Performance metrics of proposed CNN, SVM, and logistic regression models

Model	Accuracy (%)	Precision (%)	Recall (%)	F1-score (%)
CNN	95.95	96.1	96	96
SVM	97.50	97.50	97.50	97.50
Logistic regression	97.25	97.31	97.22	97.25

Fig. 4 Confusion matric of proposed CNN models with the test data

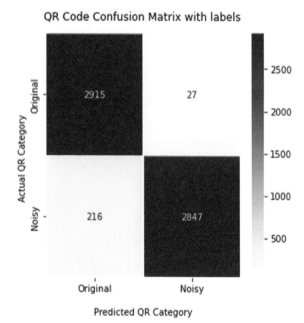

Fig. 5 Confusion matric of proposed LG model with the test data

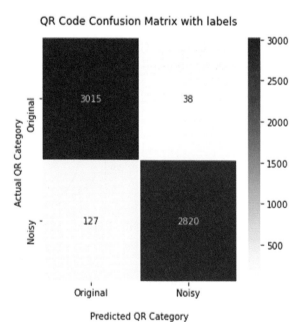

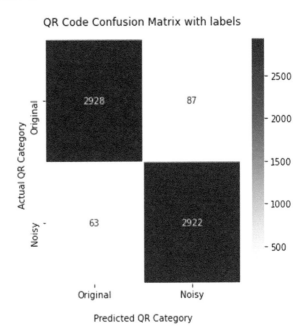

Fig. 6 Confusion matric of SVM model with the test data

4 Conclusion

When a user scans a printed visual QR code, generally it is accompanied by a noise phenomenon that interferes with recognition and causes failure. Therefore, the study aims to develop an intelligent image classification strategy to segregate the noisy and original QR code types. Although the capacity to minimize noise is crucial, it is equally important to determine if the QR code image contains noise or is an original one. This study suggested CNN, LG, and SVM-based models, to successfully identify the original and noisy QR code images to overcome this problem. To achieve this, the study developed a new dataset of QR code images that also contains various noises. Results show that SVM and LG achieved promising results by securing 97.5% and 97.25% accuracies, respectively. For proposed models, specially for CNN, all images must be of fixed (same) size. This limitation can be overcome in next study. In the future, we aim to exploit other machine learning classifiers along with other computer vision techniques to propose a classification system that can also identify the type of noise.

References

1. Chang, J. H. (2014). An introduction to using QR codes in scholarly journals. *Sci Ed, 1*, 113–117. https://doi.org/10.6087/kcse.2014.1.113

2. Chen, J., Huang, B., Mao, J., & Li, B. (2019). A novel correction algorithm for distorted QR-code image. In *2019 3rd International conference on electronic information technology and computer engineering (EITCE)* (pp 380–384). IEEE. https://doi.org/10.1109/EITCE47263. 2019.9095073

3. Hosseini, H., Hessar, F., & Marvasti, F. (2015). Real-time impulse noise suppression from images using an efficient weighted-average filtering. *IEEE Signal Processing Letters, 22,* 1050–1054. https://doi.org/10.1109/LSP.2014.2381649

4. Alzubaidi, L., Zhang, J., Humaidi, A. J., Al-Dujaili, A., Duan, Y., Al-Shamma, O., Santamaría, J., Fadhel, M. A., Al-Amidie, M., & Farhan, L. (2021). Review of deep learning: Concepts, CNN architectures, challenges, applications, future directions. *Journal of Big Data, 8,* 53.

5. Al-Saffar, A. A. M., Tao, H., & Talab, M. A. (2017). Review of deep convolution neural network in image classification. In *2017 International conference on radar, antenna, microwave, electronics, and telecommunications (ICRAMET)* (pp. 26–31). IEEE. https://doi.org/10.1109/ICR AMET.2017.8253139

6. Yim, J., & Sohn, K. -A. (2017). Enhancing the performance of convolutional neural networks on quality degraded datasets. In *2017 International conference on digital image computing: techniques and applications (DICTA)* (pp. 1–8). IEEE. https://doi.org/10.1109/DICTA.2017. 8227427

7. Rasheed, J., Hameed, A. A., Djeddi, C., Jamil, A., & Al-Turjman, F. (2021). A machine learning-based framework for diagnosis of COVID-19 from chest X-ray images. *Interdisciplinary Sciences: Computational Life Sciences, 13,* 103–117. https://doi.org/10.1007/s12539-020-00403-6

8. Rasheed, J. (2021). A sustainable deep learning based computationally intelligent seafood monitoring system for fish species screening. In *2021 International conference on artificial intelligence of things (ICAIoT)* (pp. 1–6). IEEE. https://doi.org/10.1109/ICAIoT53762.2021. 00008

9. Yahyaoui, A., Jamil, A., Rasheed, J., & Yesiltepe, M. (2019). A decision support system for diabetes prediction using machine learning and deep learning techniques. In *2019 1st International informatics and software engineering conference (UBMYK).* (pp. 1–4). IEEE. https://doi.org/10.1109/UBMYK48245.2019.8965556

10. Thai, L. H., Hai, T. S., & Thuy, N. T. (2012). Image classification using support vector machine and artificial neural network. *International Journal of Information Technology and Computer Science, 4,* 32–38. https://doi.org/10.5815/ijitcs.2012.05.05

11. Zou, X., Hu, Y., Tian, Z., & Shen, K. (2019). Logistic regression model optimization and case analysis. In *2019 IEEE 7th international conference on computer science and network technology (ICCSNT)* (pp. 135–139). IEEE. https://doi.org/10.1109/ICCSNT47585.2019.896 2457

12. Rasheed, J., Dogru, H. B., & Jamil, A. (2020). Turkish text detection system from videos using machine learning and deep learning techniques. In *2020 IEEE Third international conference on data stream mining & processing (DSMP)* (pp. 116–120). IEEE. https://doi.org/10.1109/ DSMP47368.2020.9204036

13. Sil, D., Dutta, A., & Chandra, A. (2019). Convolutional neural networks for noise classification and denoising of images. In *TENCON 2019—2019 IEEE region 10 conference (TENCON)* (pp. 447–451). IEEE. https://doi.org/10.1109/TENCON.2019.8929277

14. Khaw, H. Y., Soon, F. C., Chuah, J. H., & Chow, C. (2017). Image noise types recognition using convolutional neural network with principal components analysis. *IET Image Processing, 11,* 1238–1245. https://doi.org/10.1049/iet-ipr.2017.0374

15. Tripathi, M. (2021). Facial image noise classification and denoising using neural network. *Sustainable Engineering and Innovation, 3,* 102–111. https://doi.org/10.37868/sei.v3i2.id142

16. Geng, L., Zicheng, Z., Qian, L., Chun, L., & Jie, B. (2020). Image noise level classification technique based on image quality assessment. In *2020 IEEE international conference on power, intelligent computing and systems (ICPICS)* (pp. 651–656). IEEE. https://doi.org/10.1109/ICP ICS50287.2020.9202118

17. Momeny, M., Latif, A. M., Agha Sarram, M., Sheikhpour, R., & Zhang, Y. D. (2021). A noise robust convolutional neural network for image classification. *Results in Engineering, 10,* 100225. https://doi.org/10.1016/j.rineng.2021.100225

Voice and Text-Based Virtual Assistant for Academic Advising Using Knowledge-Based Intelligent Decision Support Expert System

D. Gnana Rajesh, G. Tamilarasi, and Mohammed Ehmer Khan

Abstract Academic advising in universities and reputed institutions serve as a roadmap in guiding the students for framing their career plan. One of the major roles and responsibilities of an academic staff is to guide the students properly in selecting the course as per the rules stipulated in bylaws. The staff members face many challenges while performing the role as an advisor. The promising challenge for an advisor is balancing the academic workload and carrying out the advisory role. One of the major constraint for the staff members is that they were not able to provide immediate answers to the queries raised by the stakeholders. The main objective of the research was intended to address this problem. Hence, the voice and text-based virtual assistant for academic advising using knowledge-based intelligent decision support expert system is used. It is a computerized program that is implemented as a conversational chatbot that is designed to handle the queries raised by the students related to academic advisement, clarification on college bylaws, progression, moving to next level, and in choosing the specialization.

Keywords Advising · Chatbot · Virtual assistant · Academic advisors

1 Introduction

Academic advising is the procedure between the student and an academic advisor for understanding the significance of overall learning, reviewing the organization's facilities and rules, discussing the educational and job ideas, and making suitable course choices [1].

Which are the best elements that encourage student achievement in higher education? [2]. Habley [3] stated that the "key contributor to the student retention and

D. Gnana Rajesh (✉) · G. Tamilarasi · M. E. Khan
Department of IT, University of Technology and Applied Sciences, Al Mussanah, Sultanate of Oman
e-mail: dgnanarajesh@gmail.com

M. E. Khan
e-mail: ehmer@act.edu.om

© The Author(s), under exclusive license to Springer Nature Singapore Pte Ltd. 2023   483
S. Shakya et al. (eds.), *Sentiment Analysis and Deep Learning*,
Advances in Intelligent Systems and Computing 1432,
https://doi.org/10.1007/978-981-19-5443-6_36

progression is based on the quality of interaction between a student and a concerned individual staff member on campus, often through academic advising. However, the influence of academic advising on student achievement has been largely overshadowed by attempts to assess student satisfaction with the advising process." While the interaction between staff and students is connected to their academic aims and results, student development is opinionated by a range of corresponding institutional hard work outside of course-related links with faculty [1].

Academic advising is considered to be the important of every academic staff member. An average of 25 advisees is assigned to each advisor (academic staff member). During their study, the students are having various queries periodically related to academic advisement, clarification on college bylaws, progression, moving to the next level, and choosing the specialization, etc., as the student and advisor (academic staff) are unable to meet regularly due to the different scheduled class hours, the students are unable to get clarified when needed.

The main objective of human–computer interaction building systems is to plan normal and instinctive linking modalities and generate safe and functional computer systems with good usability. With the latest technology, it is possible with the help of connecting things where every device is connected and communicable. In order to do so, we need to understand the variables that determine how students of University of Technology and Applied Sciences-Al Mussanah (UTAS-A) use technology and use this knowledge to develop tools and techniques that will enable us to build efficient systems.

Computer-based chatbots are becoming distinctly famous as an instinctual and successfully exposed framework between humans and types of machinery. A chatbot is a software application used to conduct an online one-to-one conversation via text or text-to-voice. It is man-made stuff that is projected to reproduce an intelligent conversation with human accomplices using their own language. Currently, chatbots are used by a lot of clients to intercede access to info or knowledge bases and furthermore to do nonspecific consultations [4].

2 Literature Review

The intention of a chatbot is to provide an effective and precise response to any question based on the dataset of frequently asked questions by using AI mark-up language and latent semantic analysis. "The general questions like welcome/greetings and general questions will be responded using AI mark-up language, and other service-based questions use latent semantic analysis to provide responses at any time that will serve user satisfaction" [4]. "The expert system is capable of advising students using prescriptive advising model and developmental advising model. The system is supported by an object-oriented database and provides a friendly graphical user interface. The system can also be used by academic advisors in their academic planning for students" [4]. Soybean pest expert system [5], a fuzzy-based expert system, asks queries in order to produce a fuzzy-based web-based answer for the user. Through the feedback received from the experts, this system is gradually becoming more accurate. Dhaya, [6] conducted works incorporating related virtual reality to mitigate low-quality images with minimal processing speed in the VR environment. Smys and Haoxiang suggested a classification system distinguish between human users and chatbot in the study. The speed and accuracy of the entropy classifier have been tested.

To handle the assessment for the families, the conversational chatbot is who needs authorized supervision regarding family cases such as child custody and legal separation. Parsing algorithms known in natural language processing is the method implemented in this research work [7]. "It is suitable for ambiguous grammar and uses the dynamic programming approach—partial hypothesized results are stored in a structure called a chart and can be re-used." The benefits and challenges of a web-based fuzzy expert system (FES) [8] are associated with the design, development, and use from a standpoint of the benefits and challenges of developing and using them. This approach is to construct a web-based fuzzy expert system for student education counselors and stays on the university grounds.

Chatbots are nowadays very popular in commercial as they can decrease customer service charge and handles any number of users at the same time with around-the-clock availability, reliability, and convenience. A smart virtual assistant for students that offers continuous instant support to the student, staff, and faculty members. Student counseling is an important activity in higher education institutes, yet it is a time-consuming effort in academic work. Through experimental exploration, this article reviews whether multiple sources of information are assisting academic advisors in providing more effective decisions for their students. Table 1 illustrated the comparison of existing chatbot and its related methodology.

Table 1 Comparison on existing chatbot system

Paper title	Chatbot for College website [9]	Q and A system for online feedable new-born chatbot	BANK CHAT BOT—an intelligent assistant system using natural language processing and machine learning [10]
Language processing technique used	Facebook messenger application programming interface	NLTK framework	NLTK framework
Chatbot system's target audience	University/college/education	Open-ended	Finance/banking
Knowledge type	Structured database	Not mentioned	Structured database
Machine learning algorithm used	Not mentioned	Not used	Support vector machine and random forest

2.1 Research Objectives

To extend the advising facilities available in UTAS-A to the fingertip of the student community and hence cater to the needs of students.

There are different types of student advising services available like advisors of the academic department and counselors of the student affairs department are maintained through email, in-person, and phone services. The current pandemic situation brought up the need for automation of all the mentioned support facilities provided by human-to-human. As there is a rise in the number of students per advisor, the waiting time for queries increases as well, hence an automated process would increase student satisfaction greatly. Microsoft Azure SQL Database: production environment will help in developing the proposed expert system.

The main aim of this project is to develop a functional chatbot with the following functionalities.

- Assistance on all basic advising-related queries from students.
- Assistance to know the grades required by students in the current level to move to the next level.
- Assistance to know the grades required by students in the current level to come out from probation.
- Assistance to predict the remaining semesters to be studied by students to complete studies.
- Assistance with specialization selection for students based on their scores in specific subjects.
- Assistance with specialization selection for students based on the current job market.
- Assistance with the prediction of grades in the current level based on their coursework marks.
- Assistance with the selection of courses that can be repeated to get the best grade improvement.
- Voice-based assistance on common FAQs and standard answers based on expert advice.

3 Research Methodology

The proposed model for constructing a knowledge-based system that wisely responds to queries enquired by students. The aim of this system is to create a worthy counsel based on information about advisement in order to decrease the delay of staff members answering student queries and to increase student satisfaction. The proposed model attempts to simulate the behavior of the staff speaking about the advisement-related issues.

The block diagram as shown in Fig. 1 illustrates the operational structure of the proposed system. The students will initialize the chatbot through the university portal

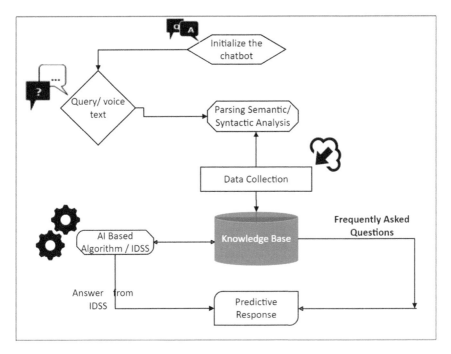

Fig. 1 Block diagram

and will seek recommendations related to advisement using text or voice messages. The system will parse the query to forecast the relative response created on the data stored in the database with the support of artificial intelligence techniques to provide a knowledge-based response.

An artificial intelligence-based algorithm is adopted to predict the response. At the initial stage, all possible data related to bylaws, amendments, and updates are converted into the form of quantitative data. The possible question and answers will be collected from the staff advisors through various data collection mechanisms like questionnaires and interviewing methods. Data will be collected from students at various levels and departments. All the collected data will be tested and stored in the database for responsive actions in the form of frequently asked questions (FAQ).

3.1 System Design Process

In this system design process section, Fig. 2 discusses the proposed classification model with the knowledge-based process and Fig. 3 describes the process diagram.

The above figure shows the proposed classification process model where the random forest classification will be used to classify the type of input to match the

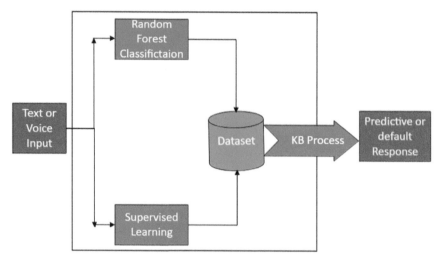

Fig. 2 Proposed classification model

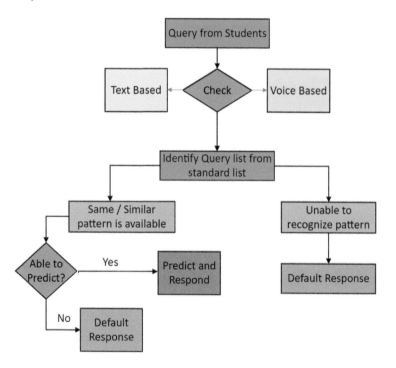

Fig. 3 Process diagram

stored dataset. Predictive or default responses will be generated using the artificial neural network-based supervised learning. To improve the response rate, the knowledge-based (KB) process will use random forest classification and supervised learning models.

The above process diagram describes the query requested from the students through a chatbot in text or voice. If it is a text query, it will find the matching from the standard list and will check for similar patterns, and the related recommendations will be given to the students. At the same time, if the request is based on voice, then it will be converted to text using voice recognition software and will find the matching from the standard list. The unsuccessful queries will be stored in the database, and the related response will be collected from experts and will be used for future purposes.

3.2 Data Accumulation

Data collection is the method of collecting and computing the data on variables of importance, in a recognized systematic manner that supports one to respond to stated research queries, test hypotheses, and calculate results. A good research design is capable of making research as efficient as possible, yielding maximal information with minimal expenditure of effort, time and money. Whenever the system responded with the default answer, the related query will be accumulated along with dataset collection for the upgrading process. The gathered information is a continuous process which will append as well as this will help the system for accuracy.

4 Research Significance

Artificial intelligence (AI) is a branch of computer science used to create intelligent machines that can work and interact with people just like intelligent human beings. AI can systematize the mission of administrative responsibilities for teachers and provide valuable responses to their students. The proposed system will help the academicians can easily deal with the academic advising rules and regulations immediately at any time. This system will decrease the common waiting time and queue size of the students as well as the academic advisors. The system also will guide the students to choose the specialization based on their earlier grades and their skills. Academic probation students and students who are academically at risk for various reasons like lack of attendance, postponement, withdrawal, etc., can also be benefited.

Artificial intelligence techniques are used to fetch the expert opinion based on the relevant request. As the system is online-based the students and even the parent's community will be benefited by getting clarifications, suggestions, and recommendations related to all academic advisement anywhere. The existing system study shows that most academic advising systems are text-based. Along with the text-based conversation, the proposed system highlights the concept of voice-based response for academic advisement queries.

5 Importance in Academics

The proposed system titled "Voice and Text-Based Virtual Assistant for Academic Advising Using Knowledge-Based Intelligent Decision Support Expert System (IDSES)" will help more than 30,000 students' population and hundreds of academic staff members/advisors so as to benefit the Oman community at large. The proposed system will help the students to know the university bylaws at any point of time. The university students will get benefited by choosing the correct specialization and the need for the GPA during their studies. The parents can also get clarified about the study plan of their children and the university rules and regulations. The academic advisor can extract the university bylaws and articles through the system 24×7. UTAS administration can expand the proposed to all other education institutions having similar bylaws.

6 Conclusion

Artificially intelligent chatbot uses AI expertise to impersonate human conversations and handle inputs from students via text and voice commands. It is used to support determinations, judgments, and courses of action by examining and analyzing massive amounts of data. The comprehensive information also helps the students to solve academic problems which can be used in decision-making. It also helps in answering all advising-related queries and responds with the best study plan/option logically possible. The proposed system provides assistance on all basic advising-related queries from students, current level to move to next level, come out from probation, score in specific subjects, specialization selection for students, prediction of grades, selection of courses for grade improvement, FAQs, and standard answers based on expert advice. Apart from answering student queries, this model also has the supporting software related to the academic advisement which helps to calculate the cumulative grade point average, and a broadcaster to update the changes of rules and regulations related to academic advisement.[11–14]

References

1. Kuh, G. (2001). Organizational culture and student persistence: Prospects and puzzles. *Journal of College Student Retention, 3*(1), 23–29.

2. Agreda, K. S. B., Fabito, E. D. B., Prado, L. C., Tebelin, M. H. G., & BenildaEleonor, V. (2013). Attorney 209: A virtual assistant adviser for family-based cases. *Journal of Automation and Control Engineering, 1*(3), 198–201.

3. Habley, W. R. (Ed.). (2004). *The status of academic advising: Findings from the ACT sixth national survey.* Monograph No. 10, National Academic Advising Association.

4. Ranoliya, B. R., Raghuwanshi, N., & Singh, S. (2017, September). Chatbot for university related FAQs. In *2017 International conference on advances in computing, communications and informatics (ICACCI)* (pp. 1525–1530). IEEE.

5. Al Ahmar, M. A. (2011). A prototype student advising expert system supported with an object-oriented database. *International Journal of Advanced Computer Science and Applications (IJACSA) Special Issue on Artificial Intelligence, 1*(3), 100–105.

6. Dhaya, R. (2020). Improved image processing techniques for user immersion problem alleviation in virtual reality environments. *Journal of Innovative Image Processing (JIIP), 2*(02), 77–84.

7. Goodarzi, M. H., & Rafe, V. (2012). Educational advisor system implemented by web-based fuzzy expert systems.

8. Saini, H. S., Kamal, R., & Sharma, A. N. (2002). Web based fuzzy expert system for integrated pest management in soybean. *International Journal of Information Technology, 8*(1), 55–74.

9. Ramesh, G. S., Nagaraju, G., Harish, V., & Kumaraswamy, P. (2021). Chatbot for college website. In *Proceedings of international conference on advances in computer engineering and communication systems* (pp. 511–521). Springer, Singapore.

10. Kulkarni, C. S., Bhavsar, A. U., Pingale, S. R., & Kumbhar, S. S. (2017). BANK CHAT BOT—An intelligent assistant system using NLP and Machine learning. *International Research Journal of Engineering and Technology (IRJET), 04*(05), 2374–2377.

11. Young-Jones, A. D., Burt, T. D., Dixon, S., & Hawthorne, M. J. (2013). Academic advising: Does it really impact student success? Quality Assurance in Education.

12. https://machias.edu/academics/faculty-handbook/section-iv-academic-pol/academic-adv-def/

13. Almutawah, K. A. (2014). A decision support system for academic advisors. *International Journal of Business Information Systems, 16*(2), 177–195.

14. Nagarhalli, T. P., Vaze, V., & Rana, N. K. (2020, March). A review of current trends in the development of chatbot systems. In *2020 6th International conference on advanced computing and communication systems (ICACCS)* (pp. 706–710). IEEE.

Breast Cancer Prediction Using Different Machine Learning Algorithm

Ujwala Ravale and Yashashree Bendale

Abstract In the Current era, due to explosion of population, early disease diagnosis has become a major concern medical research field. Due to breast cancer possibility of death is growing increasingly now days. From the previous analysis it has been disclosed that breast cancer has become second-most-severe disease as compare to other diseases. In early diagnosis and discovery of several diseases machine learning algorithm plays important role so that chance of patientâL™s survival will get increased. A reliable, effective, and quick reaction while also lowering the danger of death can be achieved using automated detection system which can be implemented using artificial intelligence or machine learning technique. In this research, we compare logistic regression, decision tree, random forest and K-nearest neighbor with the Neighbor Component Analysis (NCA) approach, which are all supervised machine learning techniques. In comparison with other algorithms, the automated system using KNN with NCA approach obtains high accuracy which is approximately 98.5%.

Keywords Machine learning · Knearest neighbor · Neighborhood component analysis · Random forest · Artificial intelligence

1 Introduction

Now a day's Breast cancer is becoming more prevalent among women. Breast cancer can be prevented if it is detected early. Women go through a variety of stages, and many of them have psychological issues such as despair, anxiety, and low self-esteem, which can lead to cancer. Even though there are effective therapies for breast cancer,

U. Ravale (✉) · Y. Bendale
SIESGST, Nerul, Navi Mumbai, India
e-mail: ujwala.ravale@siesgst.ac.in

Y. Bendale
e-mail: Yashashreebendale@gmail.com

© The Author(s), under exclusive license to Springer Nature Singapore Pte Ltd. 2023   493
S. Shakya et al. (eds.), *Sentiment Analysis and Deep Learning*,
Advances in Intelligent Systems and Computing 1432,
https://doi.org/10.1007/978-981-19-5443-6_37

the pain and suffering associated with the various treatment options is significant. Women experience depression after therapy as well.

It is a malignant or benign tumor which can grow uncontrollably in the breast. Breast cancer is caused by uncontrolled cell division in the breast [1]. Breast cancer risk is determined by age, genetic risk, and family history of the disease, as well as other variables such as obesity, lack of exercise, smoking, and eating unhealthy foods. Breast cancer is more common in women than in males.

According to experts, hormonal imbalance, lifestyle [2], and environmental variables have all been linked to an increased chance of developing BC. Gene mutations that have been passed down through the generations have been linked to about 5–6% of BC patients. Obesity, advanced age, and postmenopausal hormone imbalances are some of the other factors that cause breast cancer.

The diagnosis process becomes lengthy and time-consuming because early detection of Breast cancer has become challenging. Mammography and biopsies are used in medical diagnosis. If abnormalities are found on mammography, the patient must undergo biopsy, which is painful, time-consuming, and costly.

In this situation, an automation system would be really useful [3]. Analysis provided by the automated method can assist radiologists or other specialists in improving diagnosis accuracy. When an automated system is given a large enough data set, it can perform so effectively that a patient no longer needs to undergo unpleasant biopsies or other testing. The paper summarizes some of the most significant machine learning studies on breast cancer detection.

Data sets are classified in a variety of ways using Machine Learning algorithms. Machine learning algorithms are classified as Reinforcement Learning, Supervised, Unsupervised, Evolutionary Learning and Deep Learning Algorithms. Based on training examples supervised learning algorithm gives correct output based on this training set for all possible inputs. Supervised Learning takes the form of classification and regression. The aim of unsupervised learning technique is to identify similarities between the input data and then classifies the data based on these similarities. This research mainly focuses on to predict the accuracy of several machine learning algorithms on breast cancer data set.

Structure of this paper is arranged as explained in the following four sections. In Sect. 2, the previous work done by different authors and existing systems are presented. The multiple ML algorithms used for Breast cancer detection are explained in Sect. 3. The suggested proposed technique is explained in Sect. 4. In Sect. 5, the efficiency of proposed system are discussed in terms of different performance measurement parameters.

2   Literature Review

To design trained data model, predict or define decision rules ML algorithms should be capable to analyze any dataset. From input dataset and previous learned data the proposed system learns or knowledge will extracted in the form of rules or patterns to ensure that the ML algorithms functions intelligently without the need for human intervention.

Sajib et al. Proposed the machine learning model using XGBoost algorithm and Random Forest to identify breast cancer [4]. For analysis dataset which is used contains 275 samples and each sample has 12 features. Approximately 74.73% accuracy is achieved by applying Random forest algorithm and 73.63% is achieved using XGBoost algorithm.

Tsehay et al. introduced a model by applying support vector machine and decision tree algorithm which are important machine learning algorithm used for prediction purpose [5]. Using test dataset efficiency of the model is measured in terms of various evaluation parameters i.e. precision, recall as well as in terms of accuracy. In comparison with decision tree algorithm result had shown that SVM had performed better in the form of accuracy and precision and lowest misclassification rate. The SVM classification model shown 91.92% accuracy while decision tree model has an accuracy of 87.12%.

For prediction of breast cancer Shubham et al. Proposed NaÃ¯ve Bayes theorem, KNN and Random Forest algorithms. Wisconsin Diagnosis Breast Cancer (WBC) data set is used for comparison of various machine learning algorithms. Performance is measured using key parameters like precision and accuracy. The obtained results obtained are very useful for prediction as well as treatment [6].

Performance measurement is done using Machine learning techniques i.e. Support Vector Machine (SVM), Decision Tree (C4.5), Naive Bayes (NB) and K Nearest Neighbors (k-NN) are done by Hiba et al. [7] on Wisconsin Breast Cancer (original) dataset. From experimental results it can conclude that the highest accuracy 97.13% is achieved by SVM as well as lowest error rate.

3   Machine Learning Classifiers

The methodology of training machines (computers) to make decisions in similar cases based on data is known machine learning. Healthcare, object recognition, networks are various application areas where machine learning is used.

1. Logistic Regression: Supervised learning approach which includes larger number of dependent variables [8]. When there is a binary solution for a dependent variable, use of logistic regression is the better choice for analysis purpose. Logistic Regression, like all other forms of regression systems, is a type of predictive regression system. The association among one dependent binary variable which is dependant and one or more variables which are independent is analyzed with the help of logistic regression. The regression system which uses logistic regression generates output which varies from 0 to 1.

2. Decision Tree (DT): A recursive partition of the instance space is used to represent a decision tree. With the help of a predictive model node observations are mapped

into target value inferences. Generally, class labels are present at leaves and feature combinations that lead to class labels are present at branches in tree structure. CART, C4.5 [8] are different variations of decision tree methods.

3. Random Forest (RF): Classification as well as regression analysis both the tasks are accomplished with the help of Random Forest algorithm [5]. It is supervised machine learning algorithm. Prediction of new data samples are done based on the previous data sets so it can be considered as basic building block using machine learning algorithm.

4. Support Vector Machine: It is most popularly used classification algorithm that is supervised. For object classification, SVM builds a hyperplane. A line that differentiates between the two classes is known as hyperplane. To build a non-probabilistic binary classifier model [9], SVM train that classifier model so that new sample should be assigned to at least one of the two classes.

5. K Nearest Neighbor: In KNN classification technique, it takes a collection of labeled points and uses them to educate itself how to label more which are new. Then it finds closest point or its nearest neighbors to classify new point and asks them to vote [10].

4 Methodology

Machine learning applications are used in medical research for disease diagnosis and prediction. Machine learning tools can be used to analyze data values that can aid in disease prediction. Early prediction aids in better treatment. Using computerized computing implements and machine learning processes to promote and improve medical field investigation or diagnosis is a motivating and necessary operation. Generally, the real dataset which is used to train the proposed model is known as training dataset. The model is trained using the dataset that was used. A sample dataset used to validate the training model is referred to as test data. To process and analyze breast cancer dataset steps which are performed are as follows:

Stage 1: Preprocessing: Data missing values are handled during data preprocessing by replacing missing values with the attribute's mean value. Furthermore, the raw breast cancer data was scaled using the Standard Scaler module. Many machine learning estimators have a requirement for dataset standardization [11].

Stage 2: Features Selection: In preprocessing stage feature selection is important stage which should perform prior to actual learning. To achieve accurate results of prediction selective and important features should be targeted from dataset. To accomplish this we had implemented Neighbourhood component analysis (NCA) for feature selection. Python scikit-learn library is used for implementation of feature selection module.

Neighbourhood component analysis (NCA): (NCA) is a learning approach for a Mahalanobis distance measure that can be used in the KNN classification. NCA is a statistical method which is used improves accuracy of classification and regression algorithms [12]. This NCA method is combined with k Nearest neighbour Classifier

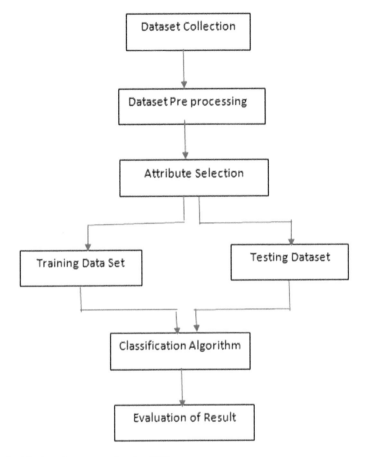

Fig. 1 Architecture for proposed system[17]

to calculate the distance measure. The proposed methodology uses NCA approach along with K Nearest Neighbour technique which is used to find nearest and similar data points within the dataset. For proposed model the block diagram is as shown in Fig. 1

5 Dataset Description

For experimental purpose, dataset form the kaggle is trained to create the supervised machine learning model for breast cancer prediction by applying the logistic regression, Decision tree, Random Forest Classifier and K Nearest Neighbor with NCA techniques are used. The dataset used for training contains benign and malignant traits. Breast radius mean, texture, perimeter mean, texture mean are few fea-

Fig. 2 Number of malignant
and benign data

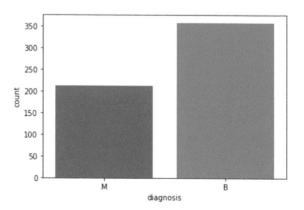

tures which are used for classification of breast cancer whether it is belonging to benign or malignant class. Dataset used for training purpose contains 569 instances of breast cancer as well as 30 attribute features for each instance. Approximately, 80% instances of dataset are used for training purpose and remaining 20% are used for testing the model. The main purpose of sklearn train_test_split function is arbitrarily partition the whole dataset into testing and training data samples which are used for analyze the model.

Dr. William H. Wolberg, a physician at the University of Wisconsin Hospital in Madison, Wisconsin, USA, prepared Wisconsin Breast Cancer Diagnostic (WBCD) dataset which are used in this study, which is freely available online. Dr. Wolberg collected samples from the patients which are in fluid samples with solid breast masses and a simple graphical computer application called Xcyt to prepare the dataset. For each cell nuclei the 10 features are calculated for WDBC dataset for each samples of the cells, which has been collected for dataset preparation along with that it computes standard error, mean and extreme value of each individual feature for the image, yielding a 30 real-valued feature vectors. In this dataset total there are 569 instances are available among this 213 are malignant and 256 are benign [13] which are shown in Fig. 2.

5.1 Attribute Explanation

Dataset which are used in this study contains different attributes like id number, Diagnosis (M = malignant, B = benign) and along with that from each imageâŁ™s cell nuclei ten attributes which are generated i.e. perimeter, symmetry, radius, concavity, texture, fractal dimension etc.

5.2 Experimental Environment

Spyder (Python), Jupyter Notebook (Python) for data analysis. Results included statistical and visualization. We analyze these results and predict the cancer disease (cancer susceptibility, recurrence, survival). We present the performance evaluation methods use to evaluate the proposed methods. Performance evaluation methods of cancer prediction or prognosis are the classification accuracy, analysis of sensitivity, specificity, TPR, FPR, TNR, FNR, confusion matrix. We analyze the accuracy and efficiency of the model and predict which model performs better for decision tree algorithms, KNN with Neighbourhood Component Analysis and Random Forest Algorithm.

6 Results and Discussions

Performance Measurement Parameters:

1. Accuracy—Accuracy is the most basic performance metric parameter. Accuracy is defined as the ratio of correctly predicted samples to the total number of samples[14].

$$\text{Accuracy} = TP + TN/TP + FP + FN + TN \qquad (1)$$

2. Precision—Precision is defined as the proportion of accurately predicted positive observations to the total predicted positive observations [15].

$$\text{Precision} = TP/TP + FP \qquad (2)$$

3. Recall (Sensitivity)—The ratio of accurately predicted positive observations to the total number of observations in the class is known as recall [15].

$$\text{Recall} = TP/TP + FN \qquad (3)$$

4. F1 score—Weighted average ratio of Precision and Recall is known as F1 Score.

$$\text{F1 Score} = 2 * (\text{Recall} * \text{Precision})/(\text{Recall} + \text{Precision}) \qquad (4)$$

where

TP True Positive
FP False Positive
TN True Negative
FN False Negative

Table 1 Algorithm comparison on precision

Algorithm	Logistic regression (%)	Decision tree (%)	Random forest classifier (%)	KNN with NCA (%)
Benign	97	98	98	99
Malignant	93	88	94	96.5
Average	95	93	96	97.75

Table 2 Algorithm comparison based on recall

Algorithm	Logistic regression (%)	Decision tree (%)	Random forest classifier (%)	KNN with NCA
Benign	96	92	97	98
Malignant	94	96	96	96
Average	95	94	96.5	97

Table 3 Algorithm comparison based on F1-score

Algorithm	Logistic regression (%)	Decision tree (%)	Random forest classifier (%)	KNN with NCA (%)
Benign	96.4	94.9	97.4	98.49
Malignant	93.4	91.82	94.9	96.24
Average	95	93.49	96.2	97.37

6.1 Implementation and Result Analysis

Tables 1 and 2 presents the precision and recall of machine learning classifiers. The highest level precision and recall is found in KNN with NCA technique as compared to other algorithms (Table 3).

6.2 Accuracy Model

Tables 4 and 2 shows accuracy for different machine learning algorithm for training and testing dataset (Fig. 3).

7 Conclusion

In the world, one of the most prevalent malignant tumours are BCS in many women, and now a days it is more commonly found in younger women. As a result, greater research scope is available in early detection of BCs; also, many of the charac-

Table 4 Accuracy of the classifiers

Algorithm	Training accuracy (%)	Testing accuracy (%)
Logistic regression	97	96
Decision tree	96	95
Random forest classifier	95	96
KNN with NCA	97	98

Fig. 3 Accuracy for different classifiers

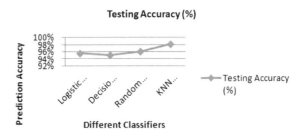

teristics of this cancer remain unknown. In this research analysis is done On the WBCD dataset by applying machine learning techniques like K Nearest Neighbor with Neighborhood Component Analysis, Logistic Regression, Decision Tree and Random Forest Classifier. Performance evaluation of various models is measured in terms of various parameters like confusion matrix, precision and recalls identifying the best machine learning algorithm. After implementation, we found that K nearest Neighbor with NCA performs better as compared to other models. As comparison with other machine algorithms the K nearest neighbor with NCA showed precision 97.5% and accuracy around 97% which shown better performance.

References

1. Priyanka, K. S. (2020). A review paper on breast cancer detection using deep learning. In *IOP Conference Series: Materials Science and Engineering1st International Conference on Computational Research and Data Analytics (ICCRDA 2020)*.
2. Vijayakumar, T. (2019). Neural network analysis for tumor investigation and cancer prediction. *Journal of Electronics and Informatics, 01*(02).
3. Monirujjaman Khan, M., Islam, S., Sarkar, S., Ayaz, F. I., Ananda, M. K., Tazin, T., Albraikan, A. A. & Almalki, F.A. (2022). Machine learning based comparative analysis for breast cancer prediction. *Journal of Healthcare Engineering, Hindwi, 2022*.
4. Kabiraj, S., Raihan, M., Alvi, N., Afrin, M., Akter, L., Sohagi, S. A., & Podder, E. (2020). Breast cancer risk prediction using XGBoost and random forest algorithm. In *11th ICCCNT, IEEE Explorer, 2020*.
5. Assegie, T. A., & Sushma, S. J. (2020). A support vector machine and decision tree-based breast cancer prediction. *International Journal of Engineering and Advanced Technology (IJEAT), 9*(3). ISSN: 2249-8958.

6. Sharma, S., Aggarwal, A., & Choudhury, T. (2018). Breast cancer detection using machine learning algorithms. In *International Conference on Computational Techniques, Electronics and Mechanical Systems (CTEMS)*, IEEE.
7. Asria, H. Mousannif, H., Al Moatassime, H., & Thomas, N. (2016). Using machine learning algorithms for breast cancer risk prediction and diagnosis. In *The 6th International Symposium on Frontiers in Ambient and Mobile Systems (FAMS 2016), Procedia Computer Science*, vol. 83.
8. Fatima, N., liu, L., Hong, S., & Ahmed, H. (2020). Prediction of breast cancer, comparative review of machine learning techniques, and their analysis. IEEE Access.
9. Rane, N., Sunny, J., Kanade, R., & Devi, S. (2020). Breast cancer classification and prediction using machine learning. *International Journal of Engineering Research and Technology (IJERT), 9*(02).
10. Amine Naji, M., El Filalib Kawtar Aarikac, S., Benlahmard, E. L. H., Abdelouhahide, R. A., & Debauchef, O. (2021). Machine learning algorithms for breast cancer prediction and diagnosis. *International Workshop on Edge IA-IoT for Smart Agriculture (SA2IOT) August 9–12*. Belgium, Leuven.
11. Rawal, R. (2020). Breast cancer prediction using machine learning. *JETIR, 7*(5).
12. Laghmati, S., Cherradi, B., Tmiri, A., Daanouni, O., & Hamida, S. (2020). Classification of patients with breast cancer using neighbourhood component analysis and supervised machine learning techniques. In *3rd International Conference on Advanced Communication Technologies and Networking (CommNet)*, IEEE.
13. Telsang, V. A., & Gurubasava, B. (2020). Breast cancer prediction analysis using machine learning algorithms. In *International Conference on Communication, Computing and Industry 4.0 (C2I4)*, IEEE.
14. Vaka, H. R., Soni, B., & Sudheer Reddy, K. (2020). Breast cancer detection by leveraging machine learning. *ICT Express, 6*(4).
15. Hamed, S., Mesleh, A., & Arabiyyat, A. (2021). Breast cancer detection using machine learning algorithms. *International Journal of Computer Science and Mobile Computing, 10*(11).
16. Dhahri, H., Al Maghayreh, E., Mahmood, A., Elkilani, W., & Faisal Nagi, M. (2019). Automated breast cancer diagnosis based on machine learning algorithms. *Hindawi Journal of Healthcare Engineering, 2019*.
17. Hammad Memon, M., Ping Li, J., Ul Haq, A., Hunain, Memon M., & Zhou, W. (2019). *Breast cancer detection in the IOT health environment using modified recursive feature selection*. Hindawi: Wireless Communications and Mobile Computing.

Printed by Printforce, the Netherlands